3D Printing of Pharmaceutical and Drug Delivery Devices

ADVANCES IN PHARMACEUTICAL TECHNOLOGY

A Wiley Book Series

Series Editors:
Dennis Douroumis, University of Greenwich, UK
Alfred Fahr, Friedrich–Schiller University of Jena, Germany
Jűrgen Siepmann, University of Lille, France
Martin Snowden, University of Greenwich, UK
Vladimir Torchilin, Northeastern University, USA

Titles in the Series

Hot-Melt Extrusion: Pharmaceutical Applications
Edited by Dionysios Douroumis

Drug Delivery Strategies for Poorly Water-Soluble Drugs
Edited by Dionysios Douroumis and Alfred Fahr

Computational Pharmaceutics: Application of Molecular Modeling in Drug Delivery
Edited by Defang Ouyang and Sean C. Smith

Pulmonary Drug Delivery: Advances and Challenges
Edited by Ali Nokhodchi and Gary P. Martin

Novel Delivery Systems for Transdermal and Intradermal Drug Delivery
Edited by Ryan Donnelly and Raj Singh

Drug Delivery Systems for Tuberculosis Prevention and Treatment
Edited by Anthony J. Hickey

Continuous Manufacturing of Pharmaceuticals
Edited by Peter Kleinebudde, Johannes Khinast, and Jukka Rantanen

Pharmaceutical Quality by Design
Edited by Walkiria S. Schlindwein and Mark Gibson

***In Vitro* Drug Release Testing of Special Dosage Forms**
Edited by Nikoletta Fotaki and Sandra Klein

Characterization of Pharmaceutical Nano- and Microsystems
Edited by Leena Peltonen

Biopharmaceutics: From Fundamentals to Industrial Practice
Edited by Hannah Batchelor

3D Printing of Pharmaceutical and Drug Delivery Devices: Progress from Bench to Bedside
Edited by Dimitrios A. Lamprou, Dennis Douroumis, and Sheng Qi

Forthcoming Titles:

Process Analytics for Pharmaceuticals
Edited by Jukka Rantanen, Clare Strachan and Thomas De Beer

Mucosal Drug Delivery
Edited by Rene Holm

3D Printing of Pharmaceutical and Drug Delivery Devices

Progress from Bench to Bedside

Edited by

DIMITRIOS A. LAMPROU
Queen's University Belfast
Liburn Road, Belfast, United Kingdom

DENNIS DOUROUMIS
School of Science, University of Greenwich
Medway Campus, Central Avenue
Chatham Maritime, United kingdom

SHENG QI
University of East Anglia
Norwich Research Park, United Kingdom

WILEY

This edition first published 2024

Registered Office(s)
John Wiley & Sons, Inc., 111 River Street, Hoboken, NJ 07030, USA
John Wiley & Sons Ltd, The Atrium, Southern Gate, Chichester, West Sussex, PO19 8SQ, UK

For details of our global editorial offices, customer services, and more information about Wiley products visit us at www.wiley.com.

Wiley also publishes its books in a variety of electronic formats and by print-on-demand. Some content that appears in standard print versions of this book may not be available in other formats.

A catalogue record for this book is available from the Library of Congress

Hardback ISBN: 9781119835974; ePub ISBN: 9781119835981; ePDF ISBN: 9781119835998; oBook ISBN: 9781119836001

Cover Image: © Pixel B/Shutterstock
Cover Design: Wiley

Set in 10/12pts Times LT Std by Integra Software Services Pvt. Ltd, Pondicherry, India
Printed and bound by CPI Group (UK) Ltd, Croydon, CR0 4YY

C9781119835974_170524

“There is nothing impossible to those who dare - (Οὐδὲν ἀνάλωτον τοῖς θαρροῦσιν ᾤετο)
- Alexander the Great”

Contents

About the Editors

Dimitrios A. Lamprou

Dimitrios A. Lamprou is Full Professor (Chair) of Biofabrication and Advanced Manufacturing at Queen's University Belfast. Dimitrios has been recognized as world leader in 3D Printing and has also been named in the Stanford University's of World's Top 2% Scientists. He is currently the author of over 150 peer-reviewed publications and of over 350 conference abstracts, and has given over 150 Invited Talks to institutions and conferences across the world. His research and academic leadership have been recognized in a range of awards, including the Royal Pharmaceutical Society of Great Britain Science Award and the Scottish Universities Life Sciences Alliance Leaders Scheme Award.

Dennis Douroumis

Dr. Dennis Douroumis is a professor in Pharmaceutical Technology and Process Engineering at the University of Greenwich, UK

His research activities focus on emerging technologies including: (a) 3D printing technologies for pharmaceutical dosage forms or novel medical devices (microneedles, bioresorbable scaffolds), (b) Continuous manufacturing processes for the development of medicinal products, and (c) Nanomaterial synthesis and surface modification for cancer treatment.

Dennis has established several national and international collaborations with world-class colleagues/researchers including industrial funded projects and several EU/UK grants. He received the prestigious award of Eminent Fellowship of the Academy of Pharmaceutical Sciences for the excellence in the pharmaceutical sciences over a prolonged period with an emphasis on advocacy and leadership. He has also received an award for his "Outstanding Scientific Contribution" in Pharmaceutical Processes and invited to deliver the Award Lecture, sponsored by AstraZeneca.

Sheng Qi

Sheng Qi is Professor of Pharmaceutical Material Science and Technology at the School of Pharmacy of the University of East Anglia (UEA). She runs a dynamic research group with most projects co-created and developed with relevant industrial partners/collaborators to address real-world clinically unmet needs. Sheng has great interests in material science and processing, and passion in innovation. By working closely with industrial partners as well as cross-discipline collaborators, her research has contributed to product development and innovations in many industrial sectors, from pharmaceutical, medical device to food, cosmetic, agri-tech and sustainable packaging. Within the field of pharmaceutical 3D printing, her efforts focus on the development of a fundamental understanding on how to adapt industrial 3D printing methods to safely process pharmaceutical materials and manufacture pharmaceutical products. She founded and leads the UEA Health and Social partner (UEAHSCP) Point of Care 3D Printing Research Group. The research group create the network for academic scientists to work closely with clinicians, pharmacists, and patients in Norfolk and Suffolk to identify and develop efficient and cost-effective uses of 3D printing in acute hospital environments for improving patient care.

List of Contributors

Sune Andersen, Research & Development Department, Janssen, Beerse, Belgium

Atheer Awad, UCL School of Pharmacy, University College London, London, UK and Department of Clinical, Pharmaceutical and Biological Sciences, University of Hertfordshire, Hatfield, UK

Abdul Basit, UCL School of Pharmacy, University College London, London, UK and FABRX Ltd., Henwood House, Henwood Ashford, Kent, UK and FABRX Artificial Intelligence, Carrete ra de Escairón, Currelos (O Saviñao), Spain

Fotios Baxevanis, Medicines & Healthcare Products Regulatory Agency, London, UK

Peter Belton, School of Chemistry, University of East Anglia, Norwich, UK

Richard Bibb, School of Design & Creative Arts, Loughborough University, Loughborough, UK

Joshua Boateng, School of Science, Faculty of Engineering and Science, University of Greenwich, Medway Campus, Kent, UK

Cecile Boudot, Evonik Corporation, Birmingham, Alabama, US

Sam Boulton, Faculty of Engineering and Science, University of Greenwich, Medway Campus, Kent, UK

Yahya E. Choonara, Wits Advanced Drug Delivery Platform Research Unit, School of Therapeutic Sciences, Faculty of Health Sciences, Department of Pharmacy and Pharmacology, University of the Witwatersrand, Johannesburg, South Africa

Marcilio Cunha-Filho, Laboratory of Food, Drugs, and Cosmetics (LTMAC), School of Health Sciences, University of Brasilia, Brasília, Brazil

Mahokh Dadsetan, Evonik Corporation, Birmingham, Alabama, US

Aikaterini Dedeloudi, School of Pharmacy, Queen's University Belfast, Belfast, UK

Dennis Douroumis, Faculty of Engineering and Science, University of Greenwich, Medway Campus, Kent, UK and Delta Pharmaceutics Ltd., Kent, UK

Andrea Engel, Evonik Corporation, Birmingham, Alabama, US

Atabak Tabriz Ghanizadeh, Faculty of Engineering and Science, University of Greenwich, Medway Campus, Kent, UK and Delta Pharmaceutics Ltd., Kent, UK

Andy Gleadall, School of Mechanical, Electrical and Manufacturing Engineering, Loughborough University, Loughborough, UK

Alvaro Goyanes, UCL School of Pharmacy, University College London, London, UK, FABRX Ltd., Henwood House, Henwood Ashford, Kent, UK, FABRX Artificial Intelligence, Carrete ra de Escairón, Currelos (O Saviñao), Spain and Departamento de Farmacología, Farmacia y Tecnología Farmacéutica, I+D Farma Group (GI-1645), Facultad de Farmacia, iMATUS and Health Research Institute of Santiago de Compostela, Universidade de Santiago de Compostela, Santiago de Compostela, Spain

Peyton Hopson, Advanced Engineering and Technology Department, Johnson & Johnson, Jacksonville, Florida, US

Scott Jones, Evonik Corporation, Birmingham, Alabama, US

Thomas Kipping, Life Science, Process Solutions and Formulation Materials, Merck Life Science KGaA, Darmstadt, Germany

Theresia Kuntz, Evonik Operations GmbH, Kirschenallee, Darmstadt, Germany

Hannah Kuofie, Faculty of Engineering and Science, University of Greenwich, Medway Campus, Kent, UK

Dimitripos A. Lamprou, School of Pharmacy, Queen's University Belfast, Belfast, UK

Ana Luiza Lima, Laboratory of Food, Drugs, and Cosmetics (LTMAC), School of Health Sciences, University of Brasilia, Brasília, Brazil

Maria Malamatari, Medicines & Healthcare Products Regulatory Agency, London, UK

Thomas McDonagh, School of Pharmacy, University of East Anglia, Norwich, UK

Sheng Qi, School of Pharmacy, University of East Anglia, Norwich, UK

Poornima Ramburrun, Wits Advanced Drug Delivery Platform Research Unit, School of Therapeutic Sciences, Faculty of Health Sciences, Department of Pharmacy and Pharmacology, University of the Witwatersrand, Johannesburg, South Africa

Lívia L. Sá-Barreto, Faculty of Ceilandia, University of Brasilia (UnB), Brasília, Brazil

James Scoble, Faculty of Engineering and Science, University of Greenwich, Medway Campus, Kent, UK

Iria Seosne-Viaño, UCL School of Pharmacy, University College London, London, UK and Department of Pharmacology, Pharmacy and Pharmaceutical Technology, Paraquasil Group (GI-2109), Faculty of Pharmacy, iMATUS and Health Research Institute of Santiago de Compostela, Universidade de Santiago de Compostela, Santiago de Compostela, Spain

John Tipton, Evonik Corporation, Birmingham, Alabama, US

Bernanbe Tucker, Evonik Corporation, Birmingham, Alabama, US

Ka-Wai Wan, Medicines & Healthcare Products Regulatory Agency, London, UK

Joey Yan, School of Mechanical, Electrical and Manufacturing Engineering, Loughborough University, Loughborough, UK

Bin Zhang, School of Pharmacy, University of East Anglia, Norwich, UK

Jian-Feng Zhang, Evonik Corporation, Birmingham, Alabama, US

Advances in Pharmaceutical Technology: Series Preface

The series *Advances in Pharmaceutical Technology* covers the principles, methods and technologies that the pharmaceutical industry uses to turn a candidate molecule or new chemical entity into a final drug form and hence a new medicine. The series will explore means of optimizing the therapeutic performance of a drug molecule by designing and manufacturing the best and most innovative of new formulations. The processes associated with the testing of new drugs, the key steps involved in the clinical trials process and the most recent approaches utilized in the manufacture of new medicinal products will all be reported. The focus of the series will very much be on new and emerging technologies and the latest methods used in the drug development process.

The topics covered by the series include the following:

Formulation: The manufacture of tablets in all forms (caplets, dispersible, fast-melting) will be described, as will capsules, suppositories, solutions, suspensions and emulsions, aerosols and sprays, injections, powders, ointments and creams, sustained release and the latest transdermal products. The developments in engineering associated with fluid, powder and solids handling, solubility enhancement, colloidal systems including the stability of emulsions and suspensions will also be reported within the series. The influence of formulation design on the bioavailability of a drug will be discussed and the importance of formulation with respect to the development of an optimal final new medicinal product will be clearly illustrated.

Drug Delivery: The use of various excipients and their role in drug delivery will be reviewed. Amongst the topics to be reported and discussed will be a critical appraisal of the current range of modified-release dosage forms currently in use and also those under development. The design and mechanism(s) of controlled release systems including macromolecular drug delivery, microparticulate controlled drug delivery, the delivery of biopharmaceuticals, delivery vehicles created for gastrointestinal tract targeted delivery, transdermal delivery and systems designed specifically for drug delivery to the lung will all be reviewed and critically appraised. Further site-specific systems used for the delivery of drugs across the blood–brain barrier including dendrimers, hydrogels and new innovative biomaterials will be reported.

***Manufacturing*:** The key elements of the manufacturing steps involved in the production of new medicines will be explored in this series. The importance of crystallisation; batch and continuous processing, seeding; and mixing including a description of the key engineering principles relevant to the manufacture of new medicines will all be reviewed and reported. The fundamental processes of quality control including good laboratory practice, good manufacturing practice, Quality by Design, the Deming Cycle, Regulatory requirements and the design of appropriate robust statistical sampling procedures for the control of raw materials will all be an integral part of this book series.

An evaluation of the current analytical methods used to determine drug stability, the quantitative identification of impurities, contaminants and adulterants in pharmaceutical materials will be described as will the production of therapeutic bio-macromolecules, bacteria, viruses, yeasts, moulds, prions and toxins through chemical synthesis and emerging synthetic/molecular biology techniques. The importance of packaging including the compatibility of materials in contact with drug products and their barrier properties will also be explored.

Advances in Pharmaceutical Technology is intended as a comprehensive one-stop shop for those interested in the development and manufacture of new medicines. The series will appeal to those working in the pharmaceutical and related industries, both large and small, and will also be valuable to those who are studying and learning about the drug development process and the translation of those drugs into new life saving and life enriching medicines.

Dennis Douroumis
Alfred Fahr
Jűrgen Siepmann
Martin Snowden
Vladimir Torchilin

Preface

Over the last 10 years the science and technological advances of three-dimensional (3D) printing (3DP) have significantly evolved. The first 3D printer was developed in early 80s by Chuck Hull, who used a technology called stereolithography (SLA) patented in 1988 by 3D Systems. Since then, we have witnessed a technological transformation where a wide range of 3DP technologies have found applications in several sectors including, art, automotive, aerospace, pharmaceutical, biomedical sciences, tissue engineering and regenerative medicine, among other sectors.

After first approval of Spritam (Apreicia Pharmaceuticals) in 2015 by the US Food and Drug Administration (FDA), 3DP technologies have attracted considerable interest among academia and industry. All three Editors are actively involved in the development of pharmaceutical dosage forms using a variety of 3DP technologies and particularly in personalised medicines, medical devices or implants, which used to edit this very important new book in this developed area. The main aim was to gather the collective experience of eminent scientists in the field to present recent trends and advances that will be of use to experienced researchers, educators or students who are actively involved in 3DPg.

It is anticipated that 3DP will revolutionise the pharmaceutical sector in the manufacturing of small batches or preclinical evaluation of new formulations, due to its unique advantages for customised dosage from in terms of dose strength, size, shape and release rates. Therefore, pharmaceutical industry will enter a new era moving to personalised dosage forms moving from the "one-size-fits-all" approach. In the near future medicines will be manufactured at the point of care (e.g., hospitals, pharmacies) to meet patient needs and improve clinical outcomes. Moreover, 3DP technologies offer a new business model through the transition from a manufacturer-centric to a consumer-centric logic and decentralised supply chain. Hence, it will offer better efficiency due to the low inventory and operating cost, less waste and print on demand personalised medicines with increased patients' access. However, the landscape in printing pharmaceuticals is not clear yet and there is a pressing need for greater flexibility both in manufacturing and regulatory matters.

The first Chapter provides an overview of key processing properties and critical attributes of materials used in a range of 3DP technologies. Excellent understanding of material properties is crucial for the development of designs with specific features but also for the selection of the appropriate printing technologies. The second chapter introduces additional 3DP technologies discussing challenges and perspectives. In addition, it covers a range of critical aspects from the computer-aided design (CAD) to the control of printing parameters and how those can be tuned to control drug release of printed dosage forms.

The third chapter introduces one of the upcoming printing methods in pharmaceutics, Selective Laser Sintering (SLS), as an emerging printing technology for the development of pharmaceutical dosage form, varying from small to large batch manufacturing or personalised dosage forms. It describes critical material properties and process parameters though a number of case studies.

Chapters 4 and 5, provide a comprehensive overview of 3DP technologies use for the fabrication of personalised medical devices and implantable drug delivery systems. The structural characteristics, such as geometry, porosity, surface properties and mechanical properties including biological aspects are among the elements someone should consider for the design of medical devices. The selection of appropriate polymers and 3DP technology, are presented in detail for a range of application such as tissue engineering, dentistry and orthopaedics. Regulatory matters related to medical devices are also discussed, briefly.

Chapters 6 and 7 are dedicated to the printing of hydrogels in various fields including wound healing, tissue engineering and drug delivery purposes. The consideration and selection of suitable materials depending on the 3DP technology are presented in detail along with limitations and future directions.

Chapter 8 presents a range of analytical techniques that are used for pre-formulation studies and particularly for the selection of materials prior to 3DP followed by advanced technologies introduced for structural characterization of the end product. In this chapter, the reader will be informed about the features and expected outcomes of analytical techniques.

The last three chapters provide the landscape on 3DP technologies for pharmaceutical product manufacturing including existing commercial applications and business models. This includes decentralised manufacturing and collaborative partnerships for the commercialization of marketed products. Emphasis is given on clinical challenges for the development of customised paediatric, geriatric or even veterinary dosages forms. Finally, a widespread overview of the existing regulatory framework and guidance for 3D printed medicinal products particularly in United Kingdom (UK) and European Union (EU) is introduced. Quality aspects embracing Quality by Design (QbD), process analytical technology (PAT) tools, and excipients safety for additive manufacturing (AM) are also discussed.

Prof. Dimitrios Lamprou (Queen's Belfast University)
Prof. Dennis Douroumis (University of Greenwich)
Prof. Sheng Qi (East Anglia University)

1

Materials for 3D Printing

Thomas McDonagh, Bin Zhang and Sheng Qi

School of Pharmacy, University of East Anglia, Norwich, UK

1.1 Introduction

Three-dimensional printing (3DP), also known as Additive Manufacturing (AM), has emerged as an exciting technology for the manufacture of pharmaceutical products for personalised patient treatment. The interindividual variability of the human population is a constant challenge when striving for effective drug delivery because patients come in different shapes, sizes, ages, genetics, and necessities and so require medication personalised to their needs for the most effective outcomes [1, 2]. 3DP has been shown to be a flexible technique that can be used to manufacture drug delivery devices (DDDs) for a wide array of applications such as tablets, implants, microneedles, and suppositories [3–5]. In addition, compared to conventional large-scale production techniques such as tablet pressing, 3DP is much more flexible with the potential to simplify supply chains and accelerate development cycles. Such traits make 3DP an ideal candidate technology to produce on-demand personalised medicines and in turn improve patient quality of life. However, there are still several challenges hindering its adoption. Limited availability of biocompatible materials, the incompatibility of the active pharmaceutical ingredient) (API), and/or polymer with the printing conditions, and regulatory hurdles present a significant challenge when designing 3DP pharmaceutical products. This is highlighted by the presence of just one 3DP pharmaceutical product currently on the market as of 2023, Spritam® [6].

3D Printing of Pharmaceutical and Drug Delivery Devices: Progress from Bench to Bedside, First Edition.
Edited by Dimitrios A. Lamprou, Dennis Douroumis and Sheng Qi.

Material choice is a fundamental consideration when designing a pharmaceutical dosage form. All drug products are comprised of an API (the bioactive component) and inactive functional excipients which facilitate the release of the API to the target location in the body. With the advance of 3D-printing medicine, API carrier materials have an increasingly important role in not only protecting the API in a convenient printable package and disguising unpalatable ingredients but also in facilitating complex release profiles. Several 3D-printing techniques are applicable for manufacturing DDD: Thermal Extrusion-based deposition systems (TE); Semi-solid extrusion (SSE); Stereolithography (SLA); and Powder Bed fusion (PBF). Successful printing of pharmaceuticals requires consideration of the nature of the 3D-printing process. Each technique has its own benefits and limitations in terms of material compatibility, print quality, and scalability and so must be considered as a whole when designing a new DDD. The objective of this chapter is to first discuss each printing technique, exploring the key material characteristics and processing parameters that influence both printability and drug delivery performance. Subsequently, the materials which have shown suitability for 3DP manufacture of DDD will be discussed with a focus on the key material attributes relevant for printing and drug release properties.

1.2 Material Processability Considerations for Pharmaceutical 3DP

1.2.1 Thermal Extrusion-Based 3D Printing

Thermal extrusion (TE)-based 3D printing is a popular solvent free 3DP technique for printing thermoplastic polymer and API formulations. In TE printing, thermal energy is applied to a print head to melt material, enabling extrusion through a nozzle onto a build plate. Thermoplastic polymers are used because they are composed of long linear chains, held together by weak attraction forces. When subjected to high temperatures in the print head, they soften or melt to enable extrusion before solidifying upon cooling. The nozzle and/or build plate are controlled by linear actuators that enable precise deposition into a pre-defined geometry in a layer-by-layer fashion according to a 3D digital model. Good print resolution is achievable with resolutions typically between 100 μm and 400 μm [7, 8], depending on the diameter of nozzle used. A benefit of TE printing is the minimal post processing requirements. Printed parts are full strength immediately after printing and unlike other 3DP techniques, no washing or drying steps are required. Furthermore, release rates of TE drug products are highly tuneable by varying printing software parameters such as infill density [9]. Due to the mechanical stability of TE drug products, this technique appears to be more suitable for controlled or sustained release applications [10, 11], although immediate release pharmaceutical products have also been explored [12].

Most current TE 3DP technologies are not compatible with the direct printing of raw powder materials. Direct TE of powders presents challenges in terms of feeding (poor flowability), homogenous mixing, and degassing (trapped air between powder particles) [13]. As such, a prior hot-melt extrusion (HME) step is commonly used to sufficiently mix reagents and process them into a more attractive feedstock for printing. In HME, heat and mechanical shear is used to mix reagents producing a homogenous melt which can be extruded into uniform filaments. Most pharmaceutical polymers require a temperature of at least 15 °C to 60 °C above their glass transition (T_g) to be sufficiently molten for processing [14]. Filaments can be printed using the widespread, low-cost technology fused deposition modelling (FDM) or can be pelletised into granules for use with direct granule TE.

1.2.1.1 Thermal Considerations

The first material processability consideration for TE 3DP concerns the thermal processing. Thermal processing presents a challenge when working with pharmaceutical formulations because some API and polymers are thermolable and will begin to degrade at high temperatures, reducing the efficacy of the pharmaceutical product. Additionally, formulation viscosity is intrinsically linked to temperature. High temperatures increase the kinetics of a formulation, reducing its apparent viscosity [15]. Therefore, successful TE operates within a temperature window that sufficiently reduces the viscosity of the polymer melt (above T_{min}) to enable good mixing and extrusion, whilst maintaining the API stability (below the thermal degradation temperature). To ensure solubility of API in the polymer melt, a third boundary temperature is often introduced, called the solubility line (T_c) [16]. This theoretical design space is shown in Figure 1.1. The design space can be seen to shrink for high drug load formulations due to the increased T_c required to solubilise the API. Additionally, increased residence time in the HME can be used to increase API dissolution but also increases the likelihood of thermal degradation. These same design constraints are also applicable when printing. Lower than optimal printing temperatures increase formulation viscosity that can cause nozzle blockages and low bond strength between printed layers [17] and high temperatures can result in further API degradation and poor print quality.

Whilst some sensitive API may appear to be incompatible with TE due to low thermal degradation temperatures (T_{deg}), strategies exist to reduce the thermal load on formulations enabling printing at lower temperatures. By using these techniques, a wide variety of polymers and API has been successfully printed, with drug loadings from 5% to 60% w/w shown to be feasible [19]. The first technique involves carefully selecting polymers and excipients which are extrudable at low temperatures. This generally involves use of

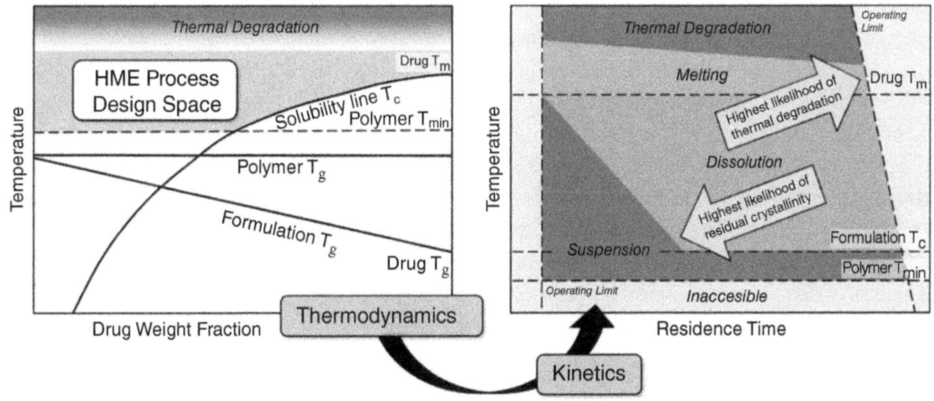

Figure 1.1 *(a) The temperature–composition phase diagram of the hot-melt extrusion process. The temperature design space falls below the thermal degradation temperature and, depending on composition, above the solubility line or polymer's minimum processing temperature T_{min}. Product phase behaviour is governed by the solubility line (formulation T_c) and formulation glass transition T_g. (b) Hot-melt extrusion process operating design space diagram. Three processing regimes (melting, dissolution, and suspension) can be delineated by temperature and kinetic considerations. (Source: Reproduced from [18] with permission from Elsevier 2018.)*

plasticisers which act as a lubricant between segments of polymer chains, thus increasing the materials flexibility and softness. Kollamaram et al. succeeded in printing Ramipril and 4-aminosalicylic acid (4-ASA) loaded tablets using a PVP PVP-VA copolymer system plasticised with PEG 1500 [20]. Ramipril is transformed into the impurity diketopiperazine upon exposure to temperatures higher than its melting point (109 °C) and 4-ASA begins to degrade at 130 °C. Filaments loaded with 3% drug were obtained by HME at 70 °C and tablets were printed at 90 °C. HPLC analysis confirmed that both drugs were stable with no signs of degradation, demonstrating how the careful selection of polymer and plasticiser can enable the printing of thermolabile API at low temperatures. Another technique involves avoiding the thermal and mechanical stress induced during HME altogether by using filament impregnation. The filament impregnation technique was first reported in 2014 [21], where API was loaded into the filament post HME by soaking in a saturated alcoholic drug solution. Whilst several authors have reported on the impregnation technique for DDD applications, drug loading is very low (<3% w/w) [22–24], limiting its use to drugs with therapeutic effects at low dose [25]. Additionally, whilst the impregnation method does avoid the thermal stress induced during HME, the API still has to pass through the heated print head. Goyanes et al. used this method to load 4-ASA into commercially produced Polyvinyl alcohol (PVA) filaments by soaking in a saturated ethanolic drug solution for 24 hours before printing [26]. After printing at 210 °C, the already limited drug loading was reduced by approximately half to 0.12% w/w, indicating that even though residence time in the print head is short (a few seconds) it is sufficient for significant degradation of thermolabile drugs.

1.2.1.2 Solubility Enhancement

One area in which thermal extrusion 3DP coupled with HME has shown great promise is in enhancing the bioavailability of poorly soluble API, opening the door to a host of molecules previously considered unviable as drugs [27]. More than 40% of new chemical entities produced during drug discovery exhibit poor solubility characteristics, greatly limiting their therapeutic effectiveness [28]. API can exist either as a solid suspension or be partially or fully dissolved in the polymer matrix. This crystallinity is highly influence by processing conditions (Figure 1.1). It is desirable that API remain in the amorphous form in DDD so that the dissolution, dispersion, or erosion of the tablets mediate the drug release rather than the solubility of the crystalline drugs. The intense mixing and heating imposed by the rotating screws in HME can cause the API molecules to dissolve in the molten polymer, transforming the crystalline drug into a more uniform, amorphous dispersion. With the drug in an amorphous form, no energy is required to break the drug crystal lattice. For this reason, relative to the crystalline form, the amorphous form of many poorly water-soluble drugs can achieve substantially higher apparent solubility and markedly faster dissolution [29]. For successful results, polymer and functional excipients should be carefully chosen to maximise API solubility. Polymer-API miscibility can be predicated using the Hansen solubility parameter [30]. For poorly miscible formulations, the drug may crystallise out during storage, resulting in physical stability issues and variability in drug release. Sometimes crystallisation inhibitors are added to formulations which can help slow this process [31].

1.2.1.3 Mechanical Considerations

In the case of FDM printing, filament strength and flexibility are a further processing consideration. For successful FDM printing, filaments must possess good mechanical strength and flexibility to endure the FDM feeding mechanism. This is a significant challenge for DDD applications because most pharmaceutical grade polymers lack such attributes, being either too brittle so that the filaments break in the motor gear or too soft so that they cannot be pushed by the drive gear, thus hindering printing [32, 33]. Recent efforts have been focused on using rheology, filament mechanical screening, and machine learning to identify useful parameters that can be used as predictive tools in pharmaceutical development [34–37]. Xu et al. found toughness to be an effective predictive parameter of filament printability using a simple stiffness test [38]. After screening over 30 in-house manufactured filaments, they found stiffness values greater than 80 g/mm^2 % to be a good indicator of filament printability. When modifying formulations to improve processability, it is important to consider the effects additional excipients will have on the function of the drug product. It is common to incorporate large quantities of additives and plasticisers into the formulation which often do not add any functionality to the main drug delivery and absorption functions, leading to the design of complex formulations with increased weight and increased potential for the adverse stability of the pharmaceutical product. Recently emerging TE techniques, termed direct powder/granule deposition TE, have focused on bypassing the need for filament forming. Using this technique, the HME extrusion and printing steps are effectively combined in a single step printing process. This not only simplifies DDD manufacture but can also be used to print formulations that would not be suitable for forming filaments. Goyanes et al. used a direct single-screw powder extruder (FabRx, UK) to prepare sustained release itraconazole printlets (3D-printed tablets) [39]. McDonagh et al. used a similar granule fed 3DP technology, Arburg Plastic Freeforming (APF), to explore the printability of non-FDM printable formulations [40]. Eudragit E PO, a very brittle polymer loaded with paracetamol, was successfully printed into immediate release tablets without the use of plasticisers. The simplified manufacturing process and more versatile material feed mechanism make direct powder/pellet thermal extrusion an exciting 3DP technology for the future. Table 1.1 summaries the typical polymers used in thermal extrusion 3D printing of solid dosage forms, which are used widely beyond the example studies listed here.

1.2.2 Semi-Solid Extrusion 3DP

Semi-solid extrusion (SSE), also known as pressure-assisted micro syringe (PAM), is a 3DP technique used to print pastes, hydrogels, and other viscous polymer systems. The working principle of SSE is similar to thermal extrusion 3DP in the sense that viscous formulations are extruded through a nozzle to form a 3D structure according to a designed geometry. However, SSE relies on use of a solvent rather than high temperature to obtain the rheological properties to enable extrusion. This allows for much milder printing conditions that enable processing of many biocompatible materials. Khaled et al. state that if a polymer can be processed into powder form, it can be printed using PAM [47]. Room temperature printing is hugely beneficial when working with thermo-sensitive API [4]. However, solvent use also presents challenges in terms of toxicity, API stability, and print quality because a drying step is required post printing that can result in significant shrinkage. SSE printing has shown promise for developing drug eluting tablets [48] and soft scaffolds for tissue engineering [49] and is currently used on the market for printing personalised gummy vitamins [50]. Very high drug loadings up to 96% have been shown to be feasible [51]. Print quality is typically not as good as FDM- and SLA-based printing methods, with resolutions around 200 μm to 400 μm being common [52, 53]. However, high precision drug delivery systems have been shown to be possible with some clever post processing.

Table 1.1 *Examples of polymers used in thermal extrusion 3D printing of pharmaceuticals.*

Polymer	Drug	Printer	Application
Polyvinyl Alcohol (PVA)	Budesonide	MakerBot Replicator 2X Desktop	Controlled release tablets [41]
Hydroxypropyl Methylcellulose (HPMC)	Ibuprofen	MakerBot Replicator 2X Desktop	Controlled release tablets [42]
Polycaprolactone (PCL)	Ibuprofen	Prusa i3 Mk3S	Controlled release tablets [9]
Eudragit® E	Paracetamol	Arburg Plastic Freeformer	Immediate release tablets [40]
Polyurethanes	Dapivirine	Arburg Plastic Freeformer	Controlled release vaginal rings [43]
Hydroxypropylcellulose	Itraconazole	FabRx M3DIMAKER™	Sustained release printlets [39]
Ethylene Vinyl Acetate (EVA)	Indomethacin	MakerBot Replicator 2X Desktop	Intrauterine implant [44]
PLA, PVA, HPMC, HPMCAS, Kollicoat® IR, PEG, Triethyl Citrate	Paracetamol	MakerBot Replicator 2X Desktop	Two pulse, oral drug delivery [45]
PVA, Sorbitol	Lisinopril Dihydrate, Indapamide, Rosuvastatin Calcium and Amlodipine Besylate	MakerBot Replicator 2X Desktop	Polypill [46]

Wu et al. manufactured microneedles patches for glucose delivery using SSE (Figure 1.2d) [54]. Pillar structures were first extruded using a sodium alginate-based paste. Subsequently, the pillars were stretched using a glass cover slide and crosslinked by spraying with a Ca^{2+} solution, producing conical microneedles with tip diameters of 24.5 μm.

1.2.2.1 Rheological Considerations

The materials extruded during SSE printing should be in a semi-solid form, also referred to as a gel or paste. These are formed by mixing polymer(s), functional excipient(s), and the drug with an appropriate solvent at a ratio that results in a paste suitable for printing. As a nozzle-based extrusion technology, optimal pastes should have suitable viscosity, yield stress under shear, compression, and viscoelastic elastic properties [55] to enable continuous printing without blockage. These rheological parameters must also be configured to the geometry of the nozzle in the print head. Depending on nozzle configuration, a wide range of paste viscosities from 30 mPa.s to $>6 \times 10^7$ mPa.s have been reported to be extrudable [56]. Of particular concern to the SSE 3D printing method is the polymer's response to being extruded, how well it is able to adhere to previously printed layers, and its ability to hold the weight of subsequent layers [57]. Shear thinning characteristics are desirable because the viscosity of the paste can be significantly reduced when a high shear rate is exerted during printing. Shear thinning properties influence not only the capability to be pushed through a narrow nozzle at a given temperature but also the ability to regain structure and shape after deposition.

1.2.2.2 Example Applications

First used to produce polypills and tablets, this technology has rapidly evolved to manufacture other types of dosage forms and medical devices. Example applications explored in the literature are chewable tablets, orodispersible films, rectal suppositories, and implantable patches [4]. Some of these applications are shown in Figure 1.2 and Table 1.2. SSE printing has been shown to be a versatile 3DP modality capable of printing a wide range of dosage forms suitable for a variety of release profiles from immediate release tablets to sustained release implants.

1.2.3 Powder Bed Fusion 3D Printing

Powder bed fusion (PBF) 3DP refers to 3D-printing technologies that use powder material as a feedstock. Many polymers and API are available as a powder, making PBF printing a versatile technique for drug delivery device manufacture. Two subset technologies are suitable for pharmaceutical applications: binder jetting (BJ) and selective laser sintering (SLS). Both technologies share the same principle of selectively binding powder in a layer-by-layer fashion to build up a 3D part. In BJ printing, powder binding is achieved by depositing a liquid binder that effectively glues adjacent particles together, whereas in SLS printing, a laser is used to partially melt and agglomerate selected particles. Subsequent layers are built up by spreading a fresh thin layer of powder on top of the bound layer

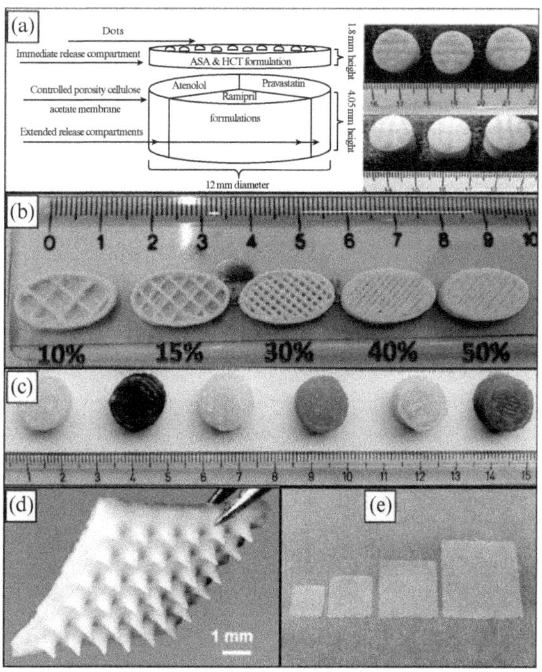

Figure 1.2 Drug delivery devices manufactured using PAM 3DP. (a) Five-in-one dose Polypill [58]. (b) Floating sustained-release printlets with different fill densities [59]. (c) Chewable isoleucine printlets prepared in different sizes, flavours, and colours [60]. (d) 3D-printed microneedle patch [54]. (e) From left to right: 25, 50, 100, and 200 mm^2 drug loaded films [61]. (Source: Reproduced with permission from [54, 58–61], Elsevier.)

Table 1.2 *Examples of solid dosage forms and devices produced by SSE 3DP.*

Polymer	Drug	Solvent	Printer	Application
HPMC Poly(Acrylic Acid)	Guaifenesin	Water	Fab@Home, USA	Tablets [57]
Croscarmellose Sodium and Hydroxypropyl Cellulose (HPC)	Levetiracetam (96%)	10% ethanol solution	Desktop semi-solid 3D extrusion printer	Immediate release tablets [51]
HPMC Cellulose Acetate	Nifedipine Captopril Glipizide	Hydro alcoholic Acetone DMSO	RegenHU, Switzerland	Polypill [62]
Cellulose Acetate, D-Mannitol, PEG 6000, Sodium Starch Glycolate, and PVP	Pravastatin (20%), atenolol (30%), ramipril (15%), aspirin (28.62%), and hydrochlorothiazide (5.86%)	Acetone and dimethyl sulfoxide (DMSO)	RegenHU 3D printer	Polypill [58]
Sucrose, Pectin, and Maltodextrin	Isoleucine (14.4%)	Water	Adapted 3D printer (The Magic Candy Factory, UK)	Chewable printlets [60]
Mesoporous Silica Nanoparticles (Msn) and Bioactive Glass Coatings in a Porous β-TCP Bioceramic	Isoniazid (INH)/ Rifampin (RFP)	Chloroform and dimethyl sulfoxide (DMSO)	3D-Bioplotter system, EnvisionTEC, Germany	Porous scaffold [63]
HPC and PVA	Warfarin (1.3%)	Ethanol and purified water	BioBots 1 (BioBot, USA) EXT 3D	Orodispersible film (ODF) [61]
Sodium Alginate and Hydroxyapatite	Insulin	Water	Allevi 2 bioprinter	Microneedle patches [54]

and repeating the fusion process. Depending on the equipment used, different mechanisms exist for this powder spreading. The most common method is the use of a metallic blade, a roller, or a rake for generating a powder layer with a precise layer height. API is typically incorporated into the powder blend but can also be added to the liquid binder in BJ printing [64]. After printing, loose powder is removed and recycled, revealing the printed structure. Further drying or curing is often necessary to further strengthen the print. Compared to other 3DP techniques, print surface finish and resolution is poor due to the rough nature of the powder feedstock and prints are fairly porous due to the loose packing of powder. Whilst less suitable for high strength, load bearing applications, the fairly weak binding and porous nature of PBF solid dosage forms can be advantageous for immediate release applications that require rapid dissolution. Porous, hygroscopic structures can be easily penetrated and broken down by liquids, making PBF prints well suited for oral delivery applications. This feature is seen in the only currently Food and Drug Administration (FDA) approved 3D-printed pharmaceutical product, Spritam, which is manufactured using a proprietary BJ technology and undergoes rapid disintegration (~10 seconds) when taken with a sip of liquid [65].

Powder is an attractive feedstock for 3DP because pharmaceutical materials are already widely available in this form factor, so can be directly used in PBF techniques, simplifying the production process. The direct use means powder properties play a key role concerning both processing performance and part properties that require careful consideration for successful printing. Figure 1.3 shows some of the relationships between powder properties and bulk powder behaviour on in-process printing performance and final print properties. Of these properties, powder flowability, packing density, and energy absorbance are the key attributes that determine formulation printability.

1.2.3.1 Powder Flowability Considerations

In PBF, good flow properties are required to ensure that a homogenous, thin layer of fresh powder is distributed during spreading at the start of each layer. Small particles (<20 μm), which are desirable for higher resolution printing, are particularly challenging to process due to their relatively large specific surface area that results in a high extent of adhesiveness to surfaces and other particles due to van der Waal interactions [67]. General methods of improving powder flowability are decreasing the width of the powder size distribution (PSD) [68, 69], increasing particle sphericity and smoothness [70], increasing particle size [71], decreasing moisture content [72], and addition of flow enhancers [73]. To enhance the particle morphology of the PBF feedstock, it can be pre-processed by grinding, milling, or

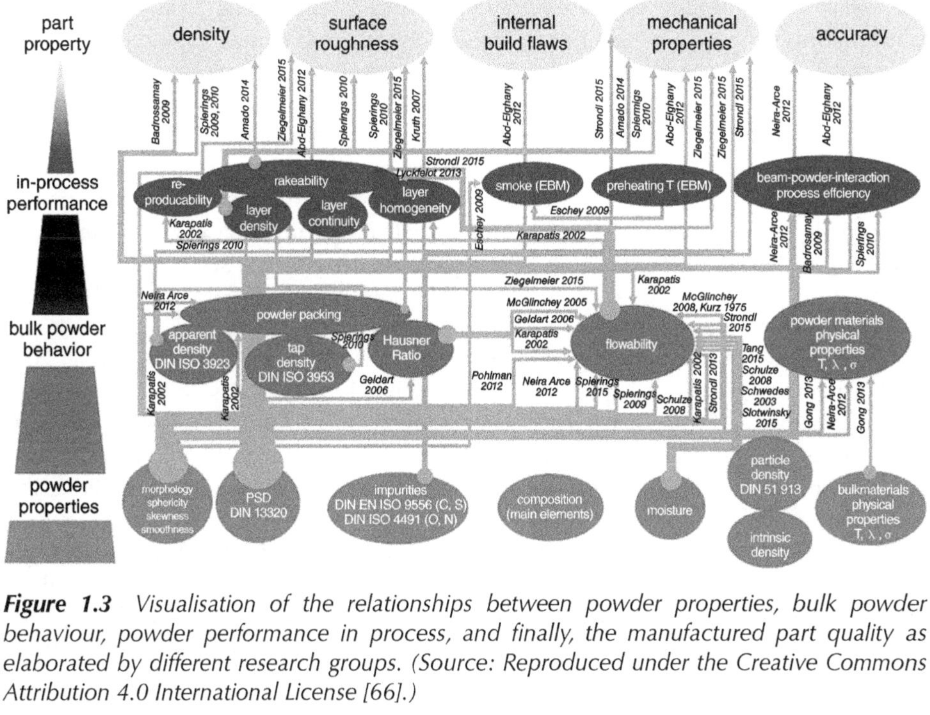

Figure 1.3 Visualisation of the relationships between powder properties, bulk powder behaviour, powder performance in process, and finally, the manufactured part quality as elaborated by different research groups. (Source: Reproduced under the Creative Commons Attribution 4.0 International License [66].)

spray drying [74, 75]. To obtain an even particle size distribution, it is recommended to sieve the feed powder prior to its use [76].

1.2.3.2 Powder Packing Density Considerations

The packing density of the powder feedstock has a strong influence on the final print density, surface roughness, and mechanical properties. A wide multimodal PSD is necessary to achieve a high powder bulk density, as the fine particles can fill the gaps between the larger ones. This is contradictory to the necessities for good powder flow and illustrates how contradictory the requirements for only one parameter can be with respect to different aspects of the PBF process. For high-strength applications such as bone tissue scaffolds, a high packing density is required to produce dense parts, whereas for pharmaceutical applications, low density prints are desirable to enhance tablet disintegration and drug release rates [77].

1.2.3.3 Powder Energy Absorbance Considerations

Powder energy absorbance describes the feedstocks' ability to absorb energy and bind to adjacent particles during printing. For the case of BJ printing, the energy absorbance refers to the interaction of the powder with the binder fluid. Binder deposition and spread is dependent on the viscosity, surface tension, and density of the binder solution and the wettability and specific surface area of the powder material [78]. In SLS printing, the energy absorbance is the feedstocks' capacity to absorb the wavelength of the laser used during printing. BJ has been shown to be effective for printing high-dose hydrophilic formulations for rapid disintegration [64, 79]. However, achieving good print quality can be a challenge when it comes to printing large doses of hydrophobic molecules which require water-based binders. Poor selection of a binder and powder bed properties may result in sliding of powder layers during printing and a decrease in the final product quality [80]. Formulating binders using a mixture of water and organic solvents can be used to overcome this but creates the problem of efficient removal of toxic solvent from the final formulation. In SLS printing, powder binding is dependent on the power and wavelength of the laser source, the laser energy absorptivity of the powder material, and the scanning speed [81, 82]. Increased laser power or decreased scanning speed increases powder particles exposure to the laser, increasing the level of sintering between particles. By tuning these parameters, part strength and porosity can be controlled which in turn mediate drug performance. For example, paracetamol loaded HPMC tablets have been printed at 100, 200, and 300 mm s^{-1} to achieve controlled drug release over 4, 3, and 2 hours, respectively [83]. Furthermore, scanning power and speed can be adjusted during a single print to create variable density dosage forms. This method was first used with biodegradable polymers by Leong et al. in 2006 [82]. The group printed PCL and poly(L-lactic acid) (PLLA) tablets with a denser outer region to act as a diffusion barrier and a more porous inner region to encapsulate the drug and achieve zero-order release. Additionally, it is important to consider the energy absorbance range of a powder feedstock. Typically, CO_2 laser diodes emit energy in the visible light region. Since most pharmaceutical polymers are white in colour, minimal energy absorption will occur naturally. As such, pharmaceutical grade colourants can be incorporated into the powder feedstock to facilitate the absorption of energy. Due to the localised heating used in SLS printing, there are concerns about the degradation of API during printing. Several studies have reported minimal drug degradation using fairly stable API such as paracetamol [84], ibuprofen [85], and progesterone [86], but there is no evidence to suggest thermolabile API can be employed with this technique. Table 1.3 provides example formulas used in powder bed fusion 3D printing of pharmaceuticals and devices.

1.2.4 Stereolithography 3D Printing

Stereolithography (SLA) was one of the first 3D-printing techniques introduced as a commercial system, when Charles W. Hull filed a patent for the photopolymerisation technology in 1986 [91]. SLA 3D parts are constructed from a vat of liquid photopolymer resin that is cured in a layer-by-layer fashion by an ultraviolet light source. Two different methods of irradiation may be applied to stereolithography, laser-based SLA, and digital light projection DLP. The laser-based method selectively cures the liquid photopolymer in a point-to-point fashion, whilst in DLP, the UV light source is projected to expose and polymerise an entire layer at once. For subsequent layers, the build plate is raised by one layer thickness and the process is repeated until the full 3D designed part is exposed. As a light-based technique, SLA printing offers the highest resolution amongst 3D-printing technologies, with XY resolutions and layer heights generally between 25 μm and 100 μm [92]. As such, SLA is commonly used for applications where a high level of detail is required. Following print completion, parts typically need to undergo two post processing steps: washing, most commonly in isoporopyl alcohol (IPA) to rinse off liquid resin, followed by UV curing to improve the mechanical properties of the printed structure, since the photopolymerisation of the resins during printing is usually incomplete. The main limitation of the SLA technique is the restrictive availability of biocompatible, photocurable materials. Only a few materials are currently available, limiting the flexibility of the technology for printing materials with different release properties.

The SLA technology is based on the phenomenon of photopolymerisation (shown in Figure 1.4). Photopolymerisation is the use of light in order initiate a polymerisation reaction, converting a liquid monomer resin into a solid polymer. The polymerisation reaction is initiated by the addition of light-sensitive compounds called photoinitiators, which become active under the appropriate wavelengths, creating free radicals due to the conversion of the absorbed light energy [93]. These free radicals are then consumed in the reaction converting the liquid monomer into a solid state, which may be referred to as a crosslinked hydrogel [94].

SLA is used to date the least studied 3DP technique in drug delivery device manufacture, which can be mainly attributed to the limited availability of biocompatible photocurable materials. As an additional drawback, the SLA printers operate primarily with single materials. Thus, the fabrication of formulations with multiple materials, such as polypills, is limited. Despite these limitations, the technique has intriguing advantages over some of the other 3DP processes as a high resolution, non-thermal process that appears promising for printing personalised implants [96], transdermal delivery systems [97, 98], and oral tablets [99]. In transdermal drug delivery, the SLA method has been shown to be suitable for preparing microneedles with various shapes, sizes, and antimicrobial coatings. The high printing precision of SLA enables great accuracy and reproducibility, making it well suited to more demanding applications in terms of both accurate dosing and geometry [100–102]. Although SLA material availability for DDD applications is limited, there is some capacity for altering the mechanical properties of the material. Goyanes et al. prepared drug-eluding topical patches with salicylic acid according to a scanned image of an individual's nose [101]. The flexibility of the nose-shaped patches was adjusted by controlling the degree of crosslinking between PEGDA chains with the addition of poly(ethylene glycol) (PEG) as a filling agent. The first application of SLA in the formulation of 3D-printed tablets was described by Wang

Table 1.3 *Example materials used in powder bed fusion 3D printing of pharmaceutical solid dosage forms and devices.*

Powder bed fusion type	Polymer	Binder ink		Drug	Printer	Application
		Binder	Solvent			
BJ	Maltitol, maltodextrin, PVP	PVP	Aqueous buffer	Captopril (in binder)	TheriForm process	Immediate release tablet [64]
	Kollidon SR and HPMC	PVP, Tween 20, triethyl citrate	Water, Ethanol	Pseudoephedrine hydrochloride (in binder)	TheriForm process	Controlled release tablet [79]
	Lactose, PVP, mannitol	PVP	Water, Ethanol	Paracetamol (in powder bed)	Fochif Mechatronics Technology (China)	Fast disintegrating tablet [87]
	D-sucrose, pregelatinised starch, povidone K30, microcrystalline cellulose, and silicon dioxide	PVP	Ethanol	Warfarin sodium (in powder bed)	Fochif Mechatronics Technology (China)	Oral disintegrating tablet [88]
	Tricalcium phosphate powder		Phosphoric acid	Vancomycin, ofloxacin, and tetracycline (post print impregnation)	3D-powder printing system (Z-Corporation, USA)	Microporous bioceramic implants [89]
SLS	Kollicoat IR and Eudragit L100–55, Candurin® Gold Sheen	N/A	N/A	Paracetamol	Sintratec Kit, AG, Brugg, Switzerland	Controlled release tablets [84]
	Hydroxypropyl methylcellulose (HPMC E5) and Kollidon® VA 64, Candurin® Gold Sheen	N/A	N/A	Paracetamol	Sintratec Kit, AG, Brugg, Switzerland	Immediate release tablets [83]
	Polyethylene	N/A	N/A	Progesterone and fluorouracil		Intrauterine drug delivery [90]

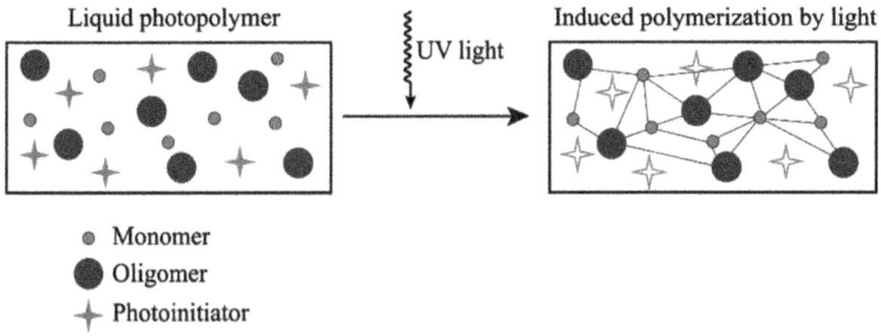

Figure 1.4 Process of photopolymerisation. Crosslinking of monomers, oligomers, and photoinitiator liquid resin upon exposure to UV light to produce a solid material. (Source: Reproduced under the Creative Commons Attribution 4.0 International License. [95].)

Table 1.4 Example materials used in SLA for 3D printing of pharmaceuticals and devices.

Polymer	Drug	Photoinitiator	Printer	Application
Polyethylene Glycol Diacrylate	Ibuprofen	DPPO	Formlabs 1 + SLA	Hydrogel tablets [104]
Poly(Ethylene Glycol) Diacrylate (PEGDA)	Acetylsalicylic acid	Darcure 2959	Home made	Tissue scaffold [105]
Polyethylene Glycol Diacrylate (PEGDA)	4-aminosalicylic acid (4-ASA) and paracetamol (acetaminophen)	DPPO	Form 1+ SLA 3D	Controlled release tablet [100]
Poly(2-Hydroxyethyl Methacrylate) (pHEMA), Poly(Ethylene Glycol) Dimethacrylate (PEGDMA)	Methylene blue hydrate	TPO		Subcutaneous implants [106]
Formlabs Elastic Resin	Lidocaine	(formulated in the commercial resin)	Form 2 SLA 3D	Bladder device for intravesical drug delivery [107]

et al., who produced dosage forms in a torus geometry using a photosensitive resin consisting of polyethylene glycol diacrylate 700 (PEGDA 700) and diphenyl (2,4,6-trimethylbenzoyl) phosphineoxide (TPO) as a photoinitiator [100]. The loading of paracetamol and 4-ASA in the printed tablets was 5.69% and 5.40%, respectively. To be able to manufacture DDD with higher drug loadings, the API must be adequately soluble in the polymer, which restricts the use of SLA for high-dose DDD. However, the manufacturing of printable objects with homogenously suspended particles (with loadings up to 53%) has been reported [103]. This indicates that the SLA printing of DDS from resins containing undissolved drug particles at higher drug content could be possible. Table 1.4 gives examples of combinations of polymer and photoinitiators used for pharmaceutical and medical device printing.

1.3 Classification of Common Materials Used in Pharmaceutical 3DP

In today's medicine, API carrier materials such as polymers and lipids form the basis of modern advanced delivery systems, having a multifaceted role on both the processability and functionality of a formulation. Suitable materials should be biocompatible, not negatively interact with the API, amenable to control API release profile, and printable. As such, the requirements are often different, depending on the printing technology used and intended application. Polymers are the most versatile category of biomaterials and have found use in the pharmaceutical industry as binders, lubricants, fillers, and solubility enhancers in solid dosage form production. Polymers are macromolecules composed of repeating units of monomers, which can either be naturally occurring, synthetically modified, or synthetically manufactured [108]. This section reviews the most common materials used in pharmaceutical 3DP as drug carriers and their key characteristics that influence processability and API function. It is important to bear in mind that the addition of API or additional excipients will have a significant effect on the thermal, rheological, and functional properties of a formulation but material choice is an important starting point that can be used to determine the necessity of additional excipients.

1.3.1 Alcohol Derived Polymers

- **Polyvinyl Alcohol (PVA) –** Polyvinyl alcohol (PVA), a water-soluble crystalline material that is easily biodegradable and considered a safe pharmaceutical excipient, is already used in multiple market drug delivery systems [109, 110]. Depending on the degree of hydrolysis of the acetate groups, the melting point of PVA may range from 180 °C (partially hydrolysed) to 228 °C (fully hydrolysed). It has a glass transition temperature of 85 °C [111] and begins to thermally degrade above 200 °C [112]. PVA is one of the most commonly used thermoplastics for 3DP and can be used with TE, PAM, and BJ [113–115].
- **Polyvinylpyrrolidone (PVP) –** Polyvinylpyrrolidone (PVP) is an amorphous non-ionic polymer made from repeating units of *N*-vinylpyrrolidone with good solubility in water and other polar solvents. PVP grades are labelled with K-values that relate to the molecular weight and degree of polymerisation and crosslinking of the material. PVP T_g is between 72 °C and 177 °C, dependent on the grade used with a degradation temperature of between 166 °C and 217 °C [116].
- **Poly(vinylpyrrolidone)/vinyl acetate (PVPVA) –** PVP can be crosslinked with vinyl acetate to form a PVPVA copolymer with is insoluble in water. It has a T_g of 105 °C and a T_{deg} of 270 °C [116]. PVPVA presents a molecular weight above 70 kDa and is widely used in pharmaceutical formulations as a direct compression, tablet disintegrant, film-forming, and taste-masking excipient, also known for its suitability for HME.

1.3.2 Eudragits

Eudragits are methacrylic polymers that were first introduced in the 1950s for enteric coatings. They are amorphous polymers that can achieve a wide range of targeted release profiles, depending on the functional groups incorporated into them. Eudragit E is suitable for immediate release in the stomach, dissolving in acidic conditions below pH 5. Alternatively, Eudragit L and S have been developed for targeted release to the gastrointestinal tract and colon at pH values of above 6 and 7, respectively, withstanding the acidic environment of the stomach [117]. Eudragit RS and RL are polymers with similar sustained release properties but suitable for pH independent release applications. Eudragit materials are commonly utilised using thermal deposition 3D-printing techniques due to their good thermal characteristics, although plasticisation is often necessary to reduce the melt viscosity of the formulations. Glass transition temperature and thermal degradation onset temperature for Eudragit E, RL, RS, L, and S are 52 °C and 250 °C [118], 63 °C and 166 °C, 64 °C and 170 °C [119], 157 °C and 206 °C, and 163 °C and 188 °C [120], respectively. Due to their versatile release characteristics, Eudragit has been used to optimise drug delivery for a variety of applications, including ophthalmic [121], buccal or sublingual [122], enteric [123], oral [40], colon [124], vaginal [125], and transdermal [126] delivery. Although Eudragit is most frequently used for thermal extrusion 3DP, it has also been used with powder bed fusion technologies [127, 128]

1.3.3 Other Polymers

- **Polycaprolactone (PCL) –** PCL is a semi-crystalline, biodegradable polyester which has received a great deal of interest for use as an implantable biomaterial. PCL undergoes slow degradation by hydrolysis of its ester linkages under physiological conditions (such as the human body), so is widely used in long-term implants and controlled drug release applications [129]. With a low melting point and glass transition temperature of 60 °C and −60 °C respectively and a high T_{deg} of 310 °C [23], PCL has great properties for HME and FDM printing and is a commonly used filament feedstock in addition to excellent solubility in organic solvents. PCL is considered a good elastic biomaterial due to its low tensile strength (~0.023 GPa) and high elongation at breakage (4700%) [130].
- **Ethylene vinyl acetate (EVA) –** EVA is the copolymer of semi-crystalline ethylene and amorphous vinyl acetate that produces materials with a 'rubber-like' feel. The weight percentage of vinyl acetate usually varies from 10% to 40%, with decreased percentage grades corresponding to increased thermal and mechanical properties. EVA has a T_g of −28 °C and a high T_{deg} of 300 °C [131].
- **Polyethylene glycol (PEG) –** PEG, also known as PEO, depending on its molecular weight, is a water soluble, highly crystalline, biocompatible, amphiphilic polymer that is widely used as an excipient in many pharmaceutical products. Polymers with Mw <100,000 are usually called PEGs, while higher molecular weight polymers are classified as PEOs. PEG has a T_g of −56 °C, a T_m of 57 °C to 63 °C, and a T_{deg} of 330 °C [132].
- **Polyglycolic acid (PGA), Polylactic acid (PLA) and their co-polymers –** PGA and PLA are both semi-crystalline, highly biocompatible polymers which have been extensively researched and utilised for long-term (months to years) medical implants and drug delivery scaffolds [133]. Both thermoplastic materials degrade by hydrolysis under physiological conditions into lactic acid and glycolic acid, which are both natural by-products of metabolic pathways in the body. They are commonly used for long-term implants such as pins, rods, plates, sutures [134], and tissue engineering scaffolds. Biodegradable implants are hugely beneficial for support structures because they gradually transfer the load to the body as it heals and does not require further surgery for removal. PLA has two pharmaceutically relevant forms, PLLA and poly(DL-lactide) (PDLLA). PLLA has a T_g of 60 °C to 65 °C, T_m of 175 °C, and strength of 4.8 GPa; whereas, PDLLA has a slightly lower T_g of 55 °C to 60 °C and a mechanical strength of 1.9 GPa [135]. Both forms thermally degrade above

300 °C [136]. PGA has a glass transition temperature between 35 °C and 40 °C and its melting point is reported to be in the range of 225 °C to 230 °C and begins to degrade at approximately 250 °C [137]. Poly(lactic acid-co-glycolic acid) (PLGA) is a copolymer of PLA and PGA that has favourable degradation mechanics, which can be tuned depending on the ratio of lactide to glycolide used for the polymerisation. The higher the content of glycolide units, the lower the time required for degradation as compared to predominantly lactide materials. An exception to this rule is the copolymer with a 50:50 monomers' ratio, which exhibits faster degradation (about two months). PLGAs typically show similar thermal properties to PLA and PGA, with a glass transition in the range of 40 °C to 60 °C.

1.3.4 Graft Polymers

- **Soluplus®** – is a swellable, non-ionic water-soluble graft polymer comprised of PCL, PVA, and PEG, which exhibits amphiphilic properties. It has a large number of hydroxyl groups that make it a great solubiliser for BCS class II drugs in aqueous media and for creating amorphous dispersions [138]. It has a low T_g of 70 °C and a high T_{deg} of 278 °C [139], making it a good candidate for HME, although plasticisation is often required to reduce the brittleness of extruded filament [140].
- **Carbopol®** – is a high molecular weight polymer composed of polyacrylic acid crosslinked with allyl sucrose or allyl pentaerythritol. The crosslinking degree determines the viscosity of the polymer. It can be incorporated into drug loaded paste formulations for PAM printing, generally used to provide sustained release [47]. Carbopol 971P, a lightly crosslinked formulation, has a viscosity range of 4,000 cP to 11,000 cP (or mPa.s) and Carbopol 974P, a highly crosslinked formulation, has a viscosity range of 29,400 cP to 39,400 cP [141].
- **Kollicoat® IR** – is a robust yet flexible PVA/polyethylene glycol graft copolymer that is rapidly water soluble and commonly used for manufacturing immediate release coatings and solid dosage forms. It has a T_g of 48 °C, a T_m of 208 °C, and a T_{deg} of 200 °C [116].

1.3.5 Photocrosslinkable

Over the past few years, a number of photocrosslinkable polymers have been developed suitable for drug delivery, such as poly(ethylene glycol) diacrylate (PEGDA), poly(2-hydroxyethyl methacrylate) (pHEMA), poly(ethylene glycol) dimethacrylate (PEGDMA), and poly(propylene fumarate)/diethyl fumarate (PPF/DEF). Typically, a photopolymer consists of a mixture of monomers and photoinitiators which, when exposed to UV light, crosslink to form a solid material. PEG-based polymers are water soluble whereas pHEMA is an insoluble but swellable material. PPF/DEF has emerged as a useful photocurable, biodegradable material for manufacturing high strength structures for applications in bone tissue engineering [142].

1.3.6 Natural Materials

Due to their abundance in nature and favourable properties (low-cost, biocompatibility, and biodegradability), polysaccharides have received increased attention in a wide range of applications, including pharmaceuticals. Their intrinsic properties can be easily tuned through chemical modification (increase solubility, change hydrophilic/hydrophobic balance, etc.) and further improved by combining them with other carefully chosen natural or synthetic polymers. Polysaccharides are susceptible to crosslinking, which can be implemented to maintain the structural integrity of the printed objects, increase their mechanical properties, or delay their degradation. Moreover, polysaccharides bearing ionic groups can exhibit a responsive behaviour, e.g., pH-responsive for chitosan, that allows for an additional level of control [143].

- **Cellulose derived polymers** – Cellulose is a highly hydrophilic polymer, although it is water insoluble due to its high degree of crystallinity and structure. For this reasons, cellulose is typically modified to form derivatives, such as Ethyl cellulose (EC), Hydroxypropyl cellulose (HPC), Hydroxypropyl methylcellulose (HPMC), and Hydroxypropyl methylcellulose Acetate Succinate (HPMCAS), with better release and processability properties. Cellulose-derived polymers are widely used in TE and SSE printing as a controlled release matrix (high molecular weight) or as a binder (low molecular weight). HPC and HPMC are amorphous, swellable, hydrophilic water-soluble polymers that exhibit pH independent release. EC is semi-crystalline and not water soluble. HPMCAS is amorphous and anionic, insoluble in gastric fluid but swells and dissolves rapidly in neutral to mildly alkaline pH. Therefore, HPMC is often used as a enteric coating material [144] and for sustained release applications [145]. Glass transition temperatures and thermal degradation onset temperatures for EC, HPC, HPMC, and HPMCAS are 130 °C and 200 °C [146], 111 °C and 227 °C [147], 96 °C to 145 °C (dependant on molecular weight) and 220 °C [148], and 115 °C and 260 °C, respectively. All but EC are amorphous, which has a melting peak at 170 °C. Cellulose-based polymers are widely used in pharmaceutical formulations because of their capability to stabilise suspensions and emulsions, form films in coating formulations, act as thickening/bonding agent, and resist hydrolysis in gastrointestinal tract [149].
- **Chitosan** – Chitosan is a natural product obtained from the shells of crustaceans such as shrimps and crabs. It is a very versatile polysaccharide with good antimicrobial and mucoadhesive properties due to the positive charge of its amino group. Pure chitosan hydrogels can be printed using SSE technologies but to improve mechanical strength an additional crosslinking step is often required. Mechanical properties can also be improved by blending with other natural polymers such as sodium alginate and pectin. The negatively charged carboxyl group of sodium alginate and pectin forms polyelectrolyte complexes with chitosan, which significantly increase the viscosity of the formulation, improving shape fidelity after printing [150].
- **Pectin** – Pectin is another anionic polysaccharide that is commonly used as a thickening agent.
- **Alginate** – Alginate (Alg) is an anionic linear polysaccharide constituted of units of β-d-mannuronic acid and its epimer α-l-guluronic acid. Similar to chitosan, Alg has good gelling properties that are appropriate for SSE, but a post processing crosslinking step is often required. Alg chains can be easily crosslinked in the presence of divalent ions, notably Ca^{2+}.
- **Collagen and Gelatin** – Collagen is an insoluble, swellable structural protein composed mainly of glycine, proline, and hydroxyproline residues. Gelatin is an swellable anionic polymer, produced by the partial degradation of collagen that is soluble at and above physiological temperatures (37 °C) [151]. Both materials form hydrogels and have shown promise in SSE tissue engineering applications [152, 153].

1.3.7 Lipid Materials

Lipid-based excipients are obtained from vegetable oils, waxes, or fatty acids. They have recently been explored for pharmaceutical 3DP due to their ability to enhance the bioavailability of poorly soluble drugs. Lipids can do this by maintaining the drugs in the solubilised state, modifying gastric emptying, and by changing the drug distribution post absorption [154–157]. In addition, the degradation products of lipids can interact with endogenous components (i.e., bile salt, phospholipid, and cholesterol) and form a variety of colloidal structures that can act as a transport medium for the absorption of lipophilic drugs [158]. Lipids typically have a low melting temperature, so can be processed using thermal extrusion 3DP, quickly solidifying after printing. Gelucire® with Koliphor®, as surfactants, have been used to print drug loaded self-microemulsifying drug delivery systems (SMEDDS) for oral drug delivery. Theobroma oil (cocoa butter), the major ingredient in chocolate, has also been used for printing paediatric friendly formulations at low temperatures (45 °C) using corn syrup as a thickening agent [159]. Besides oral dosage forms, SMEDDS have also be employed to prepare suppositories based on coconut oil [160]. Recently, a similar thermal inkjet printing technique has been using to print high resolution beeswax tablets with complex microstructure geometry [161].

It should be highlighted that in most formulations of 3D-printed pharmaceuticals and devices, functional excipients are used to aid the processability and/or drug delivery performance (i.e., drug release rate, taste masking, solubility enhancement). Table 1.5 gives examples of functional excipients used in pharmaceutical 3D printing.

Table 1.5 *Additional functional excipients used in pharmaceutical 3DP.*

Excipient	Function	Examples
Plasticiser	Plasticisers are smaller molecules that act as a lubricant between segments of polymer chains increasing the material's flexibility and softness	Triethyl citrate [23, 162], Glycerol [163], Sorbitol [164], Kolliphor [165], TPGS [165], PEG [118], PEO [118], Mannitol, and Tween 80, API also often have a plasticising effect on the formulation, which can be significant, especially at high drug loadings [166]
Ph and Release Modifiers	Speed up, slow down, or control the release of the API	• Disintegrant – Kollidon CL-F [139], Tri-calcium phosphate [167], Sodium starch glycolate [168], Croscarmellose [168] • Channelling agent – Mannitol [23] • pH dependant – tribasic phosphate sodium [162]
Processing Aids	Reduce the tackiness of the formulation	PEG 6000 [23], dibutyl sebacate [169], Talc [162], magnesium stearate [170], magnesium carbonate, Sodium [20], Lauril Sulfate [168], Calcium stearate [171]
Solubility Enhancer	Enhance the amount of API that can be dissolved in formulation	Soluplus [172], PVP [173], Tween 80 [118]
Taste Masking	Delay the release of bitter API until after swallowing	Eudragit EPO [174]
Crystallisation Inhibitors	Slow the recrystallisation of API after amorphorisation	Polycarbophil [31], (PVP) K25 [31], poloxamer 188 [31], PEG3350[31]
Colorants	Change the colour of extruded material for aesthetic reasons or to increase UV/laser energy absorbance	Candurin Orange Amber [175], Candurin Gold Sheen [175], Food Colouring [175]

1.4 Conclusions and Future Perspectives

A wide range of materials, printing technologies, and applications have been covered in this chapter, giving insight into the material considerations necessary when designing a new DDD. Traditionally, material formulation optimisation has been done using a trial-and-error approach to establish which formulations are printable. However, as the research space has matured, so has understanding on the material properties that influence printability and drug release function. Rheology has shown to be a powerful predictive tool for predicting formulation printability and even drug release profiles for nozzle-based systems (TE, SSE, BJ), whilst, for powder-based systems (BJ and SLS), good powder flowability is critical for continuous printing. Material requirements for TE, SSE, PBF, and SLA 3DP are very different due to vastly differing printing conditions. As such, there is no single 'one' technique which is best suited for all drug delivery applications. Instead, specific application requirements (release properties, mechanical stability, design complexity, etc.) should be considered along with the API stability to target the most suitable 3DP technology and API carrier materials. As pharmaceutical 3DP becomes a more mature technology, we are seeing increased research effort, both in creating new materials specifically for improved printability and API solubilising potential, as well as more versatile printing technologies with less restrictive printability windows (i.e., granule/powder direct thermal deposition systems).

3D printing of pharmaceuticals has been shown to be advantageous for manufacturing personalised medicine, offering design flexibility, high complexity, on-demand, and cost-effective production. Whilst there are still challenges associated with the limited availability of excipients, the vast quantity and variability of drug delivery systems developed within the pharmaceutical 3DP literature suggests this is not the major factor delaying the adoption of pharmaceutical 3DP in industry. Wider technical challenges around the development of printing software, scalability, and the regulatory landscape still need to be overcome before the large-scale production of 3DP pharmaceutical is widespread.

References

[1] van der Vossen, A.A.-O., Al-Hassany, L., Buljac, S. et al. (2019). Manipulation of oral medication for children by parents and nurses occurs frequently and is often not supported by instructions. *Acta Paediatrica* **108** (1651–2227): 1475–1481.

[2] Martir, J., Flanagan, T., Mann, J., and Fotaki, N. (2017). Impact of co-administration of food and drink vehicles on the solubility of poorly soluble paediatric drugs.

[3] Palo, M., Holländer, J., Suominen, J. et al. (2017). 3D printed drug delivery devices: perspectives and technical challenges. *Expert Review of Medical Devices* **14** (9): 685–696.

[4] Seoane-Viaño, I., Januskaite, P., Alvarez-Lorenzo, C. et al. (2021). Semi-solid extrusion 3D printing in drug delivery and biomedicine: personalised solutions for healthcare challenges. *Journal of Controlled Release* **332**: 367–389. https://doi.org/10.1016/j.jconrel.2021.02.027.

[5] Prasad, L.K. and Smyth, H. (2016). 3D printing technologies for drug delivery: a review. *Drug Development and Industrial Pharmacy* **42** (7): 1019–1031. https://doi.org/10.3109/03639045.2015.1120743.

[6] West, T.G. and Bradbury, T.J. (2019). 3D printing: a case of ZipDose® technology–world's first 3D printing platform to obtain FDA approval for a pharmaceutical product. *3D and 4D Printing in Biomedical Applications: Process Engineering and Additive Manufacturing* 53–79.

[7] Ghanizadeh Tabriz, A., Nandi, U., Hurt, A.P. et al. (2021). 3D printed bilayer tablet with dual controlled drug release for tuberculosis treatment. *International Journal of Pharmaceutics* **593**: 120147. https://doi.org/10.1016/j.ijpharm.2020.120147.

[8] Goole, J. and Amighi, K. (2016). 3D printing in pharmaceutics: a new tool for designing customized drug delivery systems. *International Journal of Pharmaceutics* **499** (1): 376–394. https://doi.org/10.1016/j.ijpharm.2015.12.071.

[9] Zhang, B., Gleadall, A., Belton, P. et al. (2021). New insights into the effects of porosity, pore length, pore shape and pore alignment on drug release from extrusion-based additive manufactured pharmaceuticals. *Additive Manufacturing* **46**: 102196. https://doi.org/10.1016/j.addma.2021.102196.
[10] Long, J., Gholizadeh, H., Lu, J. et al. (2017). Application of fused deposition modelling (FDM) method of 3D printing in drug delivery. *Current Pharmaceutical Design* **23** (3): 433–439.
[11] Tan, D.K., Maniruzzaman, M., and Nokhodchi, A. (2018). Advanced pharmaceutical applications of hot-melt extrusion coupled with fused deposition modelling (FDM) 3D printing for personalised drug delivery. *Pharmaceutics* **10** (4): 203.
[12] Okwuosa, T.C., Stefaniak, D., Arafat, B. et al. (2016). A lower temperature FDM 3D printing for the manufacture of patient-specific immediate release tablets. *Pharmaceutical Research* **33** (11): 2704–2712.
[13] Alshahrani, S.M., Morott, J.T., Alshetaili, A.S. et al. (2015). Influence of degassing on hot-melt extrusion process. *European Journal of Pharmaceutical Sciences* **80**: 43–52.
[14] Li, S., Tian, Y., Jones, D.S., and Andrews, G.P. (2016). Optimising drug solubilisation in amorphous polymer dispersions: rational selection of hot-melt extrusion processing parameters. *AAPS Pharmscitech* **17** (1): 200–213.
[15] Aho, J., Boetker, J.P., Baldursdottir, S., and Rantanen, J. (2015). Rheology as a tool for evaluation of melt processability of innovative dosage forms. *International Journal of Pharmaceutics* **494** (2): 623–642.
[16] Thiry, J., Krier, F., and Evrard, B. (2015). A review of pharmaceutical extrusion: critical process parameters and scaling-up. *International Journal of Pharmaceutics* **479** (1): 227–240.
[17] Yang, Y., Wang, H., Li, H. et al. (2018). 3D printed tablets with internal scaffold structure using ethyl cellulose to achieve sustained ibuprofen release. *European Journal of Pharmaceutical Sciences* **115**: 11–18.
[18] Moseson, D.E. and Taylor, L.S. (2018). The application of temperature–composition phase diagrams for hot melt extrusion processing of amorphous solid dispersions to prevent residual crystallinity. *International Journal of Pharmaceutics* **553** (1): 454–466. https://doi.org/10.1016/j.ijpharm.2018.10.055.
[19] Verstraete, G., Samaro, A., Grymonpré, W. et al. (2018). 3D printing of high drug loaded dosage forms using thermoplastic polyurethanes. *International Journal of Pharmaceutics* **536** (1): 318–325.
[20] Kollamaram, G., Croker, D.M., Walker, G.M. et al. (2018). Low temperature fused deposition modeling (FDM) 3D printing of thermolabile drugs. *International Journal of Pharmaceutics* **545** (1): 144–152. https://doi.org/10.1016/j.ijpharm.2018.04.055.
[21] Goyanes, A., Buanz, A.B.M., Basit, A.W., and Gaisford, S. (2014). Fused-filament 3D printing (3DP) for fabrication of tablets. *International Journal of Pharmaceutics* **476** (1–2): 88–92.
[22] Tagami, T., Fukushige, K., Ogawa, E. et al. (2017). 3D printing factors important for the fabrication of polyvinyl-alcohol filament-based tablets. *Biological and Pharmaceutical Bulletin* **40** (3): 357–364.
[23] Beck, R.C.R., Chaves, P.S., Goyanes, A. et al. (2017). 3D printed tablets loaded with polymeric nanocapsules: an innovative approach to produce customized drug delivery systems. *International Journal of Pharmaceutics* **528** (1): 268–279. https://doi.org/10.1016/j.ijpharm.2017.05.074.
[24] Ibrahim, M., Barnes, M., McMillin, R. et al. (2019). 3D printing of metformin HCl PVA tablets by fused deposition modeling: drug loading, tablet design, and dissolution studies. *AAPS Pharmscitech* **20** (5): 1–11.
[25] Karalia, D., Siamidi, A., Karalis, V., and Vlachou, M. (2021). 3D-printed oral dosage forms: mechanical properties, computational approaches and applications. *Pharmaceutics* **13** (9): 1401.
[26] Goyanes, A., Buanz, A.B.M., Hatton, G.B. et al. (2015). 3D printing of modified-release aminosalicylate (4-ASA and 5-ASA) tablets. *European Journal of Pharmaceutics and Biopharmaceutics* **89**: 157–162. https://doi.org/10.1016/j.ejpb.2014.12.003.
[27] Kalepu, S. and Nekkanti, V. (2015). Insoluble drug delivery strategies: review of recent advances and business prospects. *Acta Pharmaceutica Sinica B* **5** (5): 442–453. https://doi.org/10.1016/j.apsb.2015.07.003.
[28] Loftsson, T. and Brewster, M.E. (2010). Pharmaceutical applications of cyclodextrins: basic science and product development. *Journal of Pharmacy and Pharmacology* **62** (11): 1607–1621. 10.1111/j.2042-7158.2010.01030.x.

[29] Vaka, S.R.K., Bommana, M.M., Desai, D. et al. (2014). Excipients for amorphous solid dispersions. In: *Amorphous Solid Dispersions*, 123–161. Springer.

[30] Crowley, M.M., Fredersdorf, A., Schroeder, B. et al. (2004). The influence of guaifenesin and ketoprofen on the properties of hot-melt extruded polyethylene oxide films. *European Journal of Pharmaceutical Sciences* **22** (5): 409–418.

[31] Bruce, C., Fegely, K.A., Rajabi-Siahboomi, A.R., and McGinity, J.W. (2007). Crystal growth formation in melt extrudates. *International Journal of Pharmaceutics* **341** (1–2): 162–172.

[32] Alhnan, M.A., Okwuosa, T.C., Sadia, M. et al. (2016). Emergence of 3D Printed Dosage Forms: Opportunities and Challenges. *Pharmaceutical Research* **33**: 1817–1832.

[33] Yu, D.G., Zhu, L.-M., Branford-White, C.J., and Yang, X.L. (2008). Three-dimensional printing in pharmaceutics: promises and problems. *Journal of Pharmaceutical Sciences* **97** (9): 3666–3690. https://doi.org/10.1002/jps.21284.

[34] Elbadawi, M., Gustaffson, T., Gaisford, S., and Basit, A.W. (2020). 3D printing tablets: predicting printability and drug dissolution from rheological data. *International Journal of Pharmaceutics* **590**: 119868. https://doi.org/10.1016/j.ijpharm.2020.119868.

[35] Solanki, N.G., Tahsin, M., Shah, A.V., and Serajuddin, A.T.M. (2018). Formulation of 3D printed tablet for rapid drug release by fused deposition modeling: screening polymers for drug release, drug-polymer miscibility and printability. *Journal of Pharmaceutical Sciences* **107** (1): 390–401. https://doi.org/10.1016/j.xphs.2017.10.021.

[36] Nasereddin, J.M., Wellner, N., Alhijjaj, M. et al. (2018). Development of a simple mechanical screening method for predicting the feedability of a pharmaceutical FDM 3D printing filament. *Pharmaceutical Research* **35** (8): 1–13.

[37] Elbadawi, M., Muñiz Castro, B., Gavins, F.K.H. et al. (2020). M3DISEEN: a novel machine learning approach for predicting the 3D printability of medicines. *International Journal of Pharmaceutics* **590**: 119837. https://doi.org/10.1016/j.ijpharm.2020.119837.

[38] Xu, P., Li, J., Meda, A. et al. (2020). Development of a quantitative method to evaluate the printability of filaments for fused deposition modeling 3D printing. *International Journal of Pharmaceutics* **588**: 119760. https://doi.org/10.1016/j.ijpharm.2020.119760.

[39] Goyanes, A., Allahham, N., Trenfield, S.J. et al. (2019). Direct powder extrusion 3D printing: fabrication of drug products using a novel single-step process. *International Journal of Pharmaceutics* **567**: 118471. https://doi.org/10.1016/j.ijpharm.2019.118471.

[40] McDonagh, T., Belton, P., and Qi, S. (2022). Direct granule feeding of thermal droplet deposition 3D printing of porous pharmaceutical solid dosage forms free of plasticisers. *Pharmaceutical Research* **39** (3): 599–610.

[41] Goyanes, A., Chang, H., Sedough, D. et al. (2015). Fabrication of controlled-release budesonide tablets via desktop (FDM) 3D printing. *International Journal of Pharmaceutics* **496** (2): 414–420. https://doi.org/10.1016/j.ijpharm.2015.10.039.

[42] Shi, K., Salvage, J.P., Maniruzzaman, M., and Nokhodchi, A. (2021). Role of release modifiers to modulate drug release from fused deposition modelling (FDM) 3D printed tablets. *International Journal of Pharmaceutics* **597**: 120315. https://doi.org/10.1016/j.ijpharm.2021.120315.

[43] Welsh, N.R., Malcolm, R.K., Devlin, B., and Boyd, P. (2019). Dapivirine-releasing vaginal rings produced by plastic freeforming additive manufacturing. *International Journal of Pharmaceutics* **572**: 118725. https://doi.org/10.1016/j.ijpharm.2019.118725.

[44] Genina, N., Holländer, J., Jukarainen, H. et al. (2016). Ethylene vinyl acetate (EVA) as a new drug carrier for 3D printed medical drug delivery devices. *European Journal of Pharmaceutical Sciences* **90**: 53–63. https://doi.org/10.1016/j.ejps.2015.11.005.

[45] Maroni, A., Melocchi, A., Parietti, F. et al. (2017). 3D printed multi-compartment capsular devices for two-pulse oral drug delivery. *Journal of Controlled Release* **268**: 10–18. https://doi.org/10.1016/j.jconrel.2017.10.008.

[46] Pereira, B.C., Isreb, A., Forbes, R.T. et al. (2019). Temporary plasticiser: a novel solution to fabricate 3D printed patient-centred cardiovascular 'Polypill' architectures. *European Journal of Pharmaceutics and Biopharmaceutics* **135**: 94–103. https://doi.org/10.1016/j.ejpb.2018.12.009.

[47] Khaled, S.A., Burley, J.C., Alexander, M.R., and Roberts, C.J. (2014). Desktop 3D printing of controlled release pharmaceutical bilayer tablets. *International Journal of Pharmaceutics* **461** (1–2): 105–111.

[48] Pandey, M., Choudhury, H., Fern, J.L.C. et al. (2020). 3D printing for oral drug delivery: a new tool to customize drug delivery. *Drug Delivery and Translational Research* **10** (4): 986–1001.

[49] Aguilar-de-leyva, Á., Linares, V.A.-O., Casas, M.A.-O., and Caraballo, I.A.-O. (2020). 3D printed drug delivery systems based on natural products. *Pharmaceutics* **12** (7): 620.
[50] Nourished personalised gummy vitamins (2022). https://get-nourished.com/. Accessed in Jan 2022.
[51] Cui, M., Pan, H., Fang, D. et al. (2020). Fabrication of high drug loading levetiracetam tablets using semi-solid extrusion 3D printing. *Journal of Drug Delivery Science and Technology* **57**: 101683. https://doi.org/10.1016/j.jddst.2020.101683.
[52] Wen, H., He, B., Wang, H. et al. (2019). Structure-based gastro-retentive and controlled-release drug delivery with novel 3D printing. *AAPS PharmSciTech* **20** (2): 68.
[53] Elkasabgy, N.A., Mahmoud, A.A., and Maged, A. (2020). 3D printing: an appealing route for customized drug delivery systems. *International Journal of Pharmaceutics* **588**: 119732. https://doi.org/10.1016/j.ijpharm.2020.119732.
[54] Wu, M., Zhang, Y., Huang, H. et al. (2020). Assisted 3D printing of microneedle patches for minimally invasive glucose control in diabetes. *Materials Science and Engineering: C* **117**: 111299. https://doi.org/10.1016/j.msec.2020.111299.
[55] Lewis, J.A. and Gratson, G.M. (2004). Direct writing in three dimensions. *Materials Today* **7** (7–8): 32–39.
[56] Jones, N. (2012). Science in three dimensions: the print revolution. *Nature* **487** (7405): 22–23.
[57] Zidan, A., Alayoubi, A., Coburn, J. et al. (2019). Extrudability analysis of drug loaded pastes for 3D printing of modified release tablets. *International Journal of Pharmaceutics* **554**: 292–301.
[58] Khaled, S.A., Burley, J.C., Alexander, M.R. et al. (2015). 3D printing of five-in-one dose combination polypill with defined immediate and sustained release profiles. *Journal of Controlled Release* **217**: 308–314.
[59] Real, J.P., Barberis, M.E., Camacho, N.M. et al. (2020). Design of novel oral ricobendazole formulation applying melting solidification printing process (MESO-PP): an innovative solvent-free alternative method for 3D printing using a simplified concept and low temperature. *International Journal of Pharmaceutics* **587**: 119653.
[60] Goyanes, A., Madla, C.M., Umerji, A. et al. (2019). Automated therapy preparation of isoleucine formulations using 3D printing for the treatment of MSUD: first single-centre, prospective, crossover study in patients. *International Journal of Pharmaceutics* **567**: 118497. https://doi.org/10.1016/j.ijpharm.2019.118497.
[61] Sjöholm, E. and Sandler, N. (2019). Additive manufacturing of personalized orodispersible warfarin films. *International Journal of Pharmaceutics* **564**: 117–123. https://doi.org/10.1016/j.ijpharm.2019.04.018.
[62] Khaled, S.A., Burley, J.C., Alexander, M.R. et al. (2015). 3D printing of tablets containing multiple drugs with defined release profiles. *International Journal of Pharmaceutics* **494** (2): 643–650.
[63] Zhu, M., Li, K., Zhu, Y. et al. (2015). 3D-printed hierarchical scaffold for localized isoniazid/rifampin drug delivery and osteoarticular tuberculosis therapy. *Acta Biomaterialia* **16**: 145–155. https://doi.org/10.1016/j.actbio.2015.01.034.
[64] Lee, K.-J., Kang, A., Delfino, J.J. et al. (2003). Evaluation of critical formulation factors in the development of a rapidly dispersing captopril oral dosage form. *Drug Development and Industrial Pharmacy* **29** (9): 967–979.
[65] Trenfield, S.J., Madla, C.M., Basit, A.W., and Gaisford, S. (2018). Binder jet printing in pharmaceutical manufacturing. In: *3D Printing of Pharmaceuticals*, 41–54. Springer.
[66] Vock, S., Klöden, B., Kirchner, A. et al. (2019). Powders for powder bed fusion: a review. *Progress in Additive Manufacturing* **4** (4): 383–397. https://doi.org/10.1007/s40964-019-00078-6.
[67] Mullarney, M.P., Beach, L.E., Davé, R.N. et al. (2011). Applying dry powder coatings to pharmaceutical powders using a comil for improving powder flow and bulk density. *Powder Technology* **212** (3): 397–402.
[68] Schulze, D. (2008). *Powders and Bulk Solids. Behaviour, Characterization, Storage and Flow*, **22**. Springer.
[69] Pohlman, N.A., Roberts, J.A., and Gonser, M.J. (2012). Characterization of titanium powder: microscopic views and macroscopic flow. *Powder Technology* **228**: 141–148.
[70] Fu, X., Huck, D., Makein, L. et al. (2012). Effect of particle shape and size on flow properties of lactose powders. *Particuology* **10** (2): 203–208.

[71] Ziegelmeier, S., Christou, P., Wöllecke, F. et al. (2015). An experimental study into the effects of bulk and flow behaviour of laser sintering polymer powders on resulting part properties. *Journal of Materials Processing Technology* **215**: 239–250. https://doi.org/10.1016/j.jmatprotec.2014.07.029.
[72] Schwedes, J. (2003). Review on testers for measuring flow properties of bulk solids. *Granular Matter* **5** (1): 1–43.
[73] Mohamed, E.M., Barakh Ali, S.F., Rahman, Z. et al. (2020). Formulation optimization of selective laser sintering 3D-printed tablets of clindamycin palmitate hydrochloride by response surface methodology. *AAPS Pharmscitech* **21** (6): 1–15.
[74] Vehring, R. (2008). Pharmaceutical particle engineering via spray drying. *Pharmaceutical Research* **25** (5): 999–1022.
[75] Maa, Y.-F., Costantino, H.R., Nguyen, P.-A., and Hsu, C.C. (1997). The effect of operating and formulation variables on the morphology of spray-dried protein particles. *Pharmaceutical Development and Technology* **2** (3): 213–223.
[76] Awad, A., Fina, F., Trenfield, S.J. et al. (2019). 3D printed pellets (miniprintlets): a novel, multi-drug, controlled release platform technology. *Pharmaceutics* **11** (4): 148.
[77] Sen, K., Mehta, T., Sansare, S. et al. (2021). Pharmaceutical applications of powder-based binder jet 3D printing process: a review. *Advanced Drug Delivery Reviews* **177**: 113943. https://doi.org/10.1016/j.addr.2021.113943.
[78] Sen, K., Mukherjee, R., Sansare, S. et al. (2021). Impact of powder-binder interactions on 3D printability of pharmaceutical tablets using drop test methodology. *European Journal of Pharmaceutical Sciences* **160**: 105755. https://doi.org/10.1016/j.ejps.2021.105755.
[79] Wang, -C.-C., Tejwani, M.R., Roach, W.J. et al. (2006). Development of near zero-order release dosage forms using three-dimensional printing (3-DP™) technology. *Drug Development and Industrial Pharmacy* **32** (3): 367–376.
[80] Kozakiewicz-Latała, M., Nartowski, K.P., Dominik, A. et al. (2022). Binder jetting 3D printing of challenging medicines: from low dose tablets to hydrophobic molecules. *European Journal of Pharmaceutics and Biopharmaceutics* **170**: 144–159. https://doi.org/10.1016/j.ejpb.2021.11.001.
[81] Low, K.H., Leong, K.F., Chua, C.K. et al. (2001). Characterization of SLS parts for drug delivery devices. *Rapid Prototyping Journal* **7** (5): 262–268.
[82] Leong, K.F., Chua, C.K., and Gui, W.S. (2006). Building porous biopolymeric microstructures for controlled drug delivery devices using selective laser sintering. *The International Journal of Advanced Manufacturing Technology* **31** (5): 483–489.
[83] Fina, F., Madla, C.M., Goyanes, A. et al. (2018). Fabricating 3D printed orally disintegrating printlets using selective laser sintering. *International Journal of Pharmaceutics* **541** (1–2): 101–107.
[84] Fina, F., Goyanes, A., Gaisford, S., and Basit, A.W. (2017). Selective laser sintering (SLS) 3D printing of medicines. *International Journal of Pharmaceutics* **529** (1–2): 285–293.
[85] Salmoria, G.V., Cardenuto, M.R., Roesler, C.R.M. et al. (2016). PCL/ibuprofen implants fabricated by selective laser sintering for orbital repair. *Procedia CIRP* **49**: 188–192.
[86] Salmoria, G.V., Klauss, P., Zepon, K. et al. (2012). Development of functionally-graded reservoir of PCL/PG by selective laser sintering for drug delivery devices: this paper presents a selective laser sintering-fabricated drug delivery system that contains graded progesterone content. *Virtual and Physical Prototyping* **7** (2): 107–115.
[87] Yu, D.-G., Shen, X.-X., Branford-White, C. et al. (2009). Novel oral fast-disintegrating drug delivery devices with predefined inner structure fabricated by three-dimensional printing. *Journal of Pharmacy and Pharmacology* **61** (3): 323–329.
[88] Tian, P., Yang, F., Xu, Y. et al. (2018). Oral disintegrating patient-tailored tablets of warfarin sodium produced by 3D printing. *Drug Development and Industrial Pharmacy* **44** (12): 1918–1923.
[89] Gbureck, U., Vorndran, E., Müller, F.A., and Barralet, J.E. (2007). Low temperature direct 3D printed bioceramics and biocomposites as drug release matrices. *Journal of Controlled Release* **122** (2): 173–180.
[90] Salmoria, G.V., Vieira, F.E., Muenz, E.A. et al. (2018). Additive manufacturing of PE/fluorouracil/progesterone intrauterine device for endometrial and ovarian cancer treatments. *Polymer Testing* **71**: 312–317.

[91] Hull, C.W. (1986). Apparatus for production of three-dimensional objects by stereolithography. In: US patent, Editor. US 1986.
[92] Xu, X., Awad, A., Robles-Martinez, P. et al. (2021). Vat photopolymerization 3D printing for advanced drug delivery and medical device applications. *Journal of Controlled Release* **329**: 743–757.
[93] Geever, T., Killion, J., Grehan, L. et al. (2014). Effect of photoinitiator concentration on the properties of polyethylene glycol based hydrogels for potential regenerative medicine applications. *Advances in Environmental Biology* **8**: 7–17.
[94] McDermott, S., Walsh, J.E., and Howard, R.G. (2005). A novel application of UV-LEDs in the contact lens manufacturing process. *International Society for Optics and Photonics* **5826**: 119–130.
[95] Юкатан (2015). Process of photopolymerisation. In: scheme 1 P-p, Editor 2015.
[96] Guillaume, O., Geven, M.A., Sprecher, C.M. et al. (2017). Surface-enrichment with hydroxyapatite nanoparticles in stereolithography-fabricated composite polymer scaffolds promotes bone repair. *Acta biomaterialia* **54**: 386–398.
[97] Economidou, S.N., Pere, C.P.P., Reid, A. et al. (2019). 3D printed microneedle patches using stereolithography (SLA) for intradermal insulin delivery. *Materials Science and Engineering: C* **102**: 743–755. https://doi.org/10.1016/j.msec.2019.04.063.
[98] Uddin, M.J., Scoutaris, N., Economidou, S.N. et al. (2020). 3D printed microneedles for anticancer therapy of skin tumours. *Materials Science and Engineering: C* **107**: 110248. https://doi.org/10.1016/j.msec.2019.110248.
[99] Healy, A.V., Fuenmayor, E., Doran, P. et al. (2019). Additive manufacturing of personalized pharmaceutical dosage forms via stereolithography. *Pharmaceutics* **11** (12). https://doi.org/10.3390/pharmaceutics11120645.
[100] Wang, J., Goyanes, A., Gaisford, S., and Basit, A.W. (2016). Stereolithographic (SLA) 3D printing of oral modified-release dosage forms. *International Journal of Pharmaceutics* **503** (1–2): 207–212.
[101] Goyanes, A., Det-Amornrat, U., Wang, J. et al. (2016). 3D scanning and 3D printing as innovative technologies for fabricating personalized topical drug delivery systems. *Journal of Controlled Release* **234**: 41–48.
[102] Gittard, S.D., Ovsianikov, A., Akar, H. et al. (2010). Two photon polymerization-micromolding of polyethylene glycol-gentamicin sulfate microneedles. *Advanced Engineering Materials* **12** (4): B77–B82.
[103] Melchels, F.P.W., Feijen, J., and Grijpma, D.W. (2010). A review on stereolithography and its applications in biomedical engineering. *Biomaterials* **31** (24): 6121–6130.
[104] Martinez, P.R., Goyanes, A., Basit, A.W., and Gaisford, S. (2017). Fabrication of drug-loaded hydrogels with stereolithographic 3D printing. *International Journal of Pharmaceutics* **532** (1): 313–317. https://doi.org/10.1016/j.ijpharm.2017.09.003.
[105] Vehse, M., Petersen, S., Sternberg, K. et al. (2017). Drug delivery from poly(ethylene glycol) diacrylate scaffolds produced by DLC-based micro-stereolithography. *International Journal of Pharmaceutics* **519** (1–2): 186–197.
[106] Steinbach, M., Gartz, M., and Hirsch, R. (2020). Design and characterization of 3D printable photopolymer resin containing poly(2-hydroxyethyl methacrylate) for controlled drug release. *Journal of Drug Delivery Science and Technology* **59**: 101850. https://doi.org/10.1016/j.jddst.2020.101850.
[107] Xu, X., Goyanes, A., Trenfield, S.J. et al. (2021). Stereolithography (SLA) 3D printing of a bladder device for intravesical drug delivery. *Materials Science and Engineering: C* **120**: 111773. https://doi.org/10.1016/j.msec.2020.111773.
[108] Flory, P.J. (1953). *Principles of Polymer Chemistry*. Cornell University Press.
[109] Usp, C. (2008). The United States Pharmacopeia. *National Formulary* **14**.
[110] European Pharmacopoeia (2010). European Directorate for the Quality of, and Healthcare. *European Pharmacopoeia* **1**: 2010: Council of Europe.
[111] Goyanes, A., Martinez, P.R., Buanz, A. et al. (2015). Effect of geometry on drug release from 3D printed tablets. *International Journal of Pharmaceutics* **494** (2): 657–663.

[112] Holland, B.J. and Hay, J.N. (2001). The thermal degradation of poly(vinyl alcohol). *Polymer* **42** (16): 6775–6783. https://doi.org/10.1016/S0032-3861(01)00166-5.
[113] Goyanes, A., Kobayashi, M., Martínez-Pacheco, R. et al. (2016). Fused-filament 3D printing of drug products: microstructure analysis and drug release characteristics of PVA-based caplets. *International Journal of Pharmaceutics* **514** (1): 290–295.
[114] El Aita, I., Breitkreutz, J., and Quodbach, J. (2019). On-demand manufacturing of immediate release levetiracetam tablets using pressure-assisted microsyringe printing. *European Journal of Pharmaceutics and Biopharmaceutics* **134**: 29–36.
[115] Zhang, J., Allardyce, B.J., Rajkhowa, R. et al. (2021). 3D printing of silk powder by Binder Jetting technique. *Additive Manufacturing* **38**: 101820. https://doi.org/10.1016/j.addma.2020.101820.
[116] LaFountaine, J.S., McGinity, J.W., and Williams, R.O. (2016). Challenges and strategies in thermal processing of amorphous solid dispersions: a review. *AAPS PharmSciTech* **17** (1): 43–55.
[117] Bruce, L.D., Shah, N.H., Waseem Malick, A. et al. (2005). Properties of hot-melt extruded tablet formulations for the colonic delivery of 5-aminosalicylic acid. *European Journal of Pharmaceutics and Biopharmaceutics* **59** (1): 85–97. https://doi.org/10.1016/j.ejpb.2004.06.007.
[118] Alhijjaj, M., Belton, P., and Qi, S. (2016). An investigation into the use of polymer blends to improve the printability of and regulate drug release from pharmaceutical solid dispersions prepared via fused deposition modeling (FDM) 3D printing. *European Journal of Pharmaceutics and Biopharmaceutics* **108**: 111–125. https://doi.org/10.1016/j.ejpb.2016.08.016.
[119] Parikh, T., Gupta, S.S., Meena, A., and Serajuddin, A.T.M. (2016). Investigation of thermal and viscoelastic properties of polymers relevant to hot melt extrusion-III: polymethacrylates and polymethacrylic acid based polymers. *Journal of Excipients and Food Chemicals* **5** (1): 1003.
[120] Lin, S.Y. and Yu, H.L. (1999). Thermal stability of methacrylic acid copolymers of Eudragits L, S, and L30D and the acrylic acid polymer of carbopol. *Journal of Polymer Science Part A: Polymer Chemistry* **37** (13): 2061–2067.
[121] Verma, P., Gupta, R.N., Jha, A.K., and Pandey, R. (2013). Development, *in vitro* and *in vivo* characterization of Eudragit RL 100 nanoparticles for improved ocular bioavailability of acetazolamide. *Drug Delivery* **20** (7): 269–276.
[122] Diarra, M., Pourroy, G., Boymond, C., and Muster, D. (2003). Fluoride controlled release tablets for intrabuccal use. *Biomaterials* **24** (7): 1293–1300.
[123] Hao, S., Wang, B., Wang, Y. et al. (2013). Preparation of Eudragit L 100-55 enteric nanoparticles by a novel emulsion diffusion method. *Colloids and Surfaces B: Biointerfaces* **108**: 127–133.
[124] Quinteros, D.A., Manzo, R.H., and Allemandi, D.A. (2010). Design of a colonic delivery system based on cationic polymethacrylate (Eudragit E100)-mesalamine complexes. *Drug Delivery* **17** (4): 208–213.
[125] Yoo, J.-W., Giri, N., and Lee, C.H. (2011). pH-sensitive Eudragit nanoparticles for mucosal drug delivery. *International Journal of Pharmaceutics* **403** (1–2): 262–267.
[126] Verma, P.R.P. and Iyer, S.S. (2000). Transdermal delivery of propranolol using mixed grades of Eudragit: design and *in vitro* and *in vivo* evaluation. *Drug Development and Industrial Pharmacy* **26** (4): 471–476.
[127] Katstra, W.E., Palazzolo, R.D., Rowe, C.W. et al. (2000). Oral dosage forms fabricated by three dimensional printing™. *Journal of Controlled Release* **66** (1): 1–9.
[128] Rowe, C., Katstra, W., Palazzolo, R. et al. (2000). Multimechanism oral dosage forms fabricated by three dimensional printing™. *Journal of Controlled Release* **66** (1): 11–17.
[129] Holländer, J., Genina, N., Jukarainen, H. et al. (2016). Three-dimensional printed PCL-based implantable prototypes of medical devices for controlled drug delivery. *Journal of Pharmaceutical Sciences* **105** (9): 2665–2676. https://doi.org/10.1016/j.xphs.2015.12.012.
[130] Gunatillake, P., Mayadunne, R., and Adhikari, R. (2006). Recent developments in biodegradable synthetic polymers. *Biotechnology Annual Review* **12**: 301–347.

[131] Wang, K.A.-O. and Deng, Q.A.-O. (2019). The thermal and mechanical properties of poly(ethylene-co-vinyl acetate) random copolymers (PEVA) and its covalently crosslinked analogues. *Polymers* **11** (6): 1055.
[132] Vrandečić, N., Erceg, M., Jakić, M., and Klarić, I. (2010). Kinetic analysis of thermal degradation of poly(ethylene glycol) and poly(ethylene oxide)s of different molecular weight. *Thermochimica Acta* **498**: 71–80. https://doi.org/10.1016/j.tca.2009.10.005.
[133] Stewart, S.A., Domínguez-Robles, J., McIlorum, V.J. et al. (2020). Development of a biodegradable subcutaneous implant for prolonged drug delivery using 3D printing. *Pharmaceutics* **12** (2). https://doi.org/10.3390/pharmaceutics12020105.
[134] Gilding, D.K. and Reed, A.M. (1979). Biodegradable polymers for use in surgery: polyglycolic/poly(actic acid) homo- and copolymers: 1. *Polymer* **20** (12): 1459–1464. https://doi.org/10.1016/0032-3861(79)90009-0.
[135] Ulery, B.D., Nair, L.S., and Laurencin, C.T. (2011). Biomedical applications of biodegradable polymers. *Journal of Polymer Science Part B: Polymer Physics* **49** (12): 832–864. https://doi.org/10.1002/polb.22259.
[136] Wang, G. and Li, A. (2008). Thermal decomposition and kinetics of mixtures of polylactic acid and biomass during copyrolysis. *Chinese Journal of Chemical Engineering* **16** (6): 929–933. https://doi.org/10.1016/S1004-9541(09)60018-5.
[137] Samantaray, P.K., Little, A., Haddleton, D.M. et al. (2020). Poly(glycolic acid) (PGA): a versatile building block expanding high performance and sustainable bioplastic applications. *Green Chemistry* **22** (13): 4055–4081. https://doi.org/10.1039/D0GC01394C.
[138] Seoane-Viaño, I., Gómez-Lado, N., Lázare-Iglesias, H. et al. (2020). 3D printed tacrolimus rectal formulations ameliorate colitis in an experimental animal model of inflammatory bowel disease. *Biomedicines* **8** (12). https://doi.org/10.3390/biomedicines8120563.
[139] Zhang, J., Feng, X., Patil, H. et al. (2017). Coupling 3D printing with hot-melt extrusion to produce controlled-release tablets. *International Journal of Pharmaceutics* **519** (1): 186–197. https://doi.org/10.1016/j.ijpharm.2016.12.049.
[140] Yang, Y., Wang, H., Xu, X., and Yang, G. (2021). Strategies and mechanisms to improve the printability of pharmaceutical polymers Eudragit® EPO and Soluplus®. *International Journal of Pharmaceutics* **599**: 120410.
[141] Azad, M.A., Olawuni, D., Kimbell, G. et al. (2020). Polymers for extrusion-based 3D printing of pharmaceuticals: a holistic materials–process perspective. *Pharmaceutics* **12** (2): 124.
[142] Fisher, J.P., Dean, D., and Mikos, A.G. (2002). Photocrosslinking characteristics and mechanical properties of diethyl fumarate/poly(propylene fumarate) biomaterials. *Biomaterials* **23** (22): 4333–4343. https://doi.org/10.1016/S0142-9612(02)00178-3.
[143] Zhou, W., Qiao, Z., Nazarzadeh Zare, E. et al. (2020). 4D-printed dynamic materials in biomedical applications: chemistry, challenges, and their future perspectives in the clinical sector. *Journal of Medicinal Chemistry* **63** (15): 8003–8024.
[144] Brady, J., Dürig, T., Lee, P.I., and Li, J.X. (2017). Polymer properties and characterization. In: *Developing Solid Oral Dosage Forms*, 181–223. Elsevier.
[145] Zema, L., Loreti, G., Melocchi, A. et al. (2013). Gastroresistant capsular device prepared by injection molding. *International Journal of Pharmaceutics* **440** (2): 264–272.
[146] Thakkar, R., Thakkar, R., Pillai, A. et al. (2020). Systematic screening of pharmaceutical polymers for hot melt extrusion processing: a comprehensive review. *International Journal of Pharmaceutics* **576**: 118989.
[147] Chai, X., Chai, H., Wang, X. et al. (2017). Fused deposition modeling (FDM) 3D printed tablets for intragastric floating delivery of domperidone. *Scientific Reports* **7** (1): 1–9.
[148] Gómez-Carracedo, A., Alvarez-Lorenzo, C., Gómez-Amoza, J.L., and Concheiro, A. (2003). Chemical structure and glass transition temperature of non-ionic cellulose ethers. *Journal of Thermal Analysis and Calorimetry* **73** (2): 587–596.
[149] Chavan, R.B., Rathi, S., Vgss, J., and Shastri, N.R. (2019). Cellulose based polymers in development of amorphous solid dispersions. *Asian Journal of Pharmaceutical Sciences* **14** (3): 248–264.

[150] Michailidou, G., Terzopoulou, Z., Kehagia, A. et al. (2021). Preliminary evaluation of 3D printed chitosan/pectin constructs for biomedical applications. *Marine Drugs* **19** (1): 36.

[151] Farrugia, C.A. and Groves, M.J. (1999). Gelatin behaviour in dilute aqueous solution: designing a nanoparticulate formulation. *Journal of Pharmacy and Pharmacology* **51** (6): 643–649.

[152] Inzana, J.A., Olvera, D., Fuller, S.M. et al. (2014). 3D printing of composite calcium phosphate and collagen scaffolds for bone regeneration. *Biomaterials* **35** (13): 4026–4034.

[153] Billiet, T., Gevaert, E., De Schryver, T. et al. (2014). The 3D printing of gelatin methacrylamide cell-laden tissue-engineered constructs with high cell viability. *Biomaterials* **35** (1): 49–62.

[154] Porter, C.J.H., Trevaskis, N.L., and Charman, W.N. (2007). Lipids and lipid-based formulations: optimizing the oral delivery of lipophilic drugs. *Nature Reviews Drug Discovery* **6** (3): 231–248.

[155] Vithani, K., Hawley, A., Jannin, V. et al. (2017). Inclusion of digestible surfactants in solid SMEDDS formulation removes lag time and influences the formation of structured particles during digestion. *AAPS Journal* **19**: 754–764.

[156] Pouton, C.W. (2000). Lipid formulations for oral administration of drugs: non-emulsifying, self-emulsifying and 'self-microemulsifying' drug delivery systems. *European Journal of Pharmaceutical Sciences* **11**: S93–S98.

[157] Kang, B.K., Lee, J.S., Chon, S.K. et al. (2004). Development of self-microemulsifying drug delivery systems (SMEDDS) for oral bioavailability enhancement of simvastatin in Beagle dogs. *International Journal of Pharmaceutics* **274** (1–2): 65–73.

[158] Vithani, K., Goyanes, A., Jannin, V. et al. (2019). A proof of concept for 3D printing of solid lipid-based formulations of poorly water-soluble drugs to control formulation dispersion kinetics. *Pharmaceutical Research* **36** (7): 102. https://doi.org/10.1007/s11095-019-2639-y.

[159] Karavasili, C., Gkaragkounis, A., Moschakis, T. et al. (2020). Pediatric-friendly chocolate-based dosage forms for the oral administration of both hydrophilic and lipophilic drugs fabricated with extrusion-based 3D printing. *European Journal of Pharmaceutical Sciences* **147**: 105291. https://doi.org/10.1016/j.ejps.2020.105291.

[160] Seoane-Viaño, I., Ong, J.J., Luzardo-Álvarez, A. et al. (2021). 3D printed tacrolimus suppositories for the treatment of ulcerative colitis. *Asian Journal of Pharmaceutical Sciences* **16** (1): 110–119.

[161] Kyobula, M., Adedeji, A., Alexander, M.R. et al. (2017). 3D inkjet printing of tablets exploiting bespoke complex geometries for controlled and tuneable drug release. *Journal of Controlled Release* **261**: 207–215. https://doi.org/10.1016/j.jconrel.2017.06.025.

[162] Okwuosa, T.C., Pereira, B.C., Arafat, B. et al. (2017). Fabricating a shell-core delayed release tablet using dual FDM 3D printing for patient-centred therapy. *Pharmaceutical Research* **34** (2): 427–437.

[163] Melocchi, A., Uboldi, M., Inverardi, N. et al. (2019). Expandable drug delivery system for gastric retention based on shape memory polymers: development via 4D printing and extrusion. *International Journal of Pharmaceutics* **571**: 118700. https://doi.org/10.1016/j.ijpharm.2019.118700.

[164] Palekar, S., Nukala, P.K., Mishra, S.M. et al. (2019). Application of 3D printing technology and quality by design approach for development of age-appropriate pediatric formulation of baclofen. *International Journal of Pharmaceutics* **556**: 106–116. https://doi.org/10.1016/j.ijpharm.2018.11.062.

[165] Ilyés, K., Kovács, N.K., Balogh, A. et al. (2019). The applicability of pharmaceutical polymeric blends for the fused deposition modelling (FDM) 3D technique: material considerations–printability–process modulation, with consecutive effects on *in vitro* release, stability and degradation. *European Journal of Pharmaceutical Sciences* **129**: 110–123. https://doi.org/10.1016/j.ejps.2018.12.019.

[166] Prasad, E., Islam, M.T., Goodwin, D.J. et al. (2019). Development of a hot-melt extrusion (HME) process to produce drug loaded Affinisol™ 15LV filaments for fused filament fabrication (FFF) 3D printing. *Additive Manufacturing* **29**: 100776.

[167] Sadia, M., Sośnicka, A., Arafat, B. et al. (2016). Adaptation of pharmaceutical excipients to FDM 3D printing for the fabrication of patient-tailored immediate release tablets. *International Journal of Pharmaceutics* **513** (1–2): 659–668.

[168] Ehtezazi, T., Algellay, M., Islam, Y. et al. (2018). The application of 3D printing in the formulation of multilayered fast dissolving oral films. *Journal of Pharmaceutical Sciences* **107** (4): 1076–1085.

[169] Saviano, M., Aquino, R.P., Del Gaudio, P. et al. (2019). Poly(vinyl alcohol) 3D printed tablets: the effect of polymer particle size on drug loading and process efficiency. *International Journal of Pharmaceutics* **561**: 1–8. https://doi.org/10.1016/j.ijpharm.2019.02.025.

[170] Goyanes, A., Fernández-Ferreiro, A., Majeed, A. et al. (2018). PET/CT imaging of 3D printed devices in the gastrointestinal tract of rodents. *International Journal of Pharmaceutics* **536** (1): 158–164.

[171] Gioumouxouzis, C.I., Baklavaridis, A., Katsamenis, O.L. et al. (2018). A 3D printed bilayer oral solid dosage form combining metformin for prolonged and glimepiride for immediate drug delivery. *European Journal of Pharmaceutical Sciences* **120**: 40–52.

[172] Zhang, J., Yang, W., Vo, A.Q. et al. (2017). Hydroxypropyl methylcellulose-based controlled release dosage by melt extrusion and 3D printing: structure and drug release correlation. *Carbohydrate Polymers* **177**: 49–57.

[173] Kimura, S-i., Ishikawa, T., Iwao, Y. et al. (2019). Fabrication of zero-order sustained-release floating tablets via fused depositing modeling 3D printer. *Chemical and Pharmaceutical Bulletin* **67** (9): 992–999.

[174] Drašković, M., Medarević, D., Aleksić, I., and Parojčić, J. (2017). *In vitro* and *in vivo* investigation of taste-masking effectiveness of Eudragit E PO as drug particle coating agent in orally disintegrating tablets. *Drug Development and Industrial Pharmacy* **43** (5): 723–731.

[175] Goyanes, A., Scarpa, M., Kamlow, M. et al. (2017). Patient acceptability of 3D printed medicines. *International Journal of Pharmaceutics* **530** (1): 71–78. https://doi.org/10.1016/j.ijpharm.2017.07.064.

2

The Use of Microstructure Design and 3D Printing for Tailored Drug Release

Bin Zhang[1], Thomas McDonagh[1], Joey Yan[2], Andy Glendale[2], Richard Bib[3], Peter Belton[4] and Sheng Qi[1]

[1] School of Pharmacy, University of East Anglia, Norwich, UK
[2] School of Mechanical, Electrical and Manufacturing Engineering, Loughborough University, Loughborough, UK
[3] School of Design & Creative Arts, Loughborough University, Loughborough, UK
[4] School of Chemistry, University of East Anglia, Norwich, UK

2.1 Introduction

Conventional pharmaceutical products are mass produced with standardised techniques to ensure efficient and reproducible manufacturing. However, there is limited flexibility in conventional production formulations to suit variability in the population (e.g., patient age, gender, and weight) that can lead to inadequate dosing and side effects [1]. Traditional prescriptions are based on a 'one size fits all' approach, where in reality, individuals' drug dosing requirements vary significantly. Moreover, in the case of polypharmacy, in which multiple medications are prescribed per day, there is an increased risk of drug–drug interactions or adverse drug reactions. These problems are particularly common amongst the elderly population, as the average older adult takes four or more drugs each day in the USA [2, 3]. There is a large non-compliance issue in the elderly population associated with the need to take multiple tablets each day. About two-thirds of patients are non-adherent, which costs the UK National Health Service over £300 million annually [4]. Personalised

3D Printing of Pharmaceutical and Drug Delivery Devices: Progress from Bench to Bedside, First Edition.
Edited by Dimitrios A. Lamprou, Dennis Douroumis and Sheng Qi.

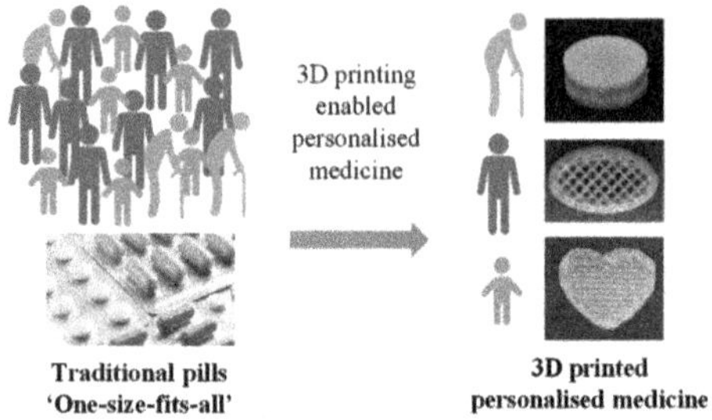

Figure 2.1 3D printing for customisation. The flexibility of 3D-printing processes enables improved customisation and small-scale production of drug-loaded devices. This can include the fabrication of tablets that are tailored to patient groups following stratification. (Source: okskaz / Adobe Stock.)

medicine can address many of the issues that we are currently facing with traditional prescriptions.

3D-printed personalised tablets can help solve the clinical problems we are facing by creating products that were not previously possible using traditional manufacturing methods [5]. The advent of 3D-printing technologies in drug product design has motivated an increased interest in fabricating personalised dosage forms that are tailored to patients and enable enhanced treatment (Figure 2.1). The influence of 3D printing is growing in the health and social care sectors, particularly in acute hospital environments, and is emerging as a critical enabler of personalised and potentially cheaper treatment.

2.2 3D-Printing Technologies

Various 3D-printing techniques have been applied to the fabrication of pharmaceutical devices, and each printing technique has its own associated advantages in terms of material and drug processibility, resolution, volume of production, and processing steps. Ragelle et al. [6] divided 3D printing into two broad categories based on the mechanism of material fabrication: deposition-based technologies and stimulus-triggered technologies. Deposition-based technologies include material extrusion (also known as fused deposition modelling), direct ink writing/semi-solid extrusion, and inkjet printing, whilst the stimulus-triggered technologies include binder jetting and vat polymerisation (such as stereolithography and two-photon polymerisation) [6–9]. Deposition-based 3D-printing techniques are commonly used for drug-related medical devices. Figure 2.2 shows the general class of deposition-based 3D printing. Table 2.1

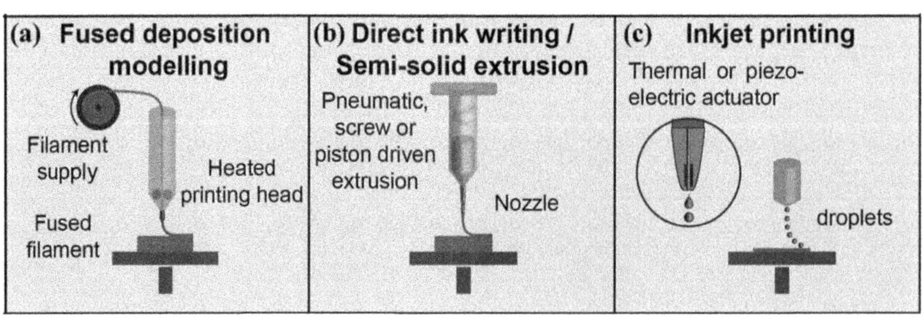

Figure 2.2 The general class of deposition-based 3D-printing approaches for the design of drug-loaded medical devices [6]. (Source: [6] with permission of Elsevier.)

Table 2.1 *Deposition-based 3D-printing technologies related to drug-loaded medical device [6, 9, 10, 12, 14].*

3D printing techniques	Resolution [mm]	RTM ratio[a]	Applied materials	Drug loading mechanism	Advantage	Disadvantage
Melt filament-based 3D printing	200	1×10^{-3} m^2 min^{-1}	Thermoplastics or drug-loaded thermoplastic polymers	Active loading through hot-melt extrusion (high drug loading and thermal stable drugs). Passive diffusion into the filament (low drug loading)	Low material costs. Low-cost printers. Simple to use	Hot-melt extrusion equipment is required for drug-loaded filament fabrication. The high temperature required during extrusion makes possible the denaturation of proteins or thermal sensitive drug molecules. The addition of the drug to filament may affect mechanical properties and flowability (printing parameters)
Melt droplet-based 3D printing	200	Circa 1×10^{-3} m^2 min^{-1}	Both hydrophilic and hydrophobic thermoplastic polymers	Active loading through drug directly mixed with powdered materials	3D printing and hot-melt extrusion is in one step	Relatively high cost. The high temperature required during extrusion and mixing stress induced destabilisation
Direct ink writing/ Semi-solid extrusion	100	0.5×10^{-3} m^2 min^{-1}	Polymer-based hydrogel	The aqueous environment of the hydrogel for hydrophilic drugs, or hydrophobic carriers	Broad materials available. Low material cost	Rheological properties of the hydrogel need to be engineered (shear-thinning, self-healing). Safety and stability of solvents
Piezoelectric inkjet 3D printing	10–100	0.1×10^{-3} m^2 min^{-1}	Hydrophilic-based polymer or drugs	Dispersed drug in the polymer solution or solvents	Good for surface modification	Limited control on drug release. Safety and stability of solvents

[a]Spatial-resolution to time-for-manufacturing (RTM) ratio compares the production efficiency of those different 3D-printing techniques. The higher the RTM ratio, the faster the fabrication.

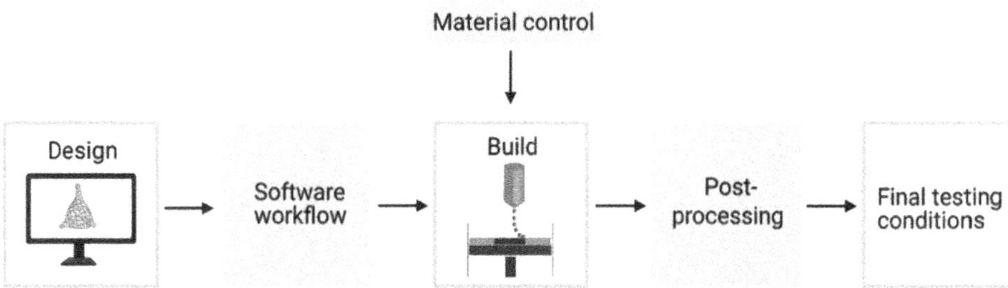

Figure 2.3 The critical steps in the deposition-based 3D-printing workflow in the development of pharmaceutical products [6]. (Source: [6] with permission of Elsevier.)

summarises the spatial resolution and other practical considerations for deposition-based 3D-printing technologies.

The use of deposition-based 3D-printing technologies for the fabrication of drug-loaded devices divides the design into macroscale and microscale. Figure 2.3 shows the basic mechanism of deposition-based 3D-printing technologies. Fused deposition modelling (FDM) or hot-melt extrusion-based 3D printing is commonly used for the fabrication of pharmaceutical products. During FDM printing, a thermoplastic filament is extruded layer-by-layer to build a digitally designed object [5, 7]. The filament is melted in the printhead and flows through a nozzle, and upon exiting the nozzle, the molten polymer cools and solidifies. The resolution of FDM printing (~200 μm) is constrained by the nozzle diameter. Moroni et al. [10] proposed spatial-resolution to time-for-manufacturing (RTM) ratio to compare the production efficiency of different 3D-printing techniques. The RTM ratio is calculated as the ratio of the spatial resolution of the technology to the time required for manufacturing. In general, the higher the RTM ratio, the faster the fabrication. The RTM ratio of FDM printing is $\sim 1 \times 10^{-3}\ m^2\ min^{-1}$ [10].

To apply FDM to the design of drug products, active pharmaceutical ingredients (APIs) need to be loaded into the filament or the final 3D construct. Two main approaches have been explored: active and passive loading [6]. In active loading, the API is mixed with the polymer to produce a printable, drug-loaded filament [5]. Passive loading is often done as a post processing step, wherein the filament or 3D construct is immersed into a drug-containing solution. allowing passive diffusion of the API into the material [11]. Recently, a complementary technique, melt droplet-based 3D printing, was introduced to circumvent the step of active loading and filament characterisation [12]. This technique uses a single-step printing process in which the API is combined with powder excipients directly into the material feed and then extruded. Importantly, this technique exploits the advantages of FDM without the need to melt the excipients to produce drug-loaded filament feedstock.

The working principle of direct ink writing or semisolid extrusion-based 3D printing is similar to the melt filament-based 3D-printing technique in that material is deposited by extrusion through a nozzle [13]. Semi-solid extrusion-based 3D printing is commonly used for robotic dispensing of shear-thinning inks such as hydrogels using pneumatic or motor-based printheads [14]. In semisolid extrusion-based 3D printing, the material is deposited layer-by-layer to form 3D objects with an RTM ratio of $0.5 \times 10^{-3}\ m^2\ min^{-1}$ [10]. Inkjet-based 3D printing uses a very small orifice to selectively deposit droplets of low viscosity material with an RTM ratio of $0.1 \times 10^{-3}\ m^2\ min^{-1}$ [10].

Binder jetting 3D printing works by ejecting a binder solution to bond specific regions of the surface of a powder bed. The process is repeated layer-by-layer for 3D printing with an RTM ratio of 0.1×10^{-3} m^2 min^{-1} [10]. There are two main types of vat photopolymerisation: stereolithography (SLA) and digital light processing (DLP). The SLA raster scans laser light to solidify a layer at the surface of a photosensitive resin [15], whilst DLP uses either a digital micromirror device or a liquid crystal display (LCD) screen to project 2D patterns across the whole resin surface to cure a whole layer at once [16], increasing the volumetric rate of fabrication. SLA remains relatively slow with an RTM ratio of 0.5×10^{-3} m^2 min^{-1}, while DLP has an accelerated RTM ratio of 2×10^{-3} m^2 min^{-1} [10]. Two-photon polymerisation is another type of light-based 3D printing, which uses ultrashort laser pulses to selectively cure resins with a very high resolution reaching sub-micron and nanoscale resolution. The fabrication process requires comparatively far more time, as the focal point of the laser is scanned through the whole object leading to an RTM ratio of 0.05×10^{-3} m^2 min^{-1}.

2.3 3D Design for Drug-Loaded Device

3D printing enables the design of new structures and features that make unconventional release profiles and the fabrication of personalised drug-loaded medical devices possible. The change in size, shape, and structure of 3D-printed medical devices directly influences drug release kinetics. Patel et al. [17] summarised different methods to design and optimise drug release patterns from 3D-printed drug dosage forms, as shown in Figure 2.4. In this section, we discuss the various design approaches of 3D-printed solid dosage forms, i.e., computer-aided design, and 3D-printing software-based and parameter-based designs, as well as the complex design of polypills.

Figure 2.4 *Different methods to design and optimise drug release patterns from 3D-printed drug dosage forms [17]. (Source: [17] with permission of Elsevier.)*

2.3.1 CAD Design-Based Design

An important tool in the design of drug-loaded medical devices is the use of computer-aided design (CAD) software. Goyanes et al. investigated the effect of CAD design of the tablet geometry on drug release for five different tablet geometries – cube, pyramid, cylinder, sphere, and torus, were printed by FDM [18]. They indicated the influence of geometrical shape on drug release due to differences in surface area to volume ratio. Additionally, CAD enables the user to import medical images for the tailored reconstruction of patient-specific 3D models. The combination of CAD software and image reconstruction has also been used to design patient-specific drug-loaded medical devices, such as anti-acne implants for topical release of salicylic acid [19].

2.3.2 Computational Software-Based Design

3D-printing software-based designs have been applied in the development of 3D-printed drug-loaded medical devices. 3D-printing software here indicates the slicing software for the conversion of a 3D CAD object model to specific instructions for the 3D printer, especially for FDM [20]. There are many open-source and free-to-use slicer software packages available on the market, such as Slic3r, Prusa Slicer, and Cura. There are many parameters related to inner structures that can be changed, such as infill percentage, infill pattern, size, with cap, base, shell, or pulsatile design, as shown in Figure 2.5. As shown in Figures 2.5

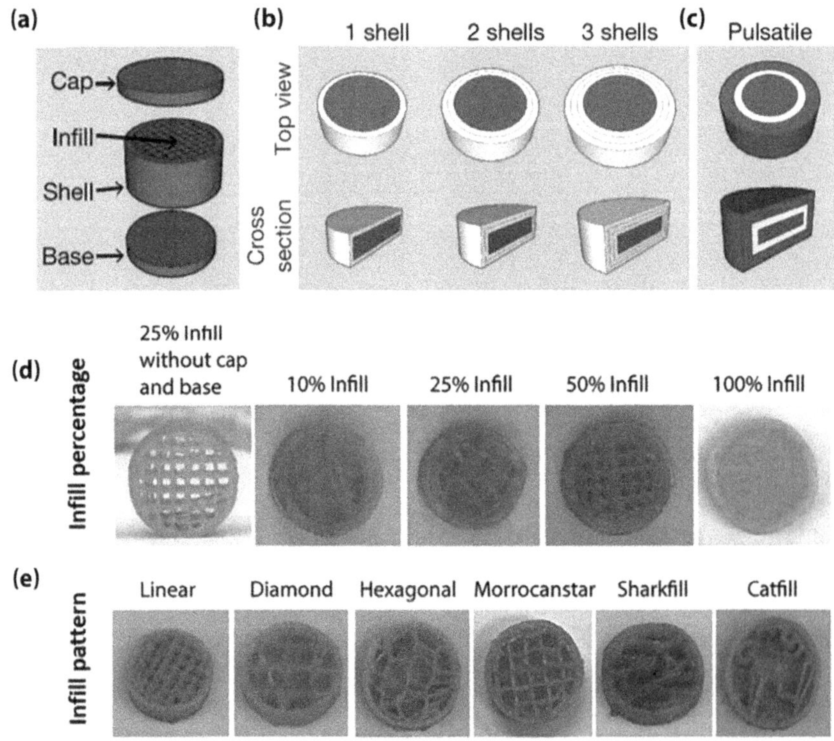

Figure 2.5 *Computational software-based designs of tablets: (a) Caps, base, shell, and infill patterns of a CAD tablet. (b) CAD drug tablet with drug-free layers is in grey and drug-loaded layers in red. (c) Pulsatile-release tablets. 3D-printed tablet with difference infill percentage (d) and infill patterns (e) [21]. (Source: [21] / with permission from Elsevier.)*

(a, d), infill percentage is the main factor governing the release rate, with slower release at larger infill percentages [5]. Kadry et al. [21] developed different inner patterns (linear, diamond, hexagonal, Moroccan star, shark, cat, hexagonal), as shown in Figure 2.5 (d). Like the tablets with increasing infill density, tablets with various infill patterns showed extended-release patterns, and the differences between the drug efficiency of various tablets were also statistically significant. Hexagonal infill showed the fastest release, while the diamond-shaped infill showed the slowest release. They explained that the drug release rate was due to the differences between various infill patterns and may have resulted from the varying perimeters or volumes of the patterns. As shown in Figures 2.5 (b, c), both shell structure and pulsatile design have both drug free and drug loaded layers. Those drug free layers can help limit or stop drug release for a period of time, because drug-free shells surrounding the tablets prevent water from penetrating into the internal tablets containing the drug.

2.3.3 3D-Printing Parameter-Based Design

Design of experiments was used to optimise 3D-printing parameters (build orientation, infill density, temperature, and print speed) to maximise the mechanical strength of the printed parts while reducing energy and material consumption [22]. The design of the experiments was leveraged to provide specific printing parameters to manufacture polyether-ether-ketone (PEEK) implants for spinal or dental applications [23]. Wang et al. used response surface methodology with three parameters (nozzle diameter, nozzle temperature, and printing speed) and with three levels for each parameter to optimise the elastic modulus and bending strength of the printed implants [23]. Daniel et al. [24] varied 3D-printing parameters (i.e., temperature and speed) and quantified the ratio of the cross-sectional area, with the sample length of the 3D-printed filament found to be within one standard deviation. Gleadall et al. [25] developed the FullControl Gcode Designer approach to enable more freedom for print-path design. Extrusion rate and speed were continuously varied in their study to achieve five different graded lattice geometries for an identical print path, as shown in Figure 2.7 (b) [26].

2.3.4 Polypills and Complex Designs

A polypill is medication in pill form that combines multiple active pharmaceutical ingredients [28]. Khaled et al. [29] used 3D-printing techniques for the fabrication of multi-active solid dosage forms. They indicated that their polypill demonstrates that complex medication regimes can be combined in a single personalised tablet, which could potentially improve adherence for those patients currently taking many separate tablets and also allow ready tailoring of a particular drug combination/drug release for the needs of an individual.

There are some complex device designs with 3D-printed components, e.g., inkjet filled capsules. Melocchi et al. (2015) successfully prepared a capsule-shaped device for the oral pulsatile release of acetaminophen [30]. The 3D-printed hollow capsule made from net hydroxypropyl cellulose filaments exhibited a typical pulsatile–release profile. They indicate that the obtained results were consistent with those from capsule shells developed using injection moulding technology with the same composition. Okwuosa et al. [27] developed a method of coordinating the use of FDM and liquid dispensing to fabricate individualised dosage forms. As shown in Figure 2.6 (c), polymethacrylate shells (Eudragit EPO and RL) for immediate and extended-release were fabricated using FDM and simultaneously filled using a computer-controlled liquid dispenser loaded with model drug solutions. Both immediate and extended drug release profiles based on polymethacrylate polymer shells were achieved.

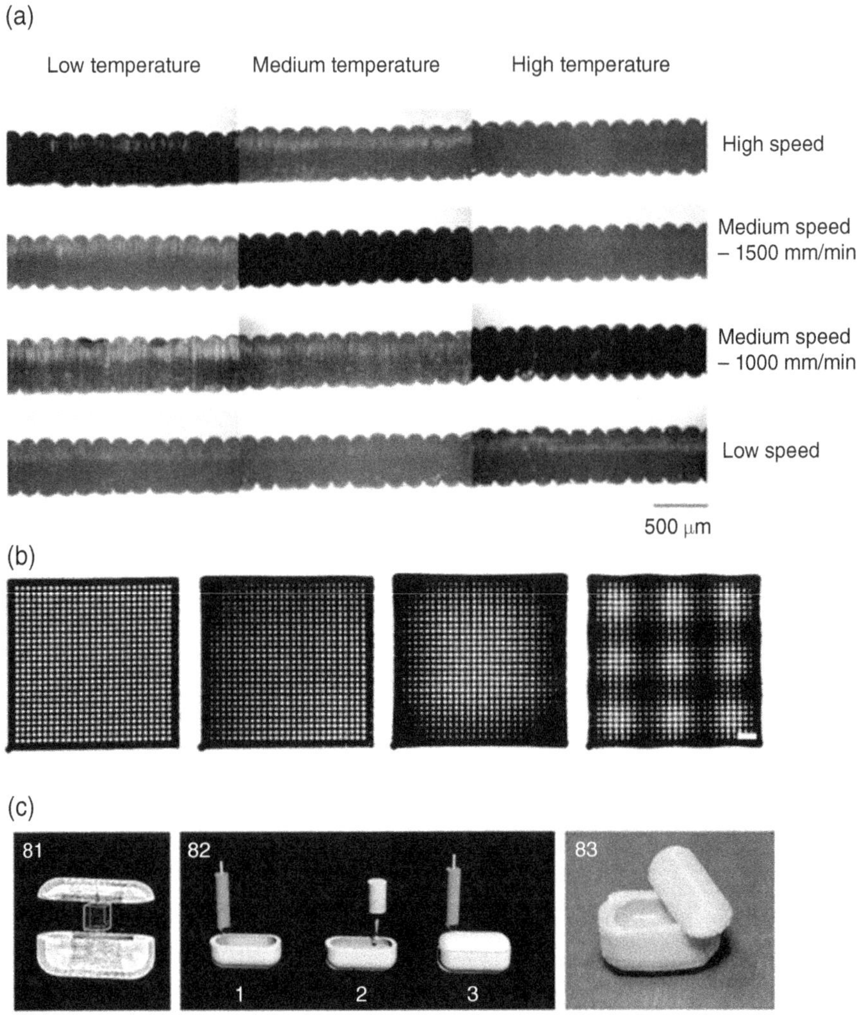

Figure 2.6 (a) Sample cross-section microscopy images of the 3D-printed filaments under different printing speeds and temperatures [24]. (b) Different graded meshes were 3D printed by designing a variable extrusion width [26]. (c) Multi-phase 3D printing: the core is located in a median level between the bottom shell (75%) and the top shell (25%); the shell is printed first followed by complete filling of the shell bottom; the printing is completed by printing the shell top [27]. (Source: (a) [24] / with permission from Elsevier; (b) [26] / with permission from Elsevier; (c) [27] / with permission from Elsevier.)

2.4 3D Designs Influence Drug Release

2.4.1 Controlling Drug Release

Controlled drug release mechanisms from pills, implants, or devices often improve the therapeutic effects. Several modes of release are possible depending on the geometry, materials chemistry, and degradation/erosion of the drug-loaded device. The classical drug release mechanisms in standard drug releases of the drug-loaded devices are the first-order and zero-order drug release mechanisms. The first-order drug release mechanism is drug release rate decreasing with time, which can be beneficial to reach a suitable drug concentration needed for rapid action followed by sustained levels of activity, while for the zero-order drug release mechanism, there is a constant rate of drug release for drugs with a narrow therapeutic window. Zhang et al. [9] developed a case study to demonstrate that the effect of pore alignment can be used to tailor the drug release rate when porosity and pore volume is kept constant, as shown in Figure 2.7. They demonstrated that it is possible to control the drug release rate whilst keeping other parameters constant, such as pore size or porosity, and crucially with no change to the feedstock materials or 3D-printing equipment. This demonstrates the ability to customise or personalise drug release rates purely through geometric design, offering enormous potential for distributed manufacture, on-demand with a single machine and single standard feedstock [5].

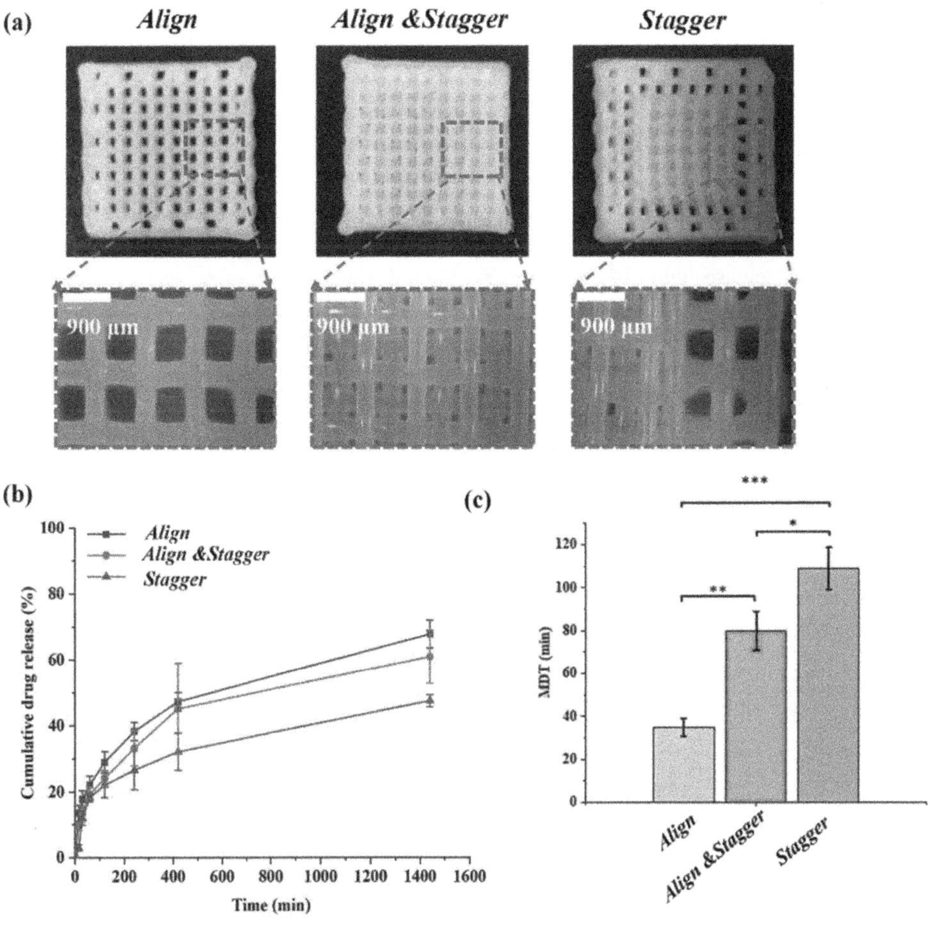

Figure 2.7 (a) The appearances of the 3D-printed constructs with constant porosity and print settings, but minor adjustments to the print path to combine aligned pores and staggered pores led to tailored drug release rate. (b) The drug release profiles and (c) the mean dissolution time (MDT) values of the constructs with aligned, staggered, and a combination (n = 3) [5]. (Source: [5] / with permission from Elsevier.)

3D printing has been used to deposit drug-loaded material with spatial control to fabricate drug products with tailored release profiles. The geometry, and more specifically the surface area-to-volume ratio and infill percentage, govern drug release in 3D-printed tablets and implants [18, 31]. One advantage of 3D printing is the ability to tailor drug product properties to enable release beyond classic release profiles. 3D printing enables more sophisticated control over the drug distribution within the product and the shape/geometry of the product through freeform designs. In addition, 3D-printing processes can improve the complexity and resolution of the drug-loaded device. 3D printing makes it possible to design and print a drug product of any geometry, architecture, and/or internal structure. Based on these concepts, 3D-printing processes have been leveraged to expand the possibilities of tailored release and produce drug-loaded devices with unconventional release profiles that are difficult (or impossible) to achieve using traditional manufacturing processes. For example, drug products with high infill density dissolved more slowly than similar constructs with low infill density due to the lower surface area in contact with the surrounding medium [12]. Goyanes et al. investigated the effect of geometry on drug release from 3D-printed tablets [18]. They indicated that drug products with increased surface-area-to-volume ratios (e.g., torus or pyramid-shaped objects) exhibited faster release rates as compared with similar devices that had smaller surface-area-to-volume ratios (e.g., cylindrical or spherical objects).

2.4.2 Modifying Drug Release

Apart from the first- and zero-order, there are some complex and modified drug release mechanisms, such as the pulsatile release mechanism. Pulsatile release can be used to synchronise drug release with biological cycles. Complex sequences of release patterns (e.g., burst, zero-order, burst) can be achieved by controlling the spatial distribution of one or multiple APIs. Another strategy to control the release profile is by using 3D printing to print tablets with a core-shell structure. As shown in Figure 2.8, Okwuosa et al. used the FDM

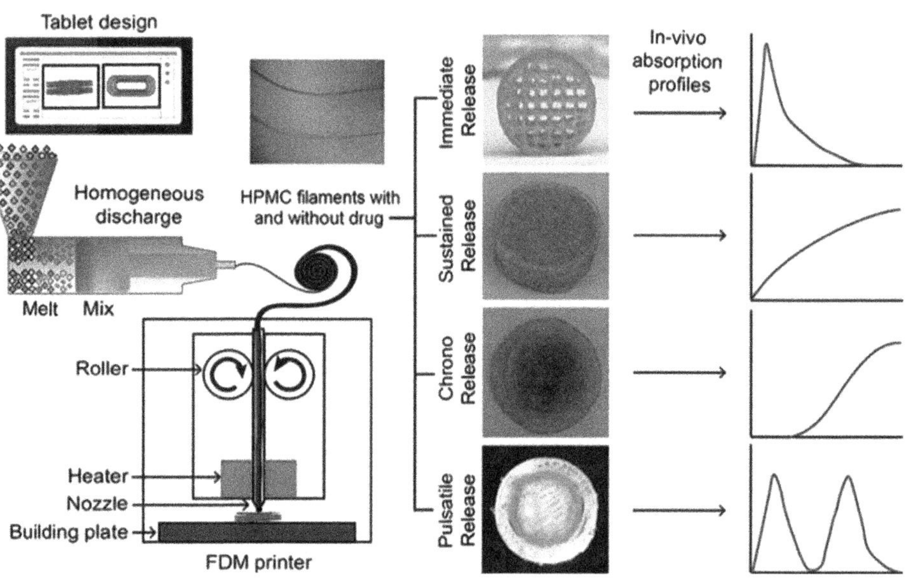

Figure 2.8 Various types of drug release mechanisms from 3D-printed tablets [21]. (Source: [21] / with permission from Elsevier.)

3D-printing technique to fabricate tablets with an external layer that restricts dissolution and drug release. In the device, the external layer prevented rapid release in the acidic environment of the stomach and was followed by a pronounced release in the intestine [32]. Xu et al. designed a hollow and solid bladder drug-loaded device containing different amounts of lidocaine hydrochloride using SLA. Hollow geometries released the API more quickly than solid geometries, and the release rate increased with drug loading [33].

Rowe et al. designed a 3D-printed tablet with two compartments for dual-stage release. In their device, each compartment contained the same API but was coated with a different polymer. The first compartment was coated with Eudragit E 100 for burst release (50% of the dose was released in 10 minutes), whilst the second compartment was coated with Eudragit RL PO for prolonged release (the remaining 50% of the dose was released over 8 hours) [34]. Similar concepts have been applied to programmed release rates using deposition-based printing of drug products, with Pluronic F127 as the main excipient [35]. Several geometries were printed to tailor drug release. A core-shell structure was used for delayed release (drug-loaded core surrounded by polymer shell), a multilayer structure enabled pulsatile release (alternating of drug and polymer layers), and a gradient distribution was prepared for constant release.

2.5 Challenges and Perspective

Spritam® is the first 3D-printing-based drug delivery system to be approved by the FDA, which is for an anti-epileptic therapy [36]. The fabrication process uses binder jetting and consists of depositing an incremental layer of levetiracetam-containing powder followed by the deposition of binding fluid onto the incremental layer to bind particles in the powder. A Spritam® conventional tablet is manufactured by compression. In comparison to the conventional tablet, the main advantage of 3D-printed Spritam® is a more convenient dosage form for patients, especially elderly and paediatric populations, than traditional tablets, as it is easier to swallow [7, 37]. Although 3D-printed drugs have advantages compared to conventional manufacturing methods, only a few pharmaceutical products have reached the market so far. One of the reasons for this is the lack of materials for building up structures. There is a crucial need for a broader range of approved materials that can be used for 3D printing to develop new drug products. The selection and variety of materials available are dictated by the 3D-printing technology itself. For example, there is a relatively low number of approved thermoplastics available for FDM 3D printing [38, 39] and there is no approved and commercialised material for selective laser sintering. There is also a lack of specifications for starting materials suitable for 3D printing with drugs, which remains a key limitation.

One of the other challenges is that some 3D-printing technologies require a post processing drying step [38], extending manufacturing time, and potential degradation of the drug. Hot-melt extrusion involving high temperatures may have detrimental impacts on the API. Increasing temperature may create API amorphous phases that will have to be evaluated for physical stability and polymorphism, which is time- and resource-intensive [40]. Additionally, in the case of deposition-based methods like FDM, filaments containing the drug are fabricated by hot-melt extrusion. Such filaments need to be monitored carefully during storage, as their physico-chemical properties may degrade with ageing, with a detrimental impact on the properties of the printed dosage form. Additionally, printing by FDM might have subjected the API to elevated temperatures twice because hot-melt extrusion is needed to fabricate the feedstock filament and then again during the printing [12]. Thus, controlling parameters in pharmaceutics is essential for drug release and developing safe products for the patient.

Another challenge is cleaning and sterilisation. 3D-printing facilitates the creation of objects with complex geometries, such as engineered porosity, channels, and internal voids or cavities that cannot be produced by traditional manufacturing methods [12, 17, 41]. These complex geometries can make it more difficult to remove material residues (cleaning) and sterilise the device due to increased surface area, generation of extensive tortuous pathways, and the creation of internal voids with limited or no access. Material removal is also a consideration in the case of 3D-printing technologies that use temporary support structures. Achieving adequate cleaning, sterility, and biocompatibility of a 3D-printed device is also an open challenge [17, 42]. Assessing and verifying sterility in porous or internally complex 3D-printed devices or drug products can present additional difficulties. Moreover, changing the paradigm of drug manufacturing is also a significant challenge [43]. On-demand 3D printing, for example, by FDM, in hospital pharmacies would present some advantages, but under current practices would require different quality controls. This modification of the traditional value chain will require effective auditing procedures at the organisational and regulatory levels [44].

Overall, 3D printing has the potential to revolutionise the production of pharmaceutical products, allowing for decentralised and customised manufacturing of therapeutics. However, the 3D-printing 'utopia' in pharmaceutics is contingent upon whether the challenges facing the market can be overcome so the technology may reach its full potential.

References

[1] Reker, D., Blum, S.M., Steiger, C. et al. (2019). "Inactive' ingredients in oral medications. *Science Translational Medicine* **11**: 483.

[2] Levy, H.B. and Barney, K.F. (2016). Pharmacology, pharmacy, and the aging adult: implications for occupational therapy. In: *Occupational Therapy with Aging Adults*, 214–234. Elsevier.

[3] Rochon, P.A. and Schmader, K. (2019). *Drug Prescribing for Older Adults*. Toronto: UpToDate.

[4] Hope, H.F., Bluett, J., Barton, A. et al. (2016). Psychological factors predict adherence to methotrexate in rheumatoid arthritis; findings from a systematic review of rates, predictors and associations with patient-reported and clinical outcomes. *RMD Open* **2** (1): e000171.

[5] GA, Z.B., Belton, P., Mcdonagh, T. et al. (2021). Data for article: new insights into the effects of porosity, pore length, pore shape and pore alignment on drug release from extrusion-based additive manufactured pharmaceuticals. *Temporary URL for peer review process*: https://figshare.com/s/d98ab20b4b1a2917bded (accessed 26 December 2021).

[6] Ragelle, H., Rahimian, S., Guzzi, E.A. et al. (2021). Additive manufacturing in drug delivery: innovative drug product design and opportunities for industrial application. *Advanced Drug Delivery Reviews* **178**: 113990.

[7] Qi, S., Zhang, B., and McDonagh, T. (2021). *Three-Dimensional Printed Implantable Products*, 252–295. Implantable Technologies.

[8] Tam, C.H., Alexander, M., Belton, P., and Qi, S. (2021). Drop-on-demand printing of personalised orodispersible films fabricated by precision micro-dispensing. *International Journal of Pharmaceutics* **610**: 121279.

[9] Zhang, B., Gleadall, A., Belton, P. et al. (2021). New insights into the effects of porosity, pore length, pore shape and pore alignment on drug release from extrusion-based additive manufactured pharmaceuticals. *Additive Manufacturing* **46**: 102196.

[10] Moroni, L., Boland, T., Burdick, J.A. et al. (2018). Biofabrication: a guide to technology and terminology. *Trends in Biotechnology* **36** (4): 384–402.

[11] Cerda, J.R., Arifi, T., Ayyoubi, S. et al. (2020). Personalised 3D printed medicines: optimising material properties for successful passive diffusion loading of filaments for fused deposition modelling of solid dosage forms. *Pharmaceutics* **12** (4): 345.

[12] Zhang, B., Nasereddin, J., McDonagh, T. et al. (2021). Effects of porosity on drug release kinetics of swellable and erodible porous pharmaceutical solid dosage forms fabricated by hot melt droplet deposition 3D printing. *International Journal of Pharmaceutics* **604**: 120626.
[13] Lewis, J.A. (2006). Direct ink writing of 3D functional materials. *Advanced Functional Materials* **16** (17): 2193–2204.
[14] Zhang, B., Teoh, X., Belton, P. et al. (2021). The development of Hypromellose based semisolid 3D printing inks for drug delivery. *Transactions on Additive Manufacturing Meets Medicine* **3** (1): 541.
[15] Hull, C.W. (1986). Apparatus for production of three-dimensional objects by stereolithography. *Google Patents*.
[16] Machekposhti, S.A., Zhang, B., Sachan, R. et al. (2021). Patterned surfaces with the controllable drug doses using inkjet printing. *Journal of Materials Research* **36**: 1–12.
[17] Patel, S.K., Khoder, M., Peak, M., and Alhnan, M.A. (2021). Controlling drug release with additive manufacturing-based solutions. *Advanced Drug Delivery Reviews* **174**: 369–386.
[18] Goyanes, A., Martinez, P.R., Buanz, A. et al. (2015). Effect of geometry on drug release from 3D printed tablets. *International Journal of Pharmaceutics* **494** (2): 657–663.
[19] Goyanes, A., Det-Amornrat, U., Wang, J. et al. (2016). 3D scanning and 3D printing as innovative technologies for fabricating personalized topical drug delivery systems. *Journal of Controlled Release* **234**: 41–48.
[20] Talalwa, L., Natour, G., Bauer, A. et al. (2020). Radiological characteristics of a new experimental rubber elastomeric polymer used in three-dimensional printing with different infill densities and patterns. *Journal of Physics Communications* **4** (12): 125006.
[21] Kadry, H., Al-Hilal, T.A., Keshavarz, A. et al. (2018). Multi-purposable filaments of HPMC for 3D printing of medications with tailored drug release and timed-absorption. *International Journal of Pharmaceutics* **544** (1): 285–296.
[22] Khalid, M. and Peng, Q. (2021). Investigation of printing parameters of Additive Manufacturing process for sustainability using Design of Experiments. *Journal of Mechanical Design* **143** (3): 032001.
[23] Wang, Y., Müller, W.-D., Rumjahn, A. et al. (2021). Mechanical properties of fused filament fabricated PEEK for biomedical applications depending on additive manufacturing parameters. *Journal of the Mechanical Behavior of Biomedical Materials* **115**: 104250.
[24] Daniel, F., Gleadall, A., and Radadia, A.D. (2021). Influence of interface in electrical properties of 3D printed structures. *Additive Manufacturing* **46**: 102206.
[25] Gleadall, A. (2021). FullControl GCode Designer: open-source software for unconstrained design in additive manufacturing. *Additive Manufacturing* **46**: 102109.
[26] Moetazedian, A., Budisuharto, A.S., Silberschmidt, V.V., and Gleadall, A. (2021). CONVEX (CONtinuously Varied EXtrusion): a new scale of design for additive manufacturing. *Additive Manufacturing* **37**: 101576.
[27] Okwuosa, T.C., Soares, C., Gollwitzer, V. et al. (2018). On demand manufacturing of patient-specific liquid capsules via co-ordinated 3D printing and liquid dispensing. *European Journal of Pharmaceutical Sciences* **118**: 134–143.
[28] Robles-Martinez, P., Xu, X., Trenfield, S.J. et al. (2019). 3D printing of a multi-layered polypill containing six drugs using a novel stereolithographic method. *Pharmaceutics* **11** (6): 274.
[29] Khaled, S.A., Burley, J.C., Alexander, M.R. et al. (2015). 3D printing of five-in-one dose combination polypill with defined immediate and sustained release profiles. *Journal of Controlled Release* **217**: 308–314.
[30] Melocchi, A., Parietti, F., Loreti, G. et al. (2015). 3D printing by fused deposition modeling (FDM) of a swellable/erodible capsular device for oral pulsatile release of drugs. *Journal of Drug Delivery Science and Technology* **30**: 360–367.
[31] H-g, Y., Choi, Y.-J., Kang, K.S. et al. (2016). A 3D-printed local drug delivery patch for pancreatic cancer growth suppression. *Journal of Controlled Release* **238**: 231–241.
[32] Okwuosa, T.C., Pereira, B.C., Arafat, B. et al. (2017). Fabricating a shell-core delayed release tablet using dual FDM 3D printing for patient-centred therapy. *Pharmaceutical Research* **34** (2): 427–437.
[33] Xu, X., Goyanes, A., Trenfield, S.J. et al. (2021). Stereolithography (SLA) 3D printing of a bladder device for intravesical drug delivery. *Materials Science and Engineering: C* **120**: 111773.

[34] Rowe, C., Katstra, W., Palazzolo, R. et al. (2000). Multimechanism oral dosage forms fabricated by three dimensional printing™. *Journal of Controlled Release* **66** (1): 11–17.
[35] Haring, A.P., Tong, Y., Halper, J., and Johnson, B.N. (2018). Programming of multicomponent temporal release profiles in 3D printed polypills via core–shell, multilayer, and gradient concentration profiles. *Advanced Healthcare Materials* **7** (16): 1800213.
[36] Konta, A.A., García-Piña, M., and Serrano, D.R. (2017). Personalised 3D printed medicines: which techniques and polymers are more successful? *Bioengineering* **4** (4): 79.
[37] Annaji, M., Ramesh, S., Poudel, I. et al. (2020). Application of extrusion-based 3D printed dosage forms in the treatment of chronic diseases. *Journal of Pharmaceutical Sciences* **109** (12): 3551–3568.
[38] Jamróz, W., Szafraniec, J., Kurek, M., and Jachowicz, R. (2018). 3D printing in pharmaceutical and medical applications: recent achievements and challenges. *Pharmaceutical Research* **35** (9): 1–22.
[39] Goole, J. and Amighi, K. (2016). 3D printing in pharmaceutics: a new tool for designing customized drug delivery systems. *International Journal of Pharmaceutics* **499** (1–2): 376–394.
[40] Alqahtani, F., Belton, P., Zhang, B. et al. (2020). An investigation into the formations of the internal microstructures of solid dispersions prepared by hot melt extrusion. *European Journal of Pharmaceutics and Biopharmaceutics* **155**: 147–161.
[41] Zhang, B., Chung, S.H., Barker, S. et al. (2021). Direct ink writing of polycaprolactone/polyethylene oxide based 3D constructs. *Progress in Natural Science: Materials International* **31** (2): 180–191.
[42] Wang, J., Zhang, Y., Aghda, N.H. et al. (2021). Emerging 3D printing technologies for drug delivery devices: current status and future perspective. *Advanced Drug Delivery Reviews* **174**: 294–316.
[43] Cui, M., Pan, H., Su, Y. et al. (2021). Opportunities and challenges of three-dimensional printing technology in pharmaceutical formulation development. *Acta Pharmaceutica Sinica B* **11** (8): 2488–2504.
[44] Ventola, C.L. (2014 October). Medical applications for 3D printing: current and projected uses. *P&T* **39** (10).

3

3D Printing of Oral Solid Dosage Forms Using Selective Laser Sintering

Atabak Tabriz Ghanizadeh[1,2], *Hannah Kuofie*[1], *James Scoble*[1], *Sam Boulton*[1] *and Dennis Douroumis*[1,2]

[1] *Faculty of Engineering and Science, University of Greenwich, Medway Campus, Kent, UK*
[2] *Delta Pharmaceutics Ltd., Kent, UK*

3.1 Introduction

Three-dimensional printing (3DP) has gained increasing popularity over the last 10 years in healthcare due to its enormous capability for resolving several of the limitations associated with conventional drug delivery and therapeutic technologies [1]. Binder jetting, fused deposition modelling (FDM), powder bed fusion (PBF), and vat polymerisation are just a few of the printing technologies that have been developed and employed in the pharmaceutical and biomedical industries throughout the years. Around 1981, Hideo Kodama developed a method for creating 3D models using photo-hardening polymers and UV radiation [2]. A few years later, Chuck W. Hull created the first 3D-printing technology, stereolithography, which was marketed by 3D systems. Initially, these technologies were limited in their application due to poor printing quality and excessive costs. Nevertheless, technological breakthroughs have resulted in the development of cost-effective, high-print-speed, and high-precision 3D printers that have already been employed in a multitude of sectors including biomedical, space, education, automotive, and art over the last two decades.

3D Printing of Pharmaceutical and Drug Delivery Devices: Progress from Bench to Bedside, First Edition.
Edited by Dimitrios A. Lamprou, Dennis Douroumis and Sheng Qi.

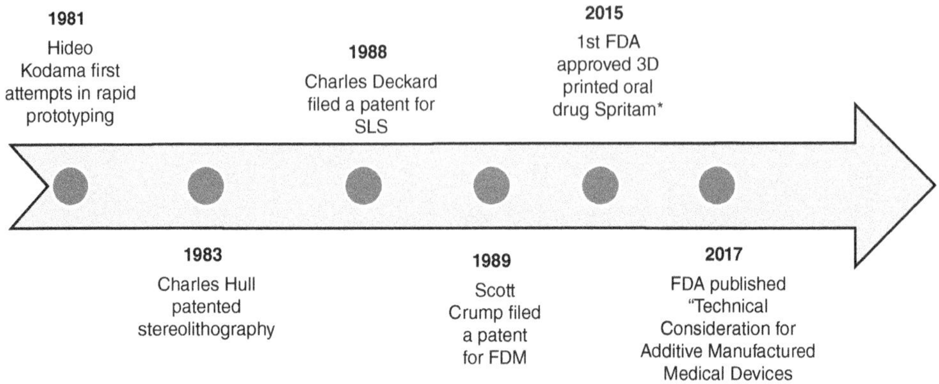

Figure 3.1 *Historic timeline of 3D printing.*

In recent years, considerable technological advancements (Figure 3.1) resulted in the development of Spritam®, an antiepileptic drug, the first commercially available 3D-printed oral medication approved by the FDA (Food and Drug Administration) in 2015. Two years later, FDA produced the 'Technical Considerations for Additive Manufactured Medical Devices,' a guide for industry and FDA employees, while research into 3D-printed medicines continues [3].

3D printing, or additive manufacturing, refers to the fabrication of a 3D object from a computer-aided design (CAD), in a layer-by-layer manner. The CAD file is converted into a stereolithography file (.stl file) that contains the necessary information for the spatial geometry of the object to be produced using CAD programs. The .stl file is divided into multiple segments after initiation, one of which is the slice file (SLI segment), which is subsequently transferred to the 3D printer for printing. Most of the 3D-printing technologies have the capacity to transform the pharmaceutical sector from one-size-fits-all tablet and capsule mass manufacturing to customised dosage forms that fit the clinical needs of the patients [4].

Since the introduction of 3D printing with stereolithography, various other techniques have been established (Figure 3.2), allowing for the processing of a broader range of materials, such as polymers, metals, and ceramics. Nanomaterials, medicines, and biological materials like cells can all be included in 3D constructions, opening a whole new world of possibilities for medical 3D printing.

Extrusion-based technologies, such as FDM, are by far the most extensively used 3D-printing technology. The technology was invented by Scott Crump, Stratasys' co-founder, and patented in 1989 [5]. An FDM printer system (Figure 3.2c) comprises of a feeder gear system that enables material to be driven through the system, a heated liquefier that is located in the print head and is a system for making the material extrudable, the print head and nozzle, the printer's axels that allow the print head to move following cartesian patterns, and finally the build platform. In this process, layers of molten or softened thermoplastic materials in the form of filaments are deposited via the printer's head at specified directions controlled by the computer software for the fabrication of the designed structure. In comparison to other 3D-printing technologies, FDM is comparatively inexpensive, and has been successfully implemented in a variety of industries [6–8].

Material jetting (MJ), also known as Poly-jetting and Drop on Demand (DoD), are two other types of 3D-printing technologies that have been used for the manufacturing of

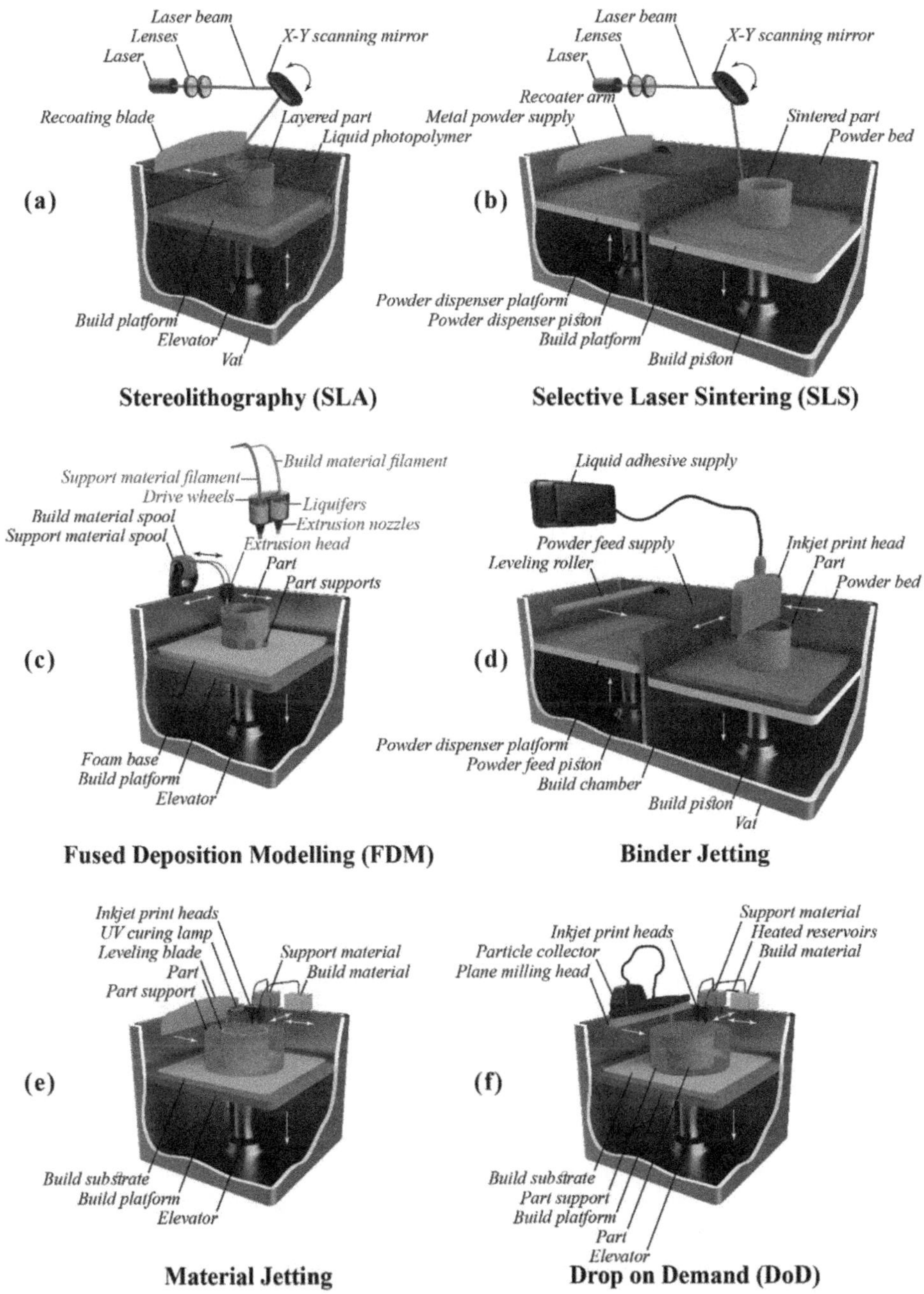

Figure 3.2 *Typical schematic diagrams of six major 3D-printing technologies: (a) Stereolithography; (b) Selective laser sintering; (c) Fused deposition modelling; (d) Binder jetting; (e) Drop on demand; and (f) Material jetting. (Source: Courtesy of Peyman Sohrabi.)*

medical devices and pharmaceutical dosage forms (Figures 3.2 e and f). These technologies deposit similarly to a typical inkjet printer, but the object is formed by stacking numerous layers of material on top of each other. Both MJ and DoD deposit very small polymer droplets, but MJ applies UV irradiation to separately cure each layer after deposition. In contrast, DoD uses wax-like substances that solidify quickly, while a fly cutter passes the build area after the deposition of each layer to minimise wastage. Binder jetting technologies (Figure 3.2d) fuse sequential layers of each cross-section layer using a low viscosity liquid binder. The layer is formed, and the next layer of powdered material is prepared for the deposition of liquid binder in a similar way to PBF processes. The post processing includes the removal of loose powder and infiltration of the layered structure with a liquid such as an epoxy resin, to help build strength and improve appearance.

3.2 Operational Principles of Selective Laser Sintering

Selective laser sintering (SLS) is a common powder fabrication technique which has found several applications within the biomedical and pharmaceutical fields, and it can be used for mass manufacturing. It was developed by Carl Deckard in 1984, at the University of Texas, and subsequently patented in 1990 [9]. The technology was using a neodymium-doped yttrium aluminium garnet (Nd:YAG) laser (100 W) and acrylonitrile butadiene styrene (ABS) powder as thermoplastic material.

SLS has been widely used for the design and fabrication of physical models through selective solidification of various types of powders [10]. Applications of SLS may include oral and maxillofacial prosthetics [11–15] and implants, tissue engineering, or for making tools for neurological surgery. In addition, it has been used for disease diagnosis planning, patient treatment, and rapid prototyping.

SLS is classified as a PBF AM technique, wherein a directed high-powered energy source, such as a laser, sinters or melts a bed of powder resin, metal, or polymer to fuse powder particles together resulting in solidification [16, 17].

An SLS 3D printer comprises of the printing chamber and powder reservoir (reserve chamber) where both are heated to a temperature just under the melting point or the glass transition of the used material (Figure 3.3). Afterwards, a high-power x–y axis laser beam is emitted on to the top surface layer of powder within the pre-heated printing chamber and begins to sinter a pre-determined 2D pattern according to the object design. After each layer is completed, another thin layer of powder is dispersed. This process is achieved by lowering the printing chamber to a pre-set height and raising the reservoir chamber to a set height. This then allows the roller to apply a fresh powder surface on top of the completed layer. This process is repeatedly performed until the final layer is printed. Eventually the powder melts down by the heat generated through the laser scanner system. After printing is complete, the sintered object, which is contained in a powder cake within the printing chamber, is extracted by shaking and sieving off the excess powder. This step can be done using an external post-processing instrument. Factors that can affect the outcome and quality of an SLS printed object include powder particle size, scan spacing, scan speed, and the power of the laser [18]. Other factors such as refresh rate, layer thickness, part bed temperature, raster angle, and hatch pattern can also have an effect on the mechanical properties of SLS printed objects [19].

The powders used for the SLS technique are freely packed or loosely arranged and they will only range by a few microns, but the powder characteristics like morphology, granulometry, density, and flowing capacity are considered over their selection. The main

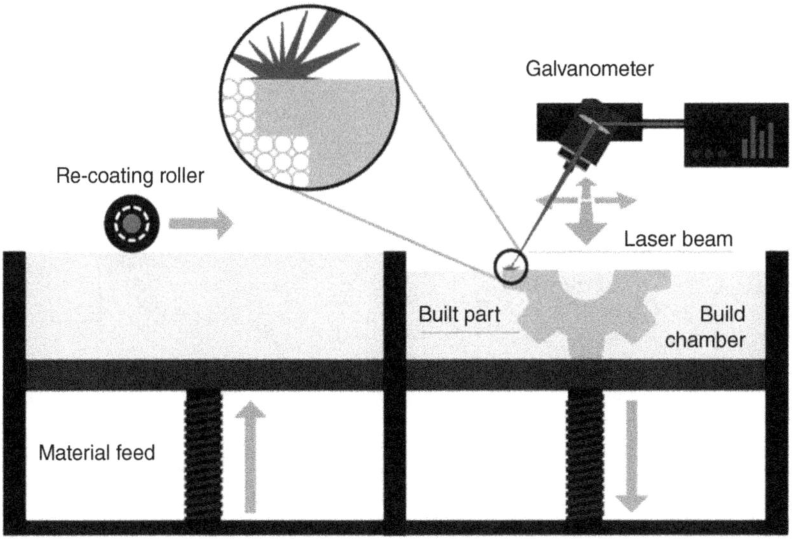

Figure 3.3 *Diagram of SLS process.*

property to be considered is the flowing capacity of the powder as it should spread uniformly while the particle size varies from 40–180 mm. The powder should also be within good sphericity and there are some agents which are used for increasing the flowing capacity and reduce cohesiveness of the powder molecules (e.g., SiO_2). An important feature of SLS is that the powder bed in the printing chamber acts as a support structure and hence overhanging layers of a design do not require the addition of supports in comparison to other 3D-printing technologies such as in SLA and FDM. Thus, SLS can be effectively used for printing complex geometries and a wide range of designs.

3.2.1 Manufacturing Challenges for SLS

SLS has been broadly used for printing metals, ceramics, and pharmaceutical products. However, for the development of oral solid dosage forms, the literature is limited and there are barely any reports in regard to the critical process parameters (CPP), powder critical material attributes (CMAs), and critical quality attributes (CQAs). There are several parameters that can affect the product quality while posing a significant challenge in process of understanding and the design of a full process control strategy. The SLS parameters can be classified into four categories [20]:

1. Laser-related and scanning parameters, including laser power, spot size, pulse duration, pulse frequency, scan speed, scan spacing, and scan pattern
2. Powder material parameters, such as particle shape, size, distribution, morphology, melting temperature, and surface roughness
3. Powder-bed and recoater parameters, such as density, layer thickness, and powder bed temperature
4. Build environment parameters, including shield gas, thermal conductivity, ambient temperature, oxygen levels, and surface free energy.

The Ishikawa diagram in Figure 3.4 illustrates the aforementioned properties that must be considered when using SLS.

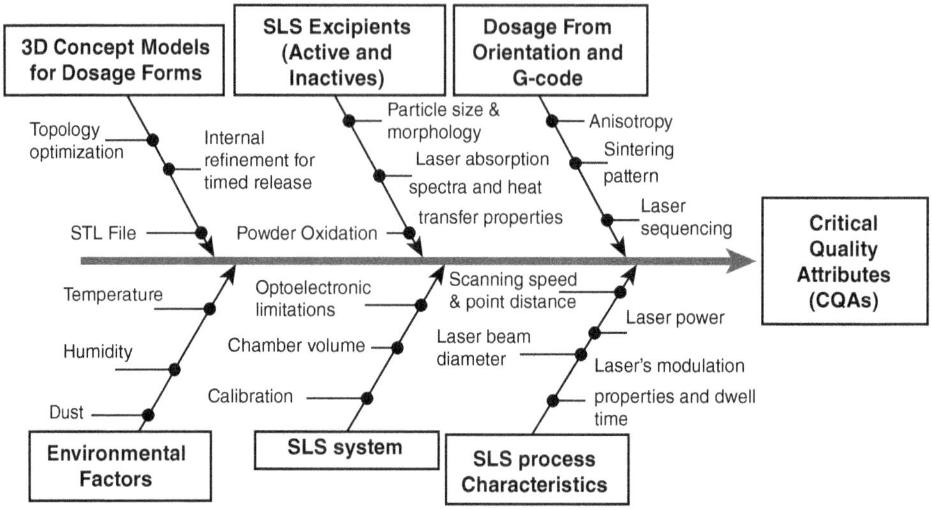

Figure 3.4 *Ishikawa diagram of critical process parameters (CPP), powder critical material attributes (CMAs), and critical quality attributes (CQAs) of SLS.*

3.2.2 Laser Selection and Scanning Speed

The interaction of the laser and the powder is crucial to the SLS process, and the laser used depends on the materials. Materials only absorb light energy from certain wavelengths based on their optical characteristics due to light–matter interactions. The CO_2 laser (10.6 m), due to its good optical absorptivity, is far more ideal for pharmaceutical polymers, whereas the Nd:YAG, Yb:YAG, or Nd:YVO4 (1.06) lasers used in industrial selective laser melting (SLM) equipment are preferable for metals and carbide ceramics [21–23]. Furthermore, CO_2 is less costly, whilst the pulsing system allows the powder to better absorb the laser. It is known that the selection of the laser beam has an impact on the end product properties such as mechanical strength, surface texture, and physical density (porosity). The laser density is defined by the following equation:

$$E = \frac{\rho}{\nu h}, \left(\frac{J}{mm^2}\right) \tag{3.1}$$

where E is the energy density, ρ is the laser power, ν is the scan speed, and h is the hatch distance.

An important process parameter, usually controlled, is the speed over which the laser is scanned. Laser scan speeds typically vary from 10–100 mm s^{-1}. Single-scan, repeated scan, or cross-scan scan patterns are available, with or without contouring. The scan pattern should be selected depending on the material qualities, as it might influence the end product's surface roughness and mechanical properties. For example, repetitive scan patterns have been shown to reduce balling, while contouring produces the same quality for all scan lines. The scan pattern directly affects the heat transfer and thermal gradient and hence the environment. The laser maximum power, the laser mode (continuous or pulsed), the region to which the beam energy is focused (spot size), and the length of time the energy is supplied to a certain area of the powder bed, are all factors that influence the total power that is delivered by the laser.

There is a limited number of studies that investigate the effect of processing parameters on the quality of the printed products. Polyamide 11 (PA11) (e.g., Duraform EX natural), polyamide 12 (PA12), and some thermoplastic elastomers are widely utilised materials. The effect of the energy density on the mechanical properties was investigated by Caulfield et al., using polyamide (DuraForm^{TM}) as model powders [24]. They introduced Eq. 3.1 for the determination of suitable mechanical properties for the printed structures. Ryse et al. further studied the impact of laser power and bed temperature using a semi-crystalline elastomeric polyester [25]. By varying the laser power and bed temperature (laser speed remained the same), it was revealed that printed structures presented better mechanical properties and higher density. Limited studies have shown that slow scanning speeds result in higher energy input due to the long interactions of the particles and the laser, while shorter hatch space supports greater energy transfer [26, 27].

3.2.3 Powder Material Parameters

Powders have a significant impact on SLS, as the laser melting effectiveness is influenced by powder size. Large particles tend to diminish the powder bed's pack density and melting consumes more energy. To the contrary, smaller particles are prone to clumping, making powder layering harder [27–29]. The powder flow ability is a key component in determining particle size distribution on the powder bed. Coarser particles capture laser energy more efficiently than finer particles because they have a larger surface area. To ensure consistent laser energy absorption, sufficient powder flow is essential to generate a smooth powder surface and a consistent height of powder layers. As flowability is affected by particle morphology, and size distribution and hence flow improves for powders with increasing sphericity and narrow size distribution (Figure 3.5). Furthermore, the packing efficiency, mechanical characteristics, and surface roughness are all affected by flowability. The selection of laser parameters and end-product qualities are also influenced by the powder composition, density, melting point (or glass transition), optical properties, heat capacity, and thermal conductivity properties. All these properties impact the light absorption that in turn affects the packing of the powder bed and the layer uniformity after the recoating deposition.

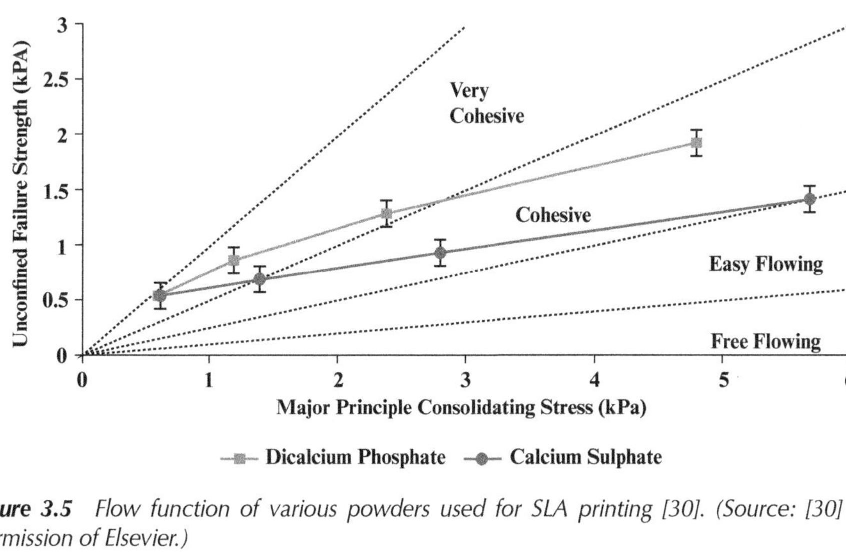

Figure 3.5 *Flow function of various powders used for SLA printing [30]. (Source: [30] with permission of Elsevier.)*

3.2.4 Powder Bed and Recoater Parameters

The powder bed properties are similar to, but distinct from, those that the bulk powder is made of. The density of the powder bed, which is influenced by powder shape, particle size, distribution, and spreading process, has a significant impact on the quality of the printed structure. Powder bed densities for most pharma grade polymers are normally between 50% and 60%; however, polymers such as EUDRAGIT EPO or Klucel ELF/EF grades can be as low as 25–35%. For high powder packing densities, the heat conductivity of the bed and the mechanical characteristics of the printed forms are greater.

3.3 3D-Printed Oral Dosages

SLS is a rapid prototyping 3D-printing technology using single powders or blends as materials. The properties of the obtained structure depend heavily on the printer parameters such as laser intensity, scan speed, bed temperature, and spot diameter of the laser beam, including the powder properties as previously discussed. SLS has gained interest for the development of oral solid dosage forms, especially after the commercialisation of Spritam® by using the ZipDose technology [31].

One of the first studies was conducted by Leong et al. [32], who investigated the potential to print structures with predetermined porosity for sustained drug release using polylactic acid and polycaprolactone blends. The laser power, scan speed, and bed temperature were found to play a critical role in the printing process. SEM analysis showed that higher porosities were obtained at temperatures of 40 °C, while at 30 °C, no necking was observed. Further increments in the bed temperature reduced the porosity of the specimens. The density of the printed tablets at low laser intensity (2W) presented the highest porosity (67.3%) with the best sintering. In contrast, while at high laser power (7W), the porosity was reduced, and curling appeared at the edge of the printed structure. High scan speeds showed uniform sintering and high porosity but relative fragile structures. By using average scan speed (5,080 mm s^{-1}), strong tablets with 50% porosity were printed.

Salmoria et al. investigated the effect of laser energy density and particle size distribution using blends of polycaprolactone/progesterone powders on the morphology of the printed structures [33]. SEM analysis showed high degree of sintering when the laser intensity was increased while particle size was reduced. High laser intensities promoted progesterone recrystallisation on the surface due to the drug melting. The T_g of PCL was affected by the plasticisation effect induced by progesterone where shifts at lower temperature were accompanied by reduced PCL crystallinity. Furthermore, low laser intensity resulted in faster tablet erosion over time and faster progesterone release as a consequence of the increased surface contact area. In addition, the smaller particle size of the powder blends led to faster dissolution rates following zero-order drug release kinetics. Hence, progesterone release rates could be tuned by altering the laser intensity and the particle size of the blends. Fatigue test analysis showed that tablets printed at lower laser intensity with large particle size presented low strength [34]. Fina et al. investigated the SLS printing of pharmaceutical dosage forms using thermoplastic polymers such as Kollicoat IR and Eudragit L100-55, which are suitable for immediate and pH dependent drug release. Paracetamol–polymer blends at 5, 20, and 35% loadings were processed at various intensities to identify potential drug degradation with increased intensities.

As the printer was supplied with a blue diode laser (445 nm), the printing was not possible without the addition of Candurin® gold sheen to facilitate the sintering process. The polymer demonstrated different porosity properties (open and closed), with Kollicoat IR having the same values irrespective of the drug loading and Eudragit with almost no pores at all due to the high sintering degree with increased paracetamol amounts. However, all printed tablets showed low friability but high crushing strength (>284 N) as a result of the complete sintering. Furthermore, only at very low drug loadings (5%), paracetamol dissolution rates were fast and all printed formulations showed sustained release varying from 2 to 12 hours. Nevertheless, the proof-of-concept study demonstrated the use of pharmaceutical polymers for the printing of drug formulations using SLS while avoiding drug degradation.

The same group investigated the feasibility of SLS for the printing of ODT formulations in the form of printlets. For the purposes of the study, hydroxypropyl methylcellulose (HPMC) or vinylpyrrolidone-vinyl acetate (Kollidon VA64) were co-processed with paracetamol (5%, PCM) to produce loaded formulations. As shown in Figure 3.6, the printlets were optimised by adjusting the applied laser intensity and varying the scanning speed (100, 200, and 300 mm s^{-1}). High scanning speeds resulted in low tablet strength (14 N) with high porosity and vice versa as a result of lower sintering extent. Interestingly, printlets comprising of Kollidon VA64 featured fast disintegrating times (<4 seconds) when sintered at very high scanning speeds.

However, the fast disintegration times were the result of the extremely low tensile strength of the printed tablets. The drug dissolution studies showed slow rates for HPMC varying from 75–100% after 2 hours, even for tablets with low tensile strength. Different results were observed for Kollidon with rapid dissolution rates of 100% in 5 minutes for tablets with high porosity but low tensile strength. It is obvious that dissolution rates were affected by the polymer nature, laser scanning speeds, and the low drug loadings of paracetamol. Hence the low printlet strength, the high friability, and the low drug loading may render the process non-viable for the design of ODTs.

A similar study was conducted by investigating the effect of various polymers such as polyethylene oxide (PEO), Eudragit (L100-55 and RL), and ethyl cellulose (EC) in

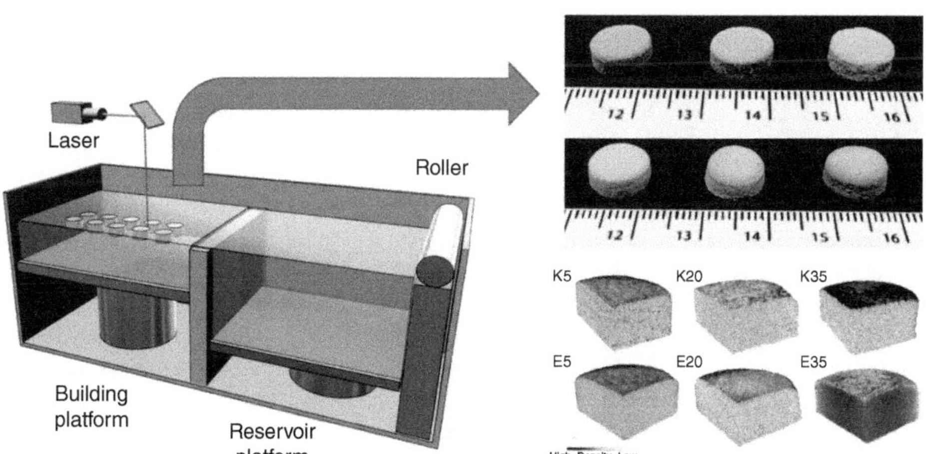

Figure 3.6 Kollidon and HPMC printlets (cm) [35]. (Source: [35] / with permission from Elsevier.)

paracetamol formulations for the printing of cylindrical gyroid lattices with customised dissolution profiles [36]. The use of gyroid lattices was an interesting approach aiming to modify the microstructures with random or periodic designs, each featuring open or closed cells [37, 38]. For such structures, small unit cell sizes present high densities as a result of the short scanning lengths, while Young's modulus increases with the decrease of the unit cell size. For successful SLS printing, it is critical to identify the optimal bed temperatures due to the different T_g of the thermoplastic polymers. By identifying the right balance between the laser speed and the bed temperature, it is the possible to print the complex gyroid lattices.

The tensile strength of the printed structures varied according to the polymer characteristic, with PEO and Eudragit RL presenting the lowest tensile strength (not detected), while the other polymers showed higher strength (280 N). The low tensile strength is attributed to the low intensity sintering process that was applied and led to greater gyroid porosity. When a more energetic sintering was applied, the observed molten areas increased on the surface of the structures. The tuning of the sintering, the polymer nature, and the selection of the polymers impacted on the porosity and tensile strength of the gyroid lattices and subsequently on the obtained dissolution rates. Despite the low paracetamol loading, the study was a good paradigm of how to develop personalised dosage forms with customised dissolution profiles (Figure 3.7).

A further application of SLS was introduced by Awad et al. for the fabrication of multi-drug miniprintlets in a two-step process [39], by using powder blend combinations of polymers with paracetamol and ibuprofen (IBU). The aim was to alter the configuration of PCM/Kollicoat and IBU/EC and vice versa in order to customise the drug dissolution rates and achieve dual release profiles (Figure 3.8).

The sintering process varied depending on the polymer properties, with Kollicoat undergoing low intensity sintering due to their round shape. The irregular shapes of EC (large and flaky) resulted in an intense sintering. The drug dissolution could be controlled through the selection of the processed blends so Kollicoat advanced fast rates, whilst for EC, sustained release was observed for both drugs. Furthermore, by varying the pellet size (1–2 mm) it was possible to partially obtain different release rates for IBU and PCM for up to 24 hours when sintered with EC. In this work, dual drug release was feasible with fast dissolution of the first drug followed by sustained release of the second from the printed minipellets.

A more detailed study was presented by Ali et al., who investigated the effect of the formulation and process variables on the quality of the printed dosage forms by applying a Quality by Design (QbD) approach [40]. The implementation of an experimental design (DoE) assisted in the investigation of critical processing and material parameters where the laser scanning speed (mm s^{-1}), surface temperature (oC), and the concentration of lactose monohydrate (LMH) were selected as the dependent variables. The DoE helped to identify the effect of the process variable on the critical quality attributes (CQA) known a dependent variable. As such, the selected CQA were the weight, disintegration time, and drug dissolution rate at 15 minutes. The main drug carriers were Kollidon VA64 and LMH, but only the latter was found to be a key material attribute. The chamber temperature had a positive effect on the tablet weight (increased), while laser scanning speed and LMH (8–12%) showed a negative impact. Higher sintering was observed for high chamber temperatures, while low sintering resulted in high scanning speeds or with increased LMH amounts.

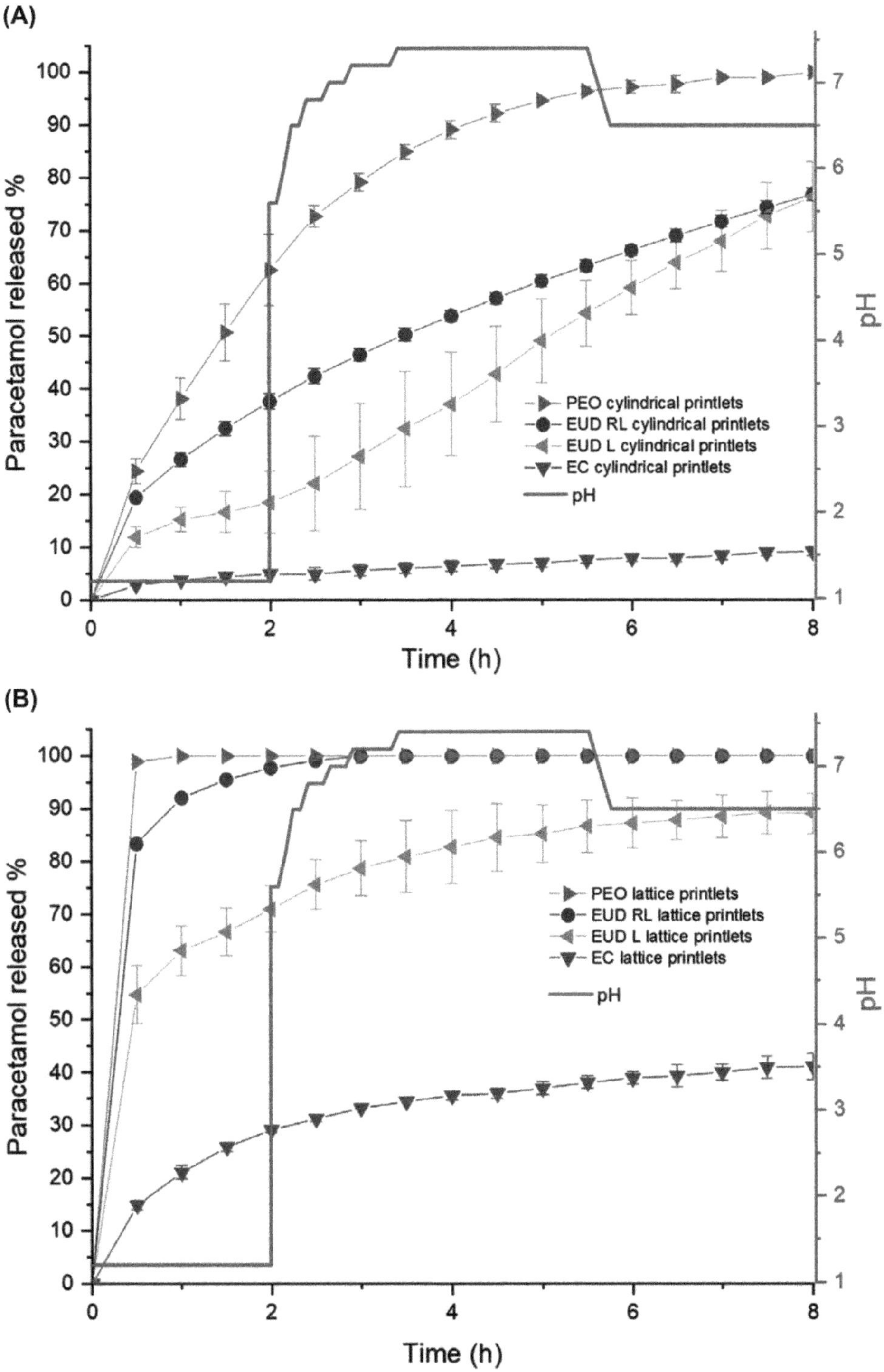

Figure 3.7 *Paracetamol dissolution profiles of (A) the cylindrical and (B) gyroid lattice constructs. The red line illustrates the pH values in acidic media (2 hours), followed by higher pH values (alkaline media).*

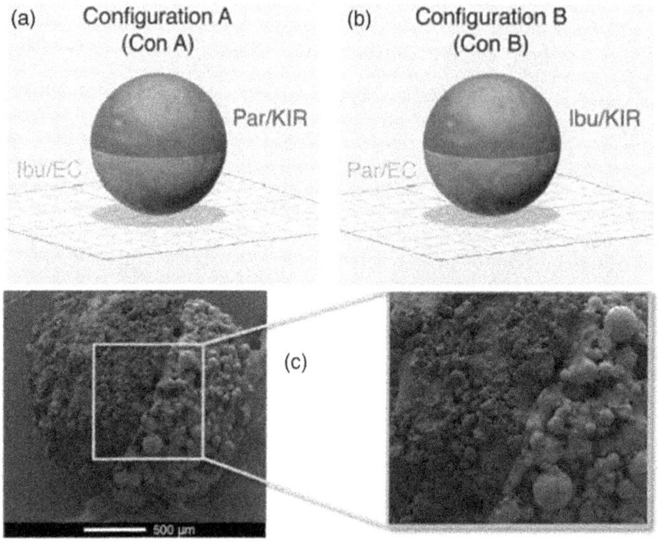

Figure 3.8 *Printed minipellets using SLS. Schematic representation of the compositions of (a) configuration A and (b) configuration B of the dual miniprintlets. (c) Scanning electron microscopy (SEM) images of the 2-mm dual minipellets.*

In a follow-up study, the same group applied the QbD approach for the development of clindamycin palmitate hydrochloride using SLS [41]. Here the concentration of lactose, microcrystalline cellulose (MCC), and the laser scanning speed were selected as critical processing parameters and tablet weight, tensile strength, disintegration times, and dissolution rates as CQAs, respectively. The integration of the Box–Bencken design showed significant effect of the laser scanning speed for all of the dependent variables. For the CMAs, the analysis revealed that MCC significantly affected the tablet tensile strength, disintegration time, and dissolution, while lactose affected the tensile strength and dissolution rates. The composition of the print formulation affected the powder flowability and thus printability with successful printing at >5% of lactose or MCC concentrations. As previously mentioned, the decrease in laser scanning speeds led to increased tablet weight due to higher sintering that is caused by the energy absorption and subsequent melting/softening. The authors observed that lactose decreased the tablet mechanical properties as it was partially melted or non-melted in the printed formulations. Despite the relatively high lactose amounts in the formulation, sintering led to the formation of tablets with low tensile strength but high disintegration times. As expected, higher chamber temperatures resulted in stronger tablets with high tensile strength due to intense sintering, while fast laser scanning produced weaker tablets and had a negative effect. Similarly, stronger tablets were produced in formulations with increased Kollidon amounts. Fast disintegration times were aligned with tablets of low tensile strength that became longer with increasing tablet hardness. The laser speed and the chamber temperature had a significant impact on both the disintegration time and dissolution rates of Na-diclofenac. High dissolution rates obtained for low tablet weight, tensile strength, or disintegration times and contrary-wise.

Gueche et al. studied the sinterability of CO_2 laser for the formation of solid oral dosage form (SODF) via SLS [42]. For the study, Kollidon® VA64, paracetamol, and DuraForm®

polyamide 12 (PA12) were used as polymeric carrier, model drug, and reference powder, respectively. The work focused on the printing of different PCM grades with one of large and plate-like particles, whilst the PCM-fine grades comprised of thin and needle-like particles. Previous studies have demonstrated that by increasing the laser wavelength, absorbance also increases for a wide range of polymers [21]. FTIR analysis showed that KVA64 absorb at a wavelength of 10.2–11 μm, which was in the range of polyamide $\lambda = 10.6$ μm and thus the addition of absorbance enhancer was not required [43]. The blending of powders with different shapes affects the powder flowability and hence the selection of PCM grades, and the PCM/polymer ratios were found to be critical for the compactness of the blends. In this study, PCM loadings varied from 10–30%, which are more representative compared to previous studies [44] and can reveal a more realistic prospect on the SLS suitability for personalised medications.

The printed SODFs showed that a high proportion of PCM was in an amorphous state without been degraded, even within SODFs printed at high laser intensities. The morphology of the printed tablets was affected by the proportion of coarse particles which form rough surfaces with irregular shapes when their content increases (Figure 3.8). The coarse particles of VA64 produced tablets with higher porosity (Figure 3.9) compared to the control PA12 which presented denser structures. Furthermore, paracetamol loading affects the tablet porosity, as non-sintered amounts prevent the formation of a continuous melting phase and hence increased porosity was observed. Overall, the implementation of a CO_2 laser resulted in SODFs with excellent tensile strength, while PCM rapid dissolution which reached 100% within 20 minutes, meets pharmacopeia specifications. The study was an outstanding paradigm for the design and printing of actual drug loaded tablets using CO_2 SLS technology.

Figure 3.9 *(a) Printed KVA64 and PA12 SODFs at optimal parameters and (b) SEM images of printed tablet surface (magnification × 30).*

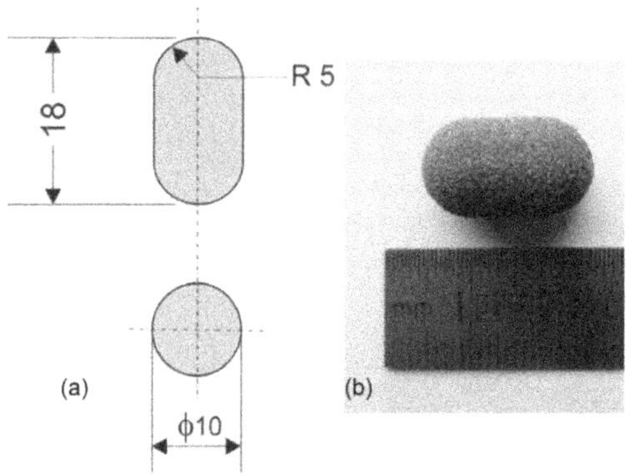

Figure 3.10 *The example of printlet: (a) Technical drawing (dimensions in mm) and (b) printed Formulation A (Met/PA12/NaCl, 80/20/0/100).*

Kulinowski et al. investigated a polymer-augmented complex formulation for the development of floating drug delivery systems (FDDS), and the use of water-insoluble nylon as an inert excipient for high metronidazole (Met) loading using SLS technology [45]. Despite the fact that carbon-stained polyamide (PA12) is not a typical pharmaceutical grade polymer, there are reports stating its biocompatibility [46]. The use of PA12 aided Met loadings of 82–92% with suitable hardness (40 N) for 800 mg tablets. As shown in Figure 3.10, the printed tablets presented a grey colour due to nylon with a rough surface.

Analysis of the phase transitions of nylon and Met during sintering showed the transformation of PA12 from the γ′ γ form [47–50], and the presence of crystalline Met with melting endothermic peaks at 179.79 °C and 159.63 °C, respectively. The drug release studies showed variable Met dissolution rates depending on the apparatus employed for the studies (USP4 or USP3) and the addition of NaCl (2% v/v) in the printed tablets. The study demonstrated another implementation of SLS technology to produce oral dosage forms with good tablet properties and customisable dissolution rates.

Madžarević et al. evaluated the effect of energy density (ED) and formulation components on the printability of Irbesartan (IRB) tablets by applying a decision tree model (Figure 3.11) as a data mining tool [51]. The decision tree modelling is an interesting approach and entails the use of input data such as energy density, particle size distribution, and the content of the drug carries. The optimal combination of processing parameters can be optimised by splitting the data between a training (70%) and test set (30%). By using more complex formulations comprising of HPMC (good printability), mannitol (good flowability, disintegrant), and Kollidon VA64 and Crospovidone (super disintegrant), the printing process was optimised by varying the chamber temperature, surface temperature, and laser speed.

The crospovidone content and the ED were the critical attributes for the printability of the designed formulations. The tablets were printable if crospovidone was >3.5%, but for lower concentrations the ED should be ≥0.615 J mm s^{-3} in order to obtain printable tablets. As shown in the decision tree (Figure 3.11), if none of the above conditions are met, then the HPMC content should be taken into account. A drawback of the approach was that for most of the printed tablets the hardness was not detectable whilst for the rest it varied from 12.5–56.0 N, which are still considered weak. Interestingly, the addition of a superdisintegrant had no impact on the disintegration times, which were very high (90 seconds to 1,510 seconds).

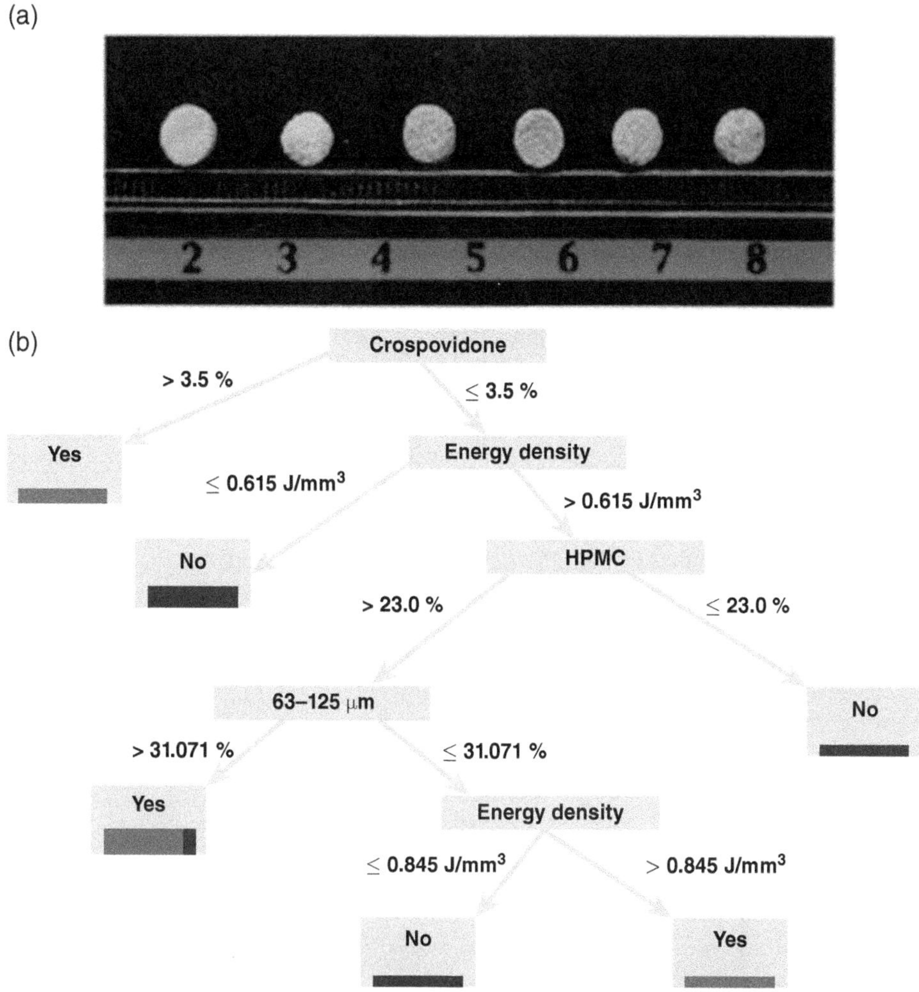

Figure 3.11 *(a) Images of selected SLS-printed tablets printed at different laser scanning speeds. From left to right: 100.00, 120.00, 140.00, 160.00, 180.00, and 220.00 mm s^{-1}. (b) Decision tree for SLS printability of powder mixtures with irbesartan.*

The authors observed that the melting point of crospovidone was close to the printing temperature, which resulted in good sintering but affected the disintegration times and subsequently the release profiles. In our opinion, crospovidone should have been combined with KollidonVA64, which presents excellent printability but most importantly a much lower glass transition temperature and thus less ED is required. This is in good agreement with the experimental findings where lower ED resulted in faster disintegration times. Furthermore, the use of superdisintegrants requires a good knowledge of the material properties. For example, the Polyplasdone XL10 grade has a particle size distribution of 30–50 μm, which is much smaller to that of HPMC which inhibited the disintegration process. However, XL grades (100–130 μm) would result in faster disintegration times due to the large particle size. Two more critical observation are that Polyplasdone grades present excellent performance at concentrations >5% and they present wicking behaviour which is not suitable for sintered tablets. For SLS printed tablets, the use of superdisintegrants with swelling properties (e.g., croscarmellose sodium, sodium starch glycolate) would be a better option for rapid disintegration.

The SLS applicability for oral solid dosages was further investigated by Yang et al. using a range of APIs for immediate and sustained release respectively combined with various polymers [52]. Some of the drugs were coloured and thus it was possible to prepare tablets through direct printing whilst the rest required the addition of a photo absorber (tartrazine lake). The amount of photo absorber is considered critical when using lasers at 450 nm and it should be investigated to achieve printing accuracy. An interesting observation was that the sintered thickness for each layer should be larger than the layer thickness to ensure print integrity between the tablet layers.

For some of the pharmaceutical grade polymers (EPO, PEG-4000), widening and warpage deformation was observed at the edges of the tablet, while carboxymethyl starch sodium presented poor printability due to the high T_g. As expected, the laser energy intensity affected the printability and low laser intensities (Figure 3.12) resulted in poor tablet properties which further improved by increasing the applied energy (>0.6 J mm s^{-2}). The addition of release modifiers (e.g., PEG, PVA) in the powder blends facilitated faster dissolution rates, especially for water-insoluble APIs such as ibuprofen. The blending of sustained release polymers such as HPMC and Eudragit RL was further investigated. The amount of HPMC was critical in the printing of the designed formulations due to its poor sinterability, but also due to the significant drug release reduction. However, RL presented superior printing features and sustained drug release was achieved for 12 hours by adjusting the HPMC:RL ratios at 1:3.

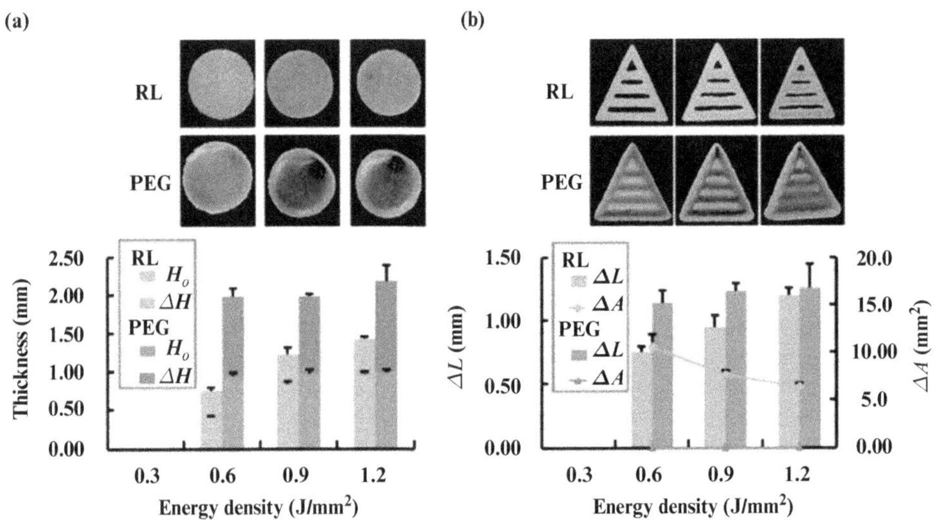

Figure 3.12 *Effect of laser energy density on the printability of Eudragit RL and PEG: (a) circle and (b) triangle samples (n = 3).*

3.4 Advantages of SLS

Conventional methods used for the manufacturing of oral dosage form incorporate direct compression, spray drying, freeze drying, granulation, or extrusion processes. However, some of these technologies present disadvantages with respect to the manufacturing costs, level of complexity, and also limitations in the required drug loadings. Whilst not necessarily getting rid of the conventional methods of manufacturing oral dosage forms, the use of SLS and its versatility in developing oral dispersion tablets has proven to be a significant advance that appears to be a promising asset to the pharmaceutical industry. As shown in this chapter, several pharmaceutical grade polymers have been introduced for printing oral dosages with flexible drug release properties. However, further studies using SLS for developing drug dosage forms with various APIs should be further conducted, including the evaluation of other excipients suited for SLS. Some of the SLS advantages are discussed below.

3.4.1 Printing Features

A great advantage of SLS is that there is no requirement for support structures. The non-sintered powder acts as the support materials to the printed tablet. As a result, the post processing is less complicated and does not damage the printed designs. In many cases the unstinted powder can be reused, which is important for expensive APIs. Several studies have shown that there is no need for powder pre-processing (e.g., extrusion) and thus SLS can be considered as a one-step printing process. Based on such features, the production cost can be significantly reduced in comparison to other printing technologies (e.g., FDM, SLA), rendering SLS a cost-effective technology.

3.4.2 Control of Surface Properties

One of the main advantages of SLS is the fabrication of various models by fusing powder particles together through sintering without the use of a solvent or the need for pre-processing of the powder blends. The sintering or melting of powders provides control over important morphological and mechanical surface features such as porosity. For many applications the control of porosity is advantageous as it allows the manipulation and alteration of the surface and its dimensions. The controls of the pore interconnection affects the tensile strength, friability, and, most importantly, the drug dissolution rates of drug formulations. The change of laser scanning speeds and the features of pharmaceutical excipients alter the structure of the produced tablets, resulting either in smooth or rough surfaces and dense content. High laser densities and slow scan speeds should be used to avoid carbonisation and burning of the tablet surface.

3.4.3 Printing of Complex Geometries

SLS is suitable for pharmaceutical applications due to its fast fabrication times and highly accurate dimensional control. When the powder blends (e.g., drug/polymer) are of narrow particle sizes below 100 μm, SLS allows the fabrication of highly precise geometrically accurate custom-made designs with complex and properties. As shown above, miniprintlets, gyroid lattices, bilayer tablets, reservoir-type designs, or triangle- and star-shape designs have been successfully printed with excellent accuracy.

3.4.4 Using a Wide Range of Materials

SLS can utilise a large variety of different materials for the fabrication of pharmaceutical dosage forms. Thermoplastic polymers are the most frequently used materials, including other excipients such as lipids and disaccharides (e.g., lactose) that can act as drug carriers or fillers and provide control over the tablet properties. For example, lactose has been shown to improve disintegration times, hardness, and dissolution. Similarly, effects have been reported when using polymer grades such as Kollidon VA64 or Povidone XL, which can promote fast disintegration times or act as superdisintigrants at low concentrations (3–5%). Hence, SLS has shown a great potential for the production of fast disintegrating tablets with relatively good mechanical properties.

As previously discussed, the selection of the polymeric excipients is crucial for the design and customisation of the drug dissolution rates. Several non-ionic or pH dependent hydrophilic polymers have been processed for the development of immediate release dosage forms by controlling the printing settings. Similarly, hydrophobic and high molecular weight polymers have been reported to provide sustained release rates of various APIs. However, hydrophilic polymers can act as sustained release matrices by increasing the sintering process and decrease the tablet porosity.

3.4.5 Drug Loading and Dose Combinations

An important feature of SLS is the capacity to co-process drug–polymer blends at various drug loadings. As the printability of the formulation solely depends on the polymer carrier, a wide range of drug loadings varying from 5–40% have been achieved when using SLS. In addition, the precise drug amount per tablet can be easily adjusted by altering the tablet dimensions and print settings without necessarily need to change the powder blend formulation. Unpublished work in our group showed accurate drug doses of 3.125, 6.25, and 12.5 mg by using carvedilol as the model substance. The tablets could be easily printed and present similar dissolution profiles without altering the drug–polymer blend composition.

Furthermore, SLS has been employed to provide dual drug release by combining two or more APIs at the required drug amounts. The dissolution rates can be adjusted using different formulations for each API. Newly designed CO_2 printers can process different powder feedstock, as they are supplied with two reservoirs.

3.4.6 Personalised Dosage Forms

A major advantage of SLS is the manufacturing of personalised dosage forms that can be fabricated at the point-of-patient care. SLS is a technology suitable for small to medium batch production and thus can be used at the point-of-care such as hospitals or even pharmacies. Currently, there are studies for fast disintegrating tablets but the customisation of palatable dosage forms by using SLS has not been proved. However, we anticipate that this will be exploited in the near future, especially for paediatric and geriatric patients. Januskaite et al. [53] demonstrate that 3D printing is an appealing technology amongst paediatric populations for the fabrication of tablets, whereas SLS was ranked as the second preferred approach. Furthermore, SLS could be used for printing polypills for geriatric patients to address issues such as polypharmacy or swallowing difficulties.

3.4.7 SLS Disadvantages

Despite the major advantages of SLS, there are a few minor drawbacks related to its use for pharmaceutical applications. The lead times for printing oral solid dosages are longer compared to other 3D-printing techniques such as FDM or SLA. This is because printing requires pre-heating of the powder bed and significant cooling times once the printing is completed. As previously mentioned, the obtained tables present a grainy surface finish, while the internal porosity of the tablets requires further processing to remove unsintered powder. There is a difficulty to print large flat surfaces and a lack of accuracy when it comes to small holes, as they are susceptive to warping and over-sintering.

3.5 Conclusions

SLS is a simple, inexpensive, and flexible 3D-printing technology that can be used for the manufacturing of personalised dosage forms or decentralised commercial production at point-of-care. The technology can be easily designed for the printing of a wide range of medicines without involving multi-step processes compared to conventional technologies. The SLS flexibility allows the accurate customisation of the active dose, design, shape, and the printing of several drugs in a single dose (polypills).

SLS has not been exploited to a great extent and its potential is not fully revealed. The existing studies are limited to simple formulations while only a few report the addition of other functional excipients. In addition, there are no clinical trials to provide evidence on the clinical efficacy of 3D-printed dosage forms and thus to support the SLS capabilities. Nevertheless, we anticipate that ongoing investigations will provide significant insights for the use of SLS in the development of novel pharmaceutical drug products.

References

[1] Curti, C., Kirby, D.J., and Russell, C.A. (2020). Current formulation approaches in design and development of solid oral dosage forms through three-dimensional printing. *Progress in Additive Manufacturing* **5**: 111–123. https://doi.org/10.1007/s40964-020-00127-5.

[2] Kodama, H. (1981). Automatic method for fabricating a three-dimensional plastic model with photo-hardening polymer. *Review of Scientific Instruments* **52**: 1770–1773. https://doi.org/10.1063/1.1136492.

[3] Gioumouxouzis, C.I., Katsamenis, O.L., Bouropoulos, N., and Fatouros, D.G. (2017). 3D printed oral solid dosage forms containing hydrochlorothiazide for controlled drug delivery. *Journal of Drug Delivery Science Technology* **40**: 164–171. https://doi.org/10.1016/j.jddst.2017.06.008.

[4] Sandler, N. and Preis, M. (2016). Printed drug-delivery systems for improved patient treatment. *Trends in Pharmacological Sciences* **37** (12): 1070–1080. https://doi.org/10.1016/j.tips.2016.10.002.

[5] Crump, S.S. (1989). Apparatus and method for creating three-dimensional objects. *Patent*, issued 1989. https://doi.org/10.2116/bunsekikagaku.28.3_195.

[6] Atabak, T.G., Fullbrook, D.H., Vilain, L. et al. (2021). Personalised tasted masked chewable 3D printed fruit-chews for paediatric patients. *Pharmaceutics* **13** (8): 1301. https://doi.org/10.3390/pharmaceutics13081301.

[7] Ghanizadeh Tabriz, A., Nandi, U. et al. (2020). 3D printed bilayer tablet with dual controlled drug release for tuberculosis treatment. *International Journal of Pharmaceutics* **593**: 120147. https://doi.org/10.1016/j.ijpharm.2020.120147.

[8] Mohamdeen, Y.M.G., Tabriz, A.G., Tighsazzadeh, M. et al. (2021). Development of 3D printed drug-eluting contact lenses. *Journal of Pharmacy and Pharmacology* **74** (10): 1467–1476. https://doi.org/10.1093/jpp/rgab173.

[9] Beamen, J.J. and Deckard, C.R. (1990). *Selective laser sintering with assisted powder handling*. Patent No. US 4938816, issued 1990.

[10] Mazzoli, A. (2013). Selective laser sintering in biomedical engineering. *Medical and Biological Engineering and Computing* **51**: 245–256. https://doi.org/10.1007/s11517-012-1001-x.

[11] Strub, J.R., Rekow, E.D., and Witkowski, S. (2006). Computer-aided design and fabrication of dental restorations: current systems and future possibilities. *Journal of the American Dental Association* **137** (9): 1289–1296. https://doi.org/10.14219/jada.archive.2006.0389.

[12] Bertrand, P., Bayle, F., Combe, C. et al. (2007). Ceramic components manufacturing by selective laser sintering. *Applied Surface Science* **254** (4): 989–992. https://doi.org/10.1016/j.apsusc.2007.08.085.

[13] Chow, L.K. and Cheung, L.K. (2007). The usefulness of stereomodels in maxillofacial surgical management. *Journal of Oral Maxillofacial Surgery* **65**: 2260–2268. https://doi.org/10.1016/j.joms.2006.11.041.

[14] Ciocca, L., Fantini, M., De Crescenzio, F. et al. (2011). Direct metal laser sintering (DMLS) of a customized titanium mesh for prosthetically guided bone regeneration of atrophic maxillary arches. *Medical and Biological Engineering and Computing* **49**: 1347–1352. https://doi.org/10.1007/s11517-011-0813-4.

[15] Drummer, D., Rietzel, D., and Kühnlein, F. (2010). Development of a characterization approach for the sintering behavior of new thermoplastics for selective laser sintering. *Physics Procedia* **5**: 533–542. https://doi.org/10.1016/j.phpro.2010.08.081.

[16] Campbell, T., Williams, C., Ivanova, O., and Garrett, B. (2011). Could 3D printing change the world? *Atlantic Council*. http://www.jstor.org/stable/resrep03564.

[17] Wang, X., Jiang, M., Zhou, Z. et al. (2017). 3D printing of polymer matrix composites: a review and prospective. *Composites, Part B: Engineering* **110** (1): 442–458. https://doi.org/10.1016/j.compositesb.2016.11.034.

[18] Gibson, I. and Shi, D. (1997). Material properties and fabrication parameters in selective laser sintering process. *Rapid Prototyping Journal* **3** (4): 129–136. https://doi.org/10.1108/13552549710191836.

[19] Jain, P.K., Pandey, P.M., and Rao, P.V.M. (2008). Experimental investigations for improving part strength in selective laser sintering. *Virtual and Physical Prototyping* **3** (3): 177–178. https://doi.org/10.1080/17452750802065893.

[20] Gibson, I., Rosen, D., and Stucker, B. (2015). Powder bed processes. In: *Additive Manufacturing Technologies: 3D Printing, Rapid Prototyping, and Direct Digital Manufacturing*, 2e. 107–145. https://doi.org/10.1007/978-1-4939-2113-3.

[21] Tolochko, N.K., Laoui, T., Khlopkov, Y.V. et al. (2000). Absorptance of powder materials suitable for laser sintering. *Rapid Prototyping Journal* **6** (3): 155–160. https://doi.org/10.1108/13552540010337029/FULL/XML.

[22] Kruth, J.P., Wang, X., Laoui, T., and Froyen, L. (2003). Lasers and materials in selective laser sintering. *Assembly Automation* **23** (4): 357–371. https://doi.org/10.1108/01445150310698652.

[23] Savalani, M., Hao, L., and Harris, R.A. (2006). Evaluation of CO_2 and Nd:YAG lasers for the selective laser sintering of HAPEX. *Proceedings of the Institution of Mechanical Engineers, Part B: Journal of Engineering Manufacturing* **220** (2): 171–182. https://doi.org/10.1243/095440505X32986.

[24] Caulfield, B., McHugh, P.E., and Lohfeld, S. (2007). Dependence of mechanical properties of polyamide components on build parameters in the SLS process. *Journal of Materials Processing Technology* **182** (1–3): 477–488. https://doi.org/10.1016/j.jmatprotec.2006.09.007.

[25] Ryse, R.V, Edeleva, M., Van Stichel, O. et al. (2022). Setting the optimal laser power for sustainable powder bed fusion processing of elastomeric polyesters: a combined experimental and theoretical study. *Materials (Basel)* **15** (1): 385. https://doi.org/10.3390/ma15010385.

[26] Goodridge, R.D., Tuck, C.J., and Hague, R.J.M. (2012). Laser sintering of polyamides and other polymers. *Progress in Material Science* **57** (2): 229–267. https://doi.org/10.1016/j.pmatsci.2011.04.001.

[27] Bai, J., Zhang, B., Song, J. et al. (2016). The effect of processing conditions on the mechanical properties of polyethylene produced by selective laser sintering. *Polymer Testing* **52**: 89–93. https://doi.org/10.1016/j.polymertesting.2016.04.004.

[28] Ziegelmeier, S., Christou, P., Wöllecke, F. et al. (2015). An experimental study into the effects of bulk and flow behaviour of laser sintering polymer powders on resulting part properties. *Journal of Materials Processing Technology* **215**: 239–250. https://doi.org/10.1016/j.jmatprotec.2014.07.029.

[29] Reesi, F., Minaiyan, M., and Taheri, A. (2018). A novel lignin-based nanofibrous dressing containing arginine for wound-healing applications. *Drug Delivery and Translational Research* **8**: 111–122. https://doi.org/10.1007/s13346-017-0441-0.

[30] Garg, V., Mallick, S.S., Garcia-Trinanes, P., and Berry, R.J. (2018). An investigation into the flowability of fine powders used in pharmaceutical industries. *Powder Technology* **336**: 375–382. https://doi.org/10.1016/j.powtec.2018.06.014.

[31] Douroumis, D. (2019). 3D printing of pharmaceutical and medical applications: a new era. *Pharmaceutical Research* **36**: 42. https://doi.org/10.1007/s11095-019-2575-x.

[32] Leong, K.F., Chua, C.K., and Gui, W.S. (2006). Building porous biopolymeric microstructures for controlled drug delivery devices using selective laser sintering. *International Journal of Advanced Manufacturing Technology* **31**: 483–489. https://doi.org/10.1007/s00170-005-0217-4.

[33] Salmoria, G.V., Klauss, P., Zepon, K.M., and Kanis, L.A. (2013). The effects of laser energy density and particle size in the selective laser sintering of polycaprolactone/progesterone specimens: morphology and drug release. *International Journal of Advanced Manufacturing Technology* **66**: 1133–1118. https://doi.org/10.1007/s00170-012-4393-8.

[34] Salmoria, G.V., Klauss, P., and Kanis, L.A. (2017). Laser printing of PCL/progesterone tablets for drug delivery applications in hormone cancer therapy. *Lasers in Manufacturing and Materials Processing* **4** (3): 108–120. https://doi.org/10.1007/s40516-017-0040-4.

[35] Fina, F., Goyanes, A., Gaisford, S., and Basit, A.W. (2017). Selective laser sintering (SLS) 3D printing of medicines. *International Journal of Pharmaceutics* **529** (1): 285–293. https://doi.org/10.1016/j.ijpharm.2017.06.082.

[36] Fina, F., Goyanes, A., Madla, C.M. et al. (2018). 3D printing of drug-loaded gyroid lattices using selective laser sintering. *International Journal of Pharmaceutics* **547** (1): 44–52. https://doi.org/10.1016/j.ijpharm.2018.05.044.

[37] Khaderi, S.N., Deshpande, V.S., and Fleck, N.A. (2014). The stiffness and strength of the gyroid lattice. *International Journal of Solids and Structures* **51** (23–24): 3866–3877. https://doi.org/10.1016/j.ijsolstr.2014.06.024.

[38] Yan, C., Hao, L., Hussein, A., and Raymont, D. (2012). Evaluations of cellular lattice structures manufactured using selective laser melting. *International Journal of Machine Tools and Manufacture* **62**: 32–33. https://doi.org/10.1016/j.ijmachtools.2012.06.002.

[39] Awad, A., Fina, F., Trenfield, S.J. et al. (2019). 3D printed pellets (Miniprintlets): a novel, multi-drug, controlled release platform technology. *Pharmaceutics* **11** (4): 148. https://doi.org/10.3390/pharmaceutics11040148.

[40] Barakh, A., Sogra, F., Mohamed, E.M. et al. (2019). Understanding the effects of formulation and process variables on the printlets quality manufactured by selective laser sintering 3D printing. *International Journal of Pharmaceutics* **30**: 118651. https://doi.org/10.1016/j.ijpharm.2019.118651.

[41] Mohamed, E.M., Barakh Ali, S.F., Rahman, Z. et al. (2020). Formulation optimization of selective laser sintering 3D-printed tablets of clindamycin palmitate hydrochloride by response surface methodology. *AAPS PharmSciTech* **21**: 232. https://doi.org/10.1208/s12249-020-01775-0.

[42] Gueche, Y.A., Sanchez-Ballester, N.M., Bataille, B. et al. (2021a). Selective laser sintering of solid oral dosage forms with copovidone and paracetamol using a CO_2 laser. *Pharmaceutics* **13** (2): 160. https://doi.org/10.3390/pharmaceutics13020160.

[43] Fina, F., Madla, C.M., Goyanes, A. et al. (2018). Fabricating 3D printed orally disintegrating printlets using selective laser sintering. *International Journal of Pharmaceutics* **541** (1): 101–107. https://doi.org/10.1016/J.IJPHARM.2018.02.015.

[44] Gueche, Y.A., Sanchez-Ballester, N.M., Bataille, B. et al. (2021b). Selective laser sintering (SLS), a new chapter in the production of solid oral dosage forms (SOFs) by 3D printing. *Pharmaceutics* **13** (8): 1212. https://doi.org/10.3390/pharmaceutics1308121.

[45] Kulinowski, P., Malczewski, P., Łaszcz, M. et al. (2022). Development of composite, reinforced, highly drug-loaded pharmaceutical printlets manufactured by selective laser sintering: in search of relevant excipients for pharmaceutical 3D printing. *Materials (Basel)* **15** (6): 2142. https://doi.org/10.3390/ma15062142.

[46] Shakiba, M., Ghomi, E.R., Khosravi, F. et al. (2021). Nylon: a material introduction and overview for biomedical applications. *Polymers for Advanced Technologies* **32** (9): 3368–3383. https://doi.org/10.1002/pat.5372.

[47] Ma, N., Liu, W., Ma, L. et al. (2020). Crystal transition and thermal behavior of Nylon 12. *E-Polymers* **20** (1): 346–352. https://doi.org/10.1515/EPOLY-2020-0039/MACHINEREADABLECITATION/RIS.

[48] Ramesh, C. (1999). Crystalline transitions in Nylon 12. *Macromolecules* **32** (17): 5704–5706. https://doi.org/10.1021/MA9904940.

[49] Dadbakhsh, S., Verbelen, L., Verkinderen, O. et al. (2017). Effect of PA12 powder reuse on coalescence behaviour and microstructure of SLS parts. *European Polymer Journal* **92**: 250–262. https://doi.org/10.1016/J.EURPOLYMJ.2017.05.014.

[50] Martynková, G.S., Slíva, A., Kratošová, G. et al. (2021). Polyamide 12 materials study of morpho-structural changes during laser sintering of 3D printing. *Polymers* **13**: 810. https://doi.org/10.3390/POLYM13050810.

[51] Madžarević, M., Medarević, Đ., Pavlović, S. et al. (2021). Understanding the effect of energy density and formulation factors on the printability and characteristics of SLS Irbesartan tablets: application of the decision tree model. *Pharmaceutics* **13** (5): 810. https://doi.org/10.3390/pharmaceutics13111969.

[52] Yang, Y., Xu, Y., Wei, S., and Shan, W. (2021). Oral preparations with tunable dissolution behavior based on selective laser sintering technique. *International Journal of Pharmaceutics* **593**: 120127. https://doi.org/10.1016/j.ijpharm.2020.120127.

[53] Januskaite, P., Xu, X., Ranmal, S.R. et al. (2020). I spy with my little eye: a paediatric visual preferences survey of 3D printed tablets. *Pharmaceutics* **12** (11): 1100. https://doi.org/10.3390/pharmaceutics12111100.

4

3D Printing for Medical Device Applications

Jian-Feng Zhang[1], *Bernabe Tucker*[1], *John Tipton*[1], *Mahrokh Dadsetan*[1], *Scott Jones*[1], *Andrea Engel*[1] *and Cecile Boudot*[2,*]

[1] *Evonik Corporation, Birmingham, Alabama, US*
[2] *Evonik Operations GmbH, Kirschenallee, Darmstadt, Germany*

4.1 Introduction

Additive manufacturing (AM) is not a new methodology for nature. Various organisms on earth have been utilising this technique for millions of years. From the honeycomb structures built by bees to the impressive reefs constructed by corals, nature has found various avenues to take base material and build up layers to form complex structures. Layer-by-layer addition of material is the basic principle of AM. Humans have been employing crude forms of AM in various endeavours such as bricklaying and pottery for millennia. However, the employment of modern industrial technology and innovations in AM is typically regarded as having its beginnings in the 1960s with the idea of layer-by-layer fabrication by machine. AM had its practical realisation in the 1980s with the advent of stereolithography (SLA) which employed photopolymers, masks, and light to create structures through layer-by-layer addition. Later, miniaturised extrusion and control equipment led to the development of fused filament fabrication (FFF), which was likely the most widely recognised form of AM and created structures via melted filament layers stacked on each other to form complex structures.

* Corresponding author: cecile.boudot@evonik.com

3D Printing of Pharmaceutical and Drug Delivery Devices: Progress from Bench to Bedside, First Edition.
Edited by Dimitrios A. Lamprou, Dennis Douroumis and Sheng Qi.

Since first being used for medical purposes such as dental implants and custom prosthetics in the 1990s, AM or 3D printing, as a disruptive technology compared to conventional manufacturing processes, has made a profound impact on the medical field, making personalised medical treatments possible and feasible, as shown in Scheme 4.1.

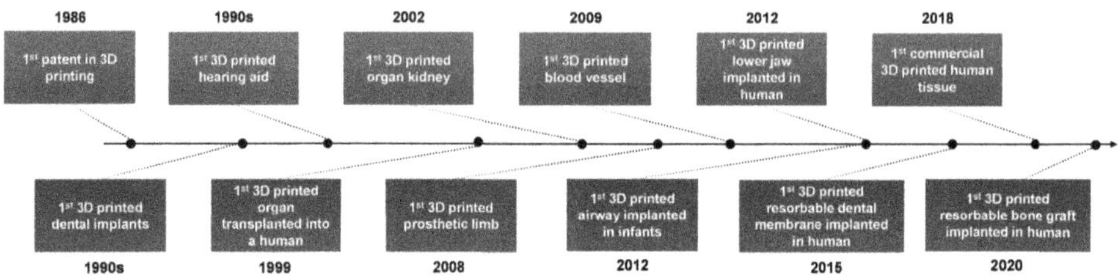

Scheme 4.1 *Evolution of 3D-printed medical devices.*

Off-the-shelf medical devices are made by conventional manufacturing processes, such as injection moulding, providing standardised structure, functions, and dimensions. Production costs can be controlled to become minimal by mass production. However, healthcare professionals must make some adjustments to the devices to better fit the intended implantation. Differences in patients with respect to their ages, genders, and sizes often make standardised devices a challenge to fit for every application. This is even true, for example, in paediatric patients, as their growth is the most important aspect of development during adolescence. The use of medical devices in these patients must consider the dynamic changes related to their growth. Personalised devices, and in certain circumstances resorption of the devices, preventing the subsequent surgical removal, provide alternative and beneficial treatment. 3D printing would provide solutions to help achieve the manufacturing of complex implants with precise internal design matching the patient's anatomy.

With these unparalleled benefits, significant advances in 3D-printing technology have been made in terms of materials, printers, and processes. Future Market Insights Global and Consulting Pvt. Ltd. forecasted in 2022 that global 3D-printed medical devices' market value would be about USD 752.5M in 2022 and USD 1,839.2M in 2028, with a stupendous growth rate at 16.1% compound annual growth rate [1].

However, this significant growth cannot underestimate the challenges and technical difficulties that 3D printing of medical implants currently encounters. From an industrial point of view, with many years of practice in materials development and optimisation for 3D printing, it is imperative to give an overview of the current landscape of 3D-printing technology with a focus on materials, printing technology, 3D-printed medical implants, challenges, and future perspectives.

4.2 3D Printers

Various 3D-printing technologies exist, and the space of available technologies is expanding. Since the original patents started to expire in the late 2000s and early 2010s, the market has seen significant growth in both the consumer and commercial/industrial sectors. The open-source movement has contributed to the current availability of machines and recent innovations in 3D-printing technology. The 3D-printing revolution that began in 2005 with the ReRap project showed the possibility of open-source technology to drive technology forwards [2]. With the expiration of the original patents, the emergence of open-source hardware, and base of enthusiasts the 3D-printing revolution is currently in full swing.

A typical process of 3D printing starts with the generation of a model *in silico* and converting it into a series of data that encode two-dimensional (2D) slices at defined thicknesses. These slices summed together approximate the original model and are converted into machine executable commands that drive the kinematics and material handling mechanisms to render these slices as stacked layers of printed material. The machine continues executing the commands to add additional layers until the entire model is fabricated from the material of choice. After printing, the model may undergo post processing to remove supports, change the surface finish, or add cosmetic modifications.

Software, called a slicer, converts the designs into layers called slices. These slices are further algorithmically converted into datasets of machine commands that drive the kinematics of the 3D printer to realise the model from material stock that is built up in layers to form the final object. These datasets are given in a machine language called gcode. These functions are the core features of 'slicers' or slicing software that is required for 3D-printing activities. Another function of slicing software is to generate gcode for printing parameters such as temperature, speed, layer height, or other parameters. Together, the combination of computer-aided design (CAD) software and slicing software make up the tools required for data generation and processing to physically realise a model from concept to machine instructions. Some popular choices for CAD software include SOLIDWORKS™, AutoCAD®, Fusion360™, and FreeCAD. Some popular choices for slicers include Simplify3D, Cura, Slic3r, PrusaSlicer, SuperSlicer, and OctoPrint, amongst others.

The workflow for 3D printing begins with design. Engineers, designers, or other application specialists will evaluate the design requirements, geometric dimensioning and tolerancing, material properties and selection, end-use applications, timeline, budget, and other factors to generate a concept that is realised with the use of CAD software. Often, the design stage will be iterative, since a range of factors related to 3D printing in general, and machine design specifically, may influence the original design to necessitate changes. Often, there is a feedback process from what the printer actually prints to what needs to be modified in the design. Next, the design file from CAD is processed with slicer software to convert the CAD file into a gcode file that the selected printer can execute. Next, the selected printer is loaded with the selected material and the gcode file is loaded onto the printer. The printer executes the gcode file to build the print from the selected material. After printing, the print is evaluated for quality and may undergo post-processing steps such as removing supports, annealing, and painting. If no additional design modification is required, the final print is released for end use.

Below are some examples of 3D printers that have been used in medical devices printing. Largely, SLA and FFF technologies make up the core of this area, but other techniques are also emerging and have a high possibility to grow as well.

4.2.1 SLA

The initial system for 3D printing was SLA and was invented by Charles Hull, founder of 3D systems, which was patented in 1984. This system used ultraviolet (UV) light to cure a photopolymer resin layer-by-layer to build a 3D object [3]. For this type of equipment, the file format known as stereolithography (STL) was developed and has since become the *de facto* standard file format for exporting CAD files to be sliced. The major benefit of the SLA technique is the ability to achieve high resolution and fine detail in the final part. This technique often necessitates the use of supports that are added to the print design to reinforce parts of the print that may be subject to movement, warping, or distortion. These supports help maintain the integrity of the print during the printing process and are removed from the part during post processing. The mechanical setup of a typical SLA printer highlights the reason support structures may be needed. First, there is usually a vat or container with a transparent floor. The vat is filled with the selected resin system. There is a build plate attached to a lead screw and stepper motor that lowers the build plate into the vat and raises the plate by the specified layer height after the layer is completed. The light system projects the pattern of each slice onto the build plate immersed in the resin at a known, specified height and cures the resin to complete the layer. In typical SLA systems, UV laser light traces the shape of the slice with an infill for solid sections as well as the shape of the supports. It is important to note that SLA with lasers cures the resin point-by-point; therefore, the maximum printing speed is limited but the achievable resolution is high. Another setup for curing is digital light processing (DLP), which utilises an array of micromirrors that project a pixelated image of the light source. The advantages are that this system is faster, requires less maintenance, and may be made quite large but at a slight loss of resolution. Liquid crystal display (LCD) technology has enabled SLA printing to be accessible at a price point that is attractive to the consumer market. The LCD is used to create a light/dark pixel map that can be illuminated with low-cost light emitting diodes (LED) to create a fast, low-cost, and minimal maintenance system, but the trade-off is resolution. This technique is sometimes called mask-SLA or MSLA technology.

4.2.2 FFF

Arguably, the most popular 3D-printing technology is FFF. Scott Crump is credited with being the founder of FFF technology which was called fused deposition modelling (FDM) by the company he founded, Stratasys [4]. FFF technology is, essentially, miniaturised extrusion with 3D control to orient the extrusion nozzle. FFF machines can achieve moderate resolution and moderate quality surface finishes. They typically have a rough and layered surface that is commonly associated with 3D-printed parts. FFF machines take a filament of selected material and push it through a heated nozzle that melts the filament. Stepper motors control the position of the nozzle as the melted filament traces out the pattern or shape dictated by the gcode onto a build plate that is often heated. The interior of the part can be adjusted for infill from solid to arbitrary percent density to balance the factors of strength and printing time. One method of directing the filament is direct drive in which the toolhead is equipped with the extruder system that uses friction to drive the filament into the heated nozzle. Another method is the Bowden tube system where the extruder system is mounted away from the toolhead holding the nozzle and pushes the filament into a Bowden tube that then guides the filament to the nozzle. There are trade-offs to each type of system. Additional considerations are that the nozzle has both a thermocouple and electric heater to precisely control the temperature of the nozzle. Furthermore, the print bed is often outfitted with a thermocouple and heater to offer independent control and temperature for the printed part. The printing chamber might be enclosed and have yet another heater/thermocouple combination to separately control the ambient temperature during printing.

4.2.3 Selective Laser Sintering (SLS)

A parallel technology to FFF and SLA printers is SLS printing, which was developed at the University of Texas in the 1980s and later sold to DTM Corporation which later sold the technology to 3D systems [5]. In SLS, a bed of powders is exposed to a laser that traces a path in the powder to sinter the powders together forming the outline and infill of the part and these are often termed the contour and hatching patterns, respectively. SLS machines have a greater versatility of materials since there are machines that can sinter polymer materials and other machines that can sinter metal materials. Thus, SLS is a contender for competing with traditional subtractive manufacturing techniques for metal components. SLS processes do not require support structures since the powder bed helps hold the shape of the printed components. As a general overview of the SLS process, a chamber is heated to a temperature near but below the sintering/melting temperature of the powders. A laser system with a movable mirror and lensing system traces the pattern imposed by the machine commands. Then, a recoating system covers the pattern with a fresh layer of powder and the process repeats to produce a part that must be excavated from the powder cake. SLS enables versatile design and printed parts with high accuracy, resolution, and detail. Typically, infrared lasers are employed with carbon dioxide laser systems being the most common. Powder processing equipment is necessary for SLS operations since the powder forms sintered agglomerates that must be sieved out prior to attempting to reuse powders. The SLS machines themselves are often complex and may use a simple radial arm with a changeable blade that recoats powders over the bed or up to a roller system on guide rails to push additional powders to recoat the next layer. Additionally, machines may include systems for inert gas delivery to prevent polymer oxidation. Currently, polyamides (PA) are the dominant polymer for SLS printing. Other powders are becoming popular such as thermoplastic polyurethane (TPU), polypropylene, bioresorbable polymers, and composites such as alumide.

4.3 Biomaterials for 3D-Printed Medical Devices

Material selection is a critical aspect in 3D printing to achieve most desirable properties. Despite the great benefit that 3D printing can offer in medical fields, there are limited suppliers of suitable medical grade materials for 3D printing. Also, there is no universal biomaterial with its properties suitable for every type of 3D printing. This has been the major bottleneck limiting the wide proliferation of 3D printing [6]. With the great demand of personalised devices by 3D printing, early stage and feasibility studies of these materials are booming both in academia and industry. It is anticipated that a number of these materials will become reality in the near future. In principle, a material suitable for medical devices 3D printing should meet the following requirements: printability, biocompatibility, and desirable mechanical properties for the intended application. Herein, we provide an overview of currently available and promising potential biomaterials for 3D printing in the medical device field.

4.3.1 Bioresorbable Polymers

Recent innovations in 3D-printing technology and resorbable polymers are redefining opportunities for 3D printing of medical devices, even with complex geometries. Bioresorbable polymers naturally can be metabolised and excreted from the human body leaving no traces behind. An immediate benefit of these polymers is the elimination of the additional surgery needed to remove implanted medical devices after completion of the intended healing mission. In terms of bioresorbable polymers origination, they can be simply divided into two groups: synthetic and natural polymers.

4.3.1.1 Synthetic Bioresorbable Polymers

Synthetic polymers can be tailor-made and fine-tuned according to polymer structure, composition, and properties during the design and synthesis processes. They offer desirable characteristics for printability and better implantation outcome prediction. Historically, synthetic resorbable polymers, for example, lactide based polymers, have been approved by the US Food and Drug Administration (FDA) for many different medical applications over the last 30 years. Below, Scheme 4.2 lists the most common synthetic resorbable polymers that have been extensively used in medical fields. These semi-crystalline homopolymers, for example, poly(L-lactic acid) (PLLA), have been exclusively produced by ring-opening polymerisation (ROP) to achieve high molecular weight compared to a polycondensation process. Flexibility of ROP enables a series of copolymers of any combination of these readily available cyclic monomers which offer various processing and medical application opportunities, such as copolymers of poly(lactic acid-*co*-glycolic acid) (PLGA), poly(lactic acid-*co*-caprolactone) (PLC), etc. These synthetic homopolymers and copolymers are a group of aliphatic polyesters that can be degraded via hydrolysis on the ester linkages. Evonik Corporation, Corbion N.V. and Poly-Med, Inc. are manufacturers of these synthetic resorbable polymers for medical applications.

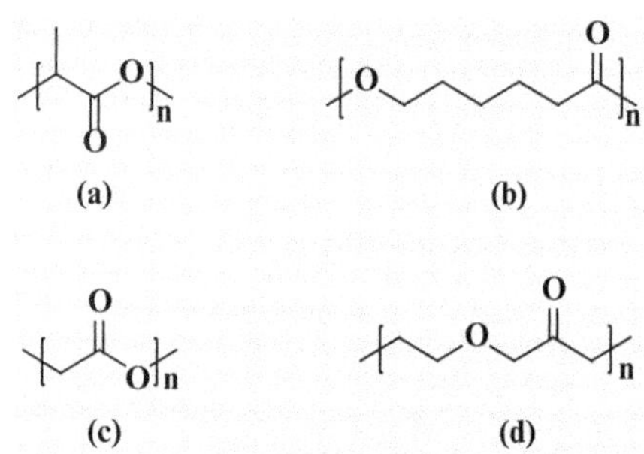

Scheme 4.2 *Typical synthetic resorbable polymers: (a) Polylactide (PLA); (b) Polycaprolactone (PCL); (c) Polyglycolide (PGA); (d) Polydioxanone (PDO).*

A semi-crystalline PLLA with a crystallinity of about 37% is obtained from L-lactide, whereas poly(DL-lactide) (PDLLA) is an amorphous polymer; however, they have different mechanical properties and degradation times. PLLA is a hard, transparent polymer with a tensile strength of 45–70 MPa. It has a melting point of 170–180 °C and a glass transition temperature (T_g) of 53 °C. PDLLA has no melting point and a similar T_g, and it has much lower tensile strength. Hydrolysis degradation starts preferably in the amorphous region and its degradation rate thus depends on crystallinity. PLLA has been arguably one of the most widely used polymers in 3D printing by FFF for medical applications, primarily due to its resorption, biocompatibility, high mechanical strength, and printability [7]. Although there are only a few studies on SLS 3D printing of PLA powder, the semi-crystalline nature fulfils important requirements for SLS printing. Sintering behaviour of SLS powders is greatly influenced by the polymer's thermal properties, melt viscosity, melt surface tension, and surface energy. PLA and carbonated hydroxyapatite (HA) nanocomposite scaffolds with controllable architecture and pore size for bone tissue engineering were 3D printed by SLS [8]. PCL, on the other hand, is a flexible resorbable polymer, having mechanical properties comparable to native scaffolds [9]. PCL is a semi-crystalline linear polymer with a T_g of around –60 °C, and melting point ranging from 60–65 °C. With these properties, PCL has attracted more attention for tissue engineering applications, in particular for bone and cartilage repairs. PCL is compatible with 3D-printing techniques such as FFF and SLS. Both PLLA and PCL homopolymers take extended time to degrade *in vitro* and *in vivo*.

PGA and PDO are two semi-crystalline polymers that usually have faster degradation rates than PLLA, PCL, and their copolymers. PGA is considered as strong and one of the first biodegradable synthetic polymers investigated for biomedical implants. It has a high crystallinity (45–50%) and therefore a high tensile modulus with very low solubility in organic solvents. Its T_g ranges from 35–40 °C and the melting point is greater than 200 °C. PDO is a flexible polymer with a T_g of about –10 °C, which is the material of choice for the first commercially developed monofilament sutures. It is known to lose its strength within 1–2 months [10] and its mass within 6–12 months by hydrolytic degradation. PGA and PDO have been utilised to adjust resorbable polymers' degradation time by blending or copolymerisation. Both polymers are suitable for 3D printing using FFF or SLS.

To meet increasing demands for 3D printing; for example, Evonik has commercialised RESOMER® Filaments based on a number of bioresorbable polymers with a diameter of 1.75 mm for FFF, which offer a variety of mechanical properties and degradation times for many medical needs. Table 4.1 lists the representative synthetic polymer properties and resultant filaments to facilitate the selection of bioresorbable polymers for FFF 3D printing. Likewise, Evonik has developed RESOMER® PrintPowder, the first of its kind, which is a free-flowing powder based on bioresorbable medical grade polymer designed for high resolution 3D printing with SLS. These powders provide optimised particle size and distribution for efficient processability by ensuring a consistent flow of the powders during the 3D-printing process [11].

Despite their biocompatibility and resorption, these synthetic polymers are hydrophobic and lack biological functions found in nature. Extensive efforts have been made to add more functions and increase the biocompatibility of these polymers; for example, by forming copolymers bearing a hydrophilic segment, like polyethylene glycol, or through blending of dextran [13] or β-tricalcium phosphate (TCP) [14], enabling FFF 3D printing

Table 4.1 *Examples of commercial bioresorbable polymer products, properties, and filaments for 3D printing [12].*

RESOMER® polymers	Tensile strength (MPa)	Elastic modulus (GPa)	Elongation at break(%)	Degradation time*	Filaments
PLLA	75–85	3.7–4.7	< 10	> 3 years	√
PLGA	80–90	3.5–4.5	< 10	1–2 years**	√
PCL	15–25	0.5–0.6	ca. 180	> 2 years	√
PDO	35–45	0.6–0.8	ca. 250	< 6 months	√

*Degradation time also depends on individual polymer's initial molecular weight.
**PLGA is an amorphous polymer; the degradation time depends on ratio between lactic/glycolic units and initial molecular weight.

to improve osteoblast response *in vivo*. A new process would be required to 3D print the modified polymers or composites. In the case of a composite filled with osteoconductive fillers, the enhanced melt viscosity, fillers particle size and its distribution, surface roughness, inter-layer bonding, as well as mechanical properties would be taken into consideration for the printing process to achieve optimal performance.

4.3.1.2 Natural Bioresorbable Polymers

Compared to most synthetic polymers, natural biopolymers comprise either functional or structural molecules, enabling them to have bioactivity or excellent performance in supporting cellular adhesion and proliferation. They have been extensively used in fabrication of scaffolds for tissue engineering applications, primarily due to their function to support the formation of new extracellular matrix after degradation [15]. Natural biopolymers, such as gelatin, hyaluronic acid, alginate, chitosan, fibrinogen, and collagen, have been formulated as the main components of bioinks in 3D bioprinting of layer-by-layer deposition processes. Instead of animal originated, recombinant natural polymers, such as fermentation-based recombinant collagen (Vecollan®, Evonik Corporation) and plant-derived recombinant collagen (rhCollagen, CollPlant Biotechnologies Ltd), eliminate the potential of pathogenic transmission from animals to human beings. The preserved collagen structure dramatically improves its acceptance for biomedical applications including 3D bioprinting. However, it is nearly impossible to print natural biopolymers directly under cell benign conditions without modification. Physical blending, physical/chemical crosslinking, and combination of both natural and appropriate synthetic biopolymers have been proved effective to resolve many challenges associated with natural biopolymer bioprinting [16].

4.3.2 Non-Bioresorbable Polymers

Rather than providing temporary or short-term function, medical implants based on non-bioresorbable polymers serve as permanent supports or when resorption is no longer desirable. For example, with artificial teeth or hearing aids, non-bioresorbable or durable polymers come into play. However, not all non-bioresorbable polymers can be used in biomedical fields, in particular for implants with respect to their biocompatibility. Table 4.2 lists non-bioresorbable polymers that are applicable to medical applications by various 3D-printing techniques.

Amongst these non-bioresorbable polymers, PEEK has emerged as the leading high-performance candidate as an alternative to metal implant components, especially in orthopaedics, trauma, and dental applications [26]. It has several advantages over metals for the use in orthopaedic applications including radiolucency and an elastic modulus of 3.6 GPa, minimising metal stress shielding effects. 3D-printing techniques have allowed the production of complex and precise implants of PEEK based composites. Recent developments are targeting to increase the adaptability of 3D-printing techniques for a wider range of applications. Newest developments on 3D printing of PEEK and PEEK-based composites are focused on improving FFF equipment design to fulfil high temperature printing requirements and improve interlayer bonding. Another approach in recent years is to develop composite filaments such as carbon fibre reinforced PEEK filaments, bioactive PEEK with enhanced microstructure, bioactivity, and osseointegration for devices produced by FFF printing [27, 28].

PAs occupy 95% of the commercial market of powder-based fusion processes. PA12 is the fastest and most affordable medical-grade material to print and is compatible

Table 4.2 *Non-bioresorbable polymers in 3D printing for medical applications.*

3D-Printing technique	Non-Resorbable biopolymers	Medical applications	Suppliers of materials
Fused Filament Fabrication (FFF)	• ABS [17, 18] • PEEK [19] • PA [20] • TPU [21] • PEBA	• Prosthetic hand • Orthopaedic implants • Surgical planning	• Evonik Corporation • Sculpteo • Lubrizol • Arkema • Initial • Oxford Performance Materials • Carbon
Selective Laser Sintering (SLS)	• PA • PEKK [22] • PEBA • TPU	• Denture • Scaffolds • Bone grafting • Artificial hip • Skull flaps	
Digital Light Synthesis [23]	• PU • Urethane methacrylate • Dental materials	• Single use medical devices • Dentures	
Stereolithography (SLA)	• Photopolymers	• Hearing aids [24] • Cranioplasty implants [25]	

Abbreviations: ABS: acrylonitrile butadiene styrene; PEEK: polyether ether ketone; PU: polyurethane; PEKK: polyether ketone ketone; PEBA: polyether block amide.

with FFF, multi-Jet Fusion printing, and SLS. PA 2200 is considered biocompatible and has stable processing along with repeatable properties of manufactured parts. Due to their biocompatibility, PAs can be used to create parts that come into contact with skin such as splints [29]. TPUs are block copolymers with soft segments consisting either of polyether or polyester chains and hard segments induced by the addition of a short-chain diol to an isocyanate. Processability of various types of ester-based TPUs for SLS focused on their melting and volume change and their powder flow and coalescence upon process cycle [30]. Unsintered TPU powders do not experience significant ageing effects in a heated build chamber and can thus be reused [31]. ABS is a biocompatible engineering thermoplastic compatible for FFF and SLS printing technologies. The application of this material is limited in tissue engineering due to its non-biodegradable nature. 3D-printed ABS displayed identical cell integration and processability similar to PLA [32].

4.3.3 Smart Polymers

Smart polymers belong to intelligent polymers that can change shape or other properties by responding to external stimuli such as heat, pH, moisture, or magnetic fields. Smart polymers have been extensively used in fabrication of medical devices [33]. Amongst them, shape memory polymers (SMP) have received substantial attention due to their capability of remembering and restoring original shape. The shape recovery triggering temperature approximately at body temperature is a key feature for biomedical applications [34]. Thermo-responsive SMPs normally consist of both hard and soft segments. Introducing soft segments inevitably reduces mechanical strength compared to the counterparts without soft segments. Maintaining adequate mechanical properties of SMPs after shape recovery in the physiological environment involving high volumes of body fluids are highly desirable as medical implants. Their surfaces also have played a prominent role in cell attachment, proliferation, and tissue ingrowth. Conventional 3D printing only provides a static construct, which finds limitation in the increasing demand for minimally invasive surgery. SMPs have made it possible in converting 3D printing to 4D printing by adding time as the fourth dimension in printing medical implants. 4D printing does not only allow for minimally invasive surgery but also allows for the tissue to regenerate between open cells in the printed architecture.

Without a reactive moiety, PLA as an SMP was printed by FFF. The printed objects self-bended when dipped in hot water at 85 °C, which was about 20 °C higher than its shape recovery triggering temperature. A geometry design enabled unbalanced pre-strain created during 3D printing and deposited through the thickness direction, which led to a mismatch in the free-strain recovery, creating curvatures and displaying a transformation from a temporary straight shape to a permanent curved shape. It demonstrated that utilising characteristics of shape memory materials, 4D FFF printing technology can manufacture self-adaptive structures through geometry design and fabrication [35]. The Evonik Corporation has developed several body heat responsive shape memory materials consisting of resorbable polymers that could be adapted for applications including 3D/4D printing of medical implants, stents, and controlled drug eluting implants [34, 36]. In PLLA-*b*-PCL multiblock copolymers, both PCL and PLLA blocks form their individually crystalline domains with an overall glass transition temperature (T_g) of around 39 °C. These multi-block copolymers produce a homogeneous melt after all components are fully melted, which is beneficial for the extrusion-based 3D-printing process to create flexibility in printing-structure-property tunability [34]. Full resorption after fulfilling the desired functions as medical implants is another important advantage over non-resorbable or permanent shape memory counterparts such as durable SMPs and shape memory alloys. In soft tissue repair application, fully resorbable composites outperformed those shape memory composites loaded with inorganic fillers. The inclusion of rigid PGA microparticles, which has a T_g of around 37 °C, into a flexible PDO matrix enables such a polymeric composite to change its shape around body temperature. The similarity of resorption time of PGA and PDO for less than 6 months allows the implants to be absorbed quickly and evenly in the human body. These polymers are suitable for a variety of 3D/4D-printing processes [36].

Several 3D-printing processes employ light sources such as UV light to cure the surface of liquid photopolymer resins that can be solidified after 3D printing. A growing number of photo-curable materials have been developed for 3D printing, such as in SLA, DLP, and volumetric printing [37]. 3D-printed static implants based on hydrogels usually do not mimic the actual dynamic movement of tissues under physiological conditions. A photocurable and temperature reversible chitosan polymer was synthesised for SLA 4D printing. The 4D structures were achieved through a shape memory effect (SME) that the polymer volume expanded upon swelling in water and contracted when the temperature was over 30 °C and recovered to its original shape [38]. Using photocurable star-shaped PCL as an ink, a stent with high mechanical properties, SME, suitable degradation rate, good biocompatibility, and sustainable drug release function was 4D printed. Due to the SME feature, the stent can be implanted with a crimped miniature size and restore to the originally 3D-printed shape after deployment under thermal stimulus. The self-expansion after implantation could significantly reduce the occurrence of surgical approaches and damage to the tissue [39].

Commonly employed SMPs process inappropriate shape transition temperatures leading to challenges in implantation operation. Major efforts in development of SMPs have long focused on materials that have quick or even instant shape memory properties when exposed to a stimulus. A slow recovery, which takes days to weeks, would benefit certain medical applications as well. A DLP printed light cured PDLLA dimethacylate SMP with a broad range of T_g took two weeks to recover at a temperature close to onset of T_g. The broad T_g range gave this SMP a unique staged recovery, with different parts of the samples recovering at different times. Potential applications for slow recovery SMPs include tissue expanders that would gradually increase in size and volume over weeks and slow deploying stents for gradual dilation of vascular or urethras, giving surrounding tissue time to adjust and grow [40].

Moving away from acrylate-based systems may improve the biocompatibility of printed devices. A mixture of four-arm star-shaped poly(propylene maleate) copolymer and reactive diethyl fumarate as low viscosity ink was 4D printed to scaffolds by DLP. The printed strut dimension impacted shape memory properties with respect to the fact that the UV light was not able to penetrate the specimens, leaving unreacted monomer residuals as a diluent of the ink. Extending post-cure exposure increasing the SME and mechanical properties further confirmed the unreacted monomer in the system [41]. Ring opening copolymerisation of allyl glycidol ether and cyclohexene dicarboxylic anhydride produced a library of polyesters of thiol-ene photopolymer resins. The thiol-ene click chemistry crosslinking process is within an effective timeframe for 3D printing by SLA or DLP. These materials displayed shape memory properties after photo crosslinking during the 3D-printing process due to the network rigidity. The thiol-ene process displays a significant advancement in the need for 4D printing materials [42]. A soft, elastomeric aliphatic polycarbonate-based material exhibiting shape memory properties enables minimally invasive 4D printed self-fitting scaffolds to fill model soft tissue voids without deforming or applying pressure to the surrounding tissue [43]. On the basis of this technology, 4D Biomaterials, a spin-off company from University of Birmingham and University of Warwick in the UK, commercialises 4Degra® resorbable liquid resin-inks suitable for DLP 3D-printed medical implants ranging from implantable micro-devices to tibial fracture repair devices [44].

4.3.4 Metal and Ceramic

Medical metal materials are mainly applied to prepare permanent implants such as orthopaedic or dental implants containing stainless steel, cobalt-chromium alloy, titanium alloy, or tantalum alloy [45]. These compositions of metal materials possess good biocompatibility that meets medical standards. 3D-printed biomedical metal materials have advantages over traditional implants. 3D-printed metal implants tend to have small grains with better mechanical performance. In addition, highly controlled printing environments ensure high purity of the printed parts, thus maintaining the desired characteristics of the material. Moreover, the design complexity of 3D-printed personalised products is reduced, allowing customisation of implants with mechanical behaviours that are similar to those of bone or teeth. The preciseness and efficiency of high-energy 3D-printing equipment such as electron beam and laser beam can meet the requirements for manufacturing small parts or achieving high-scale production. Under the computer-aided design, 3D printing can rapidly manufacture shaped implants; meanwhile, it can also produce size-controllable micropore structures. These micro-pore structures can lower the elastic modulus of metal materials and decrease the stress shielding at the solid parts of the implants and can promote the integration between metal and bones at the surface of the implants. Surface treatment, such as electrochemical deposition, chemical modification, and alkali-heat treatment, is usually carried out to enhance the bioactivity of porous metal implants. As part of the metal 3D-printing process chain, heat treatment can be used to tailor biocompatibility and mechanical properties simultaneously [46].

In addition to extensive research on the 3D printing of polymers and metals, 3D printing of ceramics is now actively under development. The ability to fabricate ceramic components of arbitrarily complex shapes has been extremely challenging without 3D printing. In many cases, the 3D-printing step utilises a ceramic slurry or blend of material to build a part that can be sintered like the process for bound metal deposition. Promising explorations have been made in the biomedical field, including the fabrication of components for tissue engineering, which generally require less precision in resolution and surface finish, as well as porous features of the printed parts for cultivation purposes. Biocompatible ceramics such as HA, calcium phosphates (CP), and TCP are often used in 3D printing of scaffolds for bone replacement. A pre-ceramic polymer, i.e., silicon resin as a binder, which also reacted with fillers to form desired ceramic phases, was 3D printed to $CaSiO_3$-based biocompatible ceramic parts with a porosity of approximately 64% by volume [47].

Ceramic SLS has also become increasingly popular in biomedical applications, particularly in the fabrication of customised complex and highly cellular biocompatible scaffolds for tissue engineering. These fabrications generally involve high volume fractions of binder phases of up to 60 vol%, and so geometrical accuracy and surface roughness are not strictly required for these applications, although macro-porous structures are often customised in a controllable manner [48]. Examples include bone implants made by ceramic–polymer blends, such as HA-TCP [49], PCL-HA [50], and PEEK-HA [51]. Ceramic–glass composites have also been investigated for the fabrication of biocompatible scaffolds, such as HA–phosphate glass [52]. In these applications, the low-melting point polymers and glasses serve as liquid-phase binders during SLS to facilitate densification. For 3D printing of ceramics using FFF, composite filaments are prepared by densely loading ceramic particles into thermoplastic binders [53]. After printing, the printed ceramic part is subjected to binder removal and sintering to achieve densification.

Photo-polymerisation methods have shown greater potential over powder bed fusion in the manufacture of 3D ceramics parts [54]. Slurry-based photo-polymerisation methods such as SLA, DLP, and two-photon polymerisation for ceramics fabrication have shown promise. A controllable feature resolution and surface finish with desirable mechanical performance can be achieved for the fabricated parts. A variety of commercial machines are now available for the fabrication of complex 3D ceramic components. DLP is more favourable for widespread use due to lower price [55]. Aerospace and medical industries are the two most promising markets for ceramic 3D printing. However, high standards and compliance challenges result in complex and long regulatory pathways for commercialisation.

4.4 3D-Printed Personalised Medical Devices

Medical implants fabricated by 3D printing have demonstrated unique features and/or functions that were not possible or too costly to fabricate by traditional manufacturing techniques such as injection moulding. It is imperative to understand the core mechanisms relating to the materials in 3D printing; for example, the key characteristics of materials suitable for 3D printing include thermal stability, flowability, printing temperature, and reproducibility, etc. Additional material features such as biocompatibility, biosorption, self-healing, shape memory, self-assembly, self-deployment, etc., can add extra advantages in the design and manufacturing of medical implants [56]. Probably the greatest advantage of 3D printing is related to the rapid fabrication of patient-specific or personalised designs of medical implants. Surgical treatments involving 3D-printed medical implants have successfully demonstrated their profound benefits. In this section, selected medical implants enabled by 3D printing are reviewed.

4.4.1 Vascular Repair Devices

Coronary and other vascular diseases have been the leading causes of morbidity and mortality worldwide. Autografts, xenografts, and bioengineered extracellular matrix have been explored as treatments. 3D printing offers design freedom for stents that has not been possible before. SMPs have shown promise as polymeric stents owing to their self-deployment capabilities without auxiliary devices, reducing catheter size for delivery and vascular biocompatibility. Resorbable SMPs could prevent the negative effects of strain induced hardening in conventional polymers impacting resorption uniformity. Rather than using a product with predetermined size, 3D-printed personalised stents with desirable final diameter were crimped onto a much smaller diameter balloon or into a smaller sized catheter sheath to enable it for easy implantation and correct placement. With time as an additional dimension in the 3D printing, 4D printing better reflects this unique advantage. After being implanted and placed at the right site in the anatomy, the stent then recovers to its original diameter triggered by body heat without the need of balloon expansion and limit stent migration [57, 58]. High-resolution projection micro-stereolithography (PμSLA) has been demonstrated to be efficient in fabricating stents utilising SMPs [59]. Direct-write printing of photo-curable PLA-based inks is another viable fabrication technique for 4D printing stents with magneto-responsive materials, allowing them to be magnetically remotely guided to the placement destination [60]. A highly stretchable elastomer with a semi-interpenetrating polymer network (IPN) structure was 4D printed through UV light assisted direct-ink-writing. This semi-IPN elastomer features a urethane diacrylate for UV crosslinking and a crystalline PCL component in the network. The self-healing capability of this material can repair vascular tissue in 3–5 minutes [61]. The world's smallest stent, just 50 μm wide and 0.5 mm long, 40 times smaller than any other stents produced, was only achievable by means of 4D printing. This type of small stent could help widen life-threatening constrictions of the urinary tract in foetuses in the womb. The shape memory properties give the stent the fourth dimension that once the stent is deformed, it remembers and returns to the original expanded shape when facing body temperature over time [62]. Resorbable and personalised occluders enabling remote control were 4D printed through PLA. Introducing microgrooves to polymeric stent surfaces to form micropatterns has promoted endothelial cell attachment and alignment along stent grooves, due to the improved surface hydrophobicity. It is suggested that SMPs with improved surface and material characteristic could be possible for next-generation blood-contacting devices [63].

4.4.2 Splints

For medical implants, a critical aspect is the geometry that becomes even more critical and challenging for paediatric patients to allow the implants' proper function. Traditionally, prefabricated medical implants have difficulties in accommodating for the growth over time in paediatric patients. This even becomes more critical and challenging for infants; for example, with tracheobronchomalacia (TBM). TBM is a life-threating condition where the airway of the patient excessively collapses during respiration. Under provisions for emergency clearance from the FDA, a 3-month-old infant with TBM became the first baby to receive treatment by a 3D-printed resorbable tracheal splint created by Professor Hollister and Dr Glenn at the University of Michigan [64]. In another case, an infant with localised TBM was treated with a 3D-printed resorbable airway splint that resulted in improved ventilation [65]. Patient-specific resorbable airway splints based on PCL, with 4% of hydroxyapatite 3D printed by SLS, demonstrated initial clinical efficacy in treatment of critically ill children with severe TBM. No child required removal or replacement of the splints. The first four children to receive splints are all alive with sustained significant clinical benefit for more than two years after splint implantation [66]. 4D printed implants could provide an answer for the accommodation of shape change fitting to growth. In a clinical study, personalised 4D printed external airway splints were successfully implanted in infant patients. A splint based on biocompatible and resorbable polymer blend was designed to accommodate the growth of the airway, consequently preventing unwanted compression [67]. Until further data is presented, PCL appears to be the most suitable material for splint manufacture and the corresponding applications [68].

4.4.3 Nerve Guidance Conduits

Nerve guided conduits (NGCs), as alternatives to nerve autografts and allografts, are tubular devices that facilitate nerve regeneration by meeting physical, chemical, and biological requirements. The emerging 3D-printing technology has led to remarkable advances in fabrication and functionalisation of personalised NGCs in peripheral nerve repair and regeneration. A recent overview [69] summarised materials that have been used in 3D printing of NGCs including PLA, PCL, PLGA, collagen, and photo-curable monomers like gelatin-methacrylate, silk-methacrylate, alginate-methacrylate, etc. These materials exhibited both advantages and disadvantages for their application in NGCs. 3D-printed NGS based on PLA, PLA/PCL, and PLGA from inkjet printing displayed low mechanical strength. High resolution SLA printing has been shown as a better alternative to fabricate NGCs of different complexity and size using photocuring biomaterials.

SMPs represent the possibility to produce minimally invasive medical devices including NGCs. Soybean oil epoxidised acrylate was used as an ink for SLA 4D printing for an NGC with several desirable functions for nerve tissue regeneration [70]. The cured reactive polymer by UV light allowed for easy readjustments of the NGC through thermal stimulation during implantation. The axial tension force generated from this stimuli-responsive polymer can guide tissue regrowth. However, possible cytotoxicity of residual photo-initiator and uncured resin may limit their clinical applications. There is a great need for safe bio-inks for 3D printing.

4.4.4 Tissue Engineering

Tissue engineering bridges material engineering and life science to create scaffolds or organs to regenerate, restore, replace, improve, and maintain tissues damaged by injury, disease, or congenital disability [71]. An ideal scaffold must provide a surface to allow cell attachment and a continuous 3D inter-connected porous structure to promote extracellular matrix formation and vascularisation. 3D printing allows fabrication of scaffolds with more controlled and precise structures than any other techniques so far [72].

Bone regeneration has been one of the focus areas of tissue engineering. Several works have demonstrated that inclusion of osteoconductive fillers to polymers or composites can promote osteosis inclusion. Porous scaffolds with biomimetic architecture are highly promising for bone tissue engineering applications. Bioresorbable porous composite scaffolds constituting PLA and HA were 3D printed by FFF. Ceramic particles inhibited the growth of cracks during compression-heating-compression cycles, whilst porous PLA/HA scaffolds recovered their initial shape. In the subsequent reheating process, the self-healing effect of the scaffold narrowed the cracks. With 15% HA, the 3D-printed porous scaffold was able to recover 98% of its original shape, which merited it as a candidate for self-fitting implants for small bone defect replacement [73]. Incorporation of methacrylates bearing 2-ureido-4[1H]-pyrimidinone units to PCL dimethacrylate macro-monomers provided 4D printed structures with self-healing properties. Even after the healing process, the printed objects still exhibited remarkable shape memory properties [74]. Similar results were observed in FFF 3D-printed microporous osteoconductive bone grafts based on amphiphilic resorbable PLGA-*b*-PEG-*b*-PLGA copolymer with 25% of HA. The SME and PEG hydration-induced stiffening and swelling enabled a facile surgical delivery and stable fixation in critical long bone defects in rats [75]. Magnetically induced SMPs enable remote non-contact control and selective actuation. 4D printing of these materials to biomimetic scaffolds with pre-determined appropriate morphology and microstructure provide better compatibility to the geometry and better supporting functions. PLA with 20% of ferromagnetic nanoparticles (Fe_3O_4) was 3D printed into porous scaffolds through FFF. The scaffolds not only had excellent shape memory properties but also processed adequate mechanical strength to resist external load generated during motion. Water borne resorbable polyurethane containing iron oxide nanoparticles as 3D-printing ink was developed as customised bone substitutes for bone tissue engineering applications. The ink containing 500 ppm of iron oxide nanoparticles was designed to not only promote osteogenic induction but also to improve shape fixity capability. The scaffolds were printed by low-temperature FFF and exhibited high shape recovery at 37 °C. The iron oxide nanoparticles could be released gradually from the scaffolds and further improved the osteogenesis [76]. A 3D-printed PCL mesh serving as inner core, for patient-specific ear-shaped cartilage, was used clinically for the first time for auricle reconstruction in five patients and achieved satisfactory aesthetical outcome with mature cartilage formation for 2.5 years [77].

Natural polymers have been successfully employed in 3D printing of tissue engineering scaffolds. In 1999, the first 3D bio-printed bladder using the patient's own cells and a collagen-PGA composite as the scaffold was implanted into a human [78]. Poieskin®, a product from Poietis, became the first commercially available bio-printed human tissue based on laser-assisted 4D bioprinting in 2018. This product consists of a dermal compartment composed of primary human fibroblasts embedded in a collagen I matrix [79]. BellaSeno has developed a 3D-printed porous resorbable breast scaffold based on Evonik's RESOMER polymers to avoid the use of silicone implants and respective side effects [80]. In 2019, Organovo announced a successful automated production of kidney organoids with detailed kidney tissues through its advanced bioprinting technique, which revealed the world's first commercial 3D bioprinter to print human tissues [81].

4.4.5 3D Printing in Dentistry

Dental implants were the pioneer in adapting 3D-printing technologies in the medical field in the 1990s. Advances in 3D printing and imaging technologies have fostered applications in clinical and experimental dentistry. Even now, nearly all structures that dental professionals produce for patients can be made by 3D printing [82], such as in oral surgery, prosthodontics, periodontics, endodontics, and temporomandibular joints (TMJ). 3D printing has resolved many long-standing challenges in dentistry, for example, replacing damaged TMJ discs. An artificial TMJ disc was fabricated using polyvinyl alcohol (PVA) hydrogel crosslinked by cyclic freeze-thaw and reinforced by 3D-printed PCL, which induced disc defect repair for 12 weeks. Such a disc demonstrated comparable mechanical strength to a natural disc and providing additional advantages regarding better fatigue resistance, hydrophilicity, less creep, and improved cell adhesion [83]. The first human case of treatment of a large periodontal osseous defect with a patient-specific scaffold, based on 3D-printed PCL powder by SLS with signalling growth factor was reported in 2015. The scaffold remained covered for 12 months, exhibiting a 3 mm gain of clinical attachment and partial root coverage [84]. Separately, 3D-printed PLA/HA scaffolds loaded with dental pulp stem cells were explored in an animal study for the feasibility of tooth regeneration. HA promotes a high degree of mineralisation; however, after nine months of initial surgery, the scaffolds were not completely absorbed [85]. These studies revealed that a more rapidly resorbing matrix with a healing timeframe shorter than 1 year together with a less bulky design would be more suitable for oral tissue regeneration around teeth. Fast bioresorbable polymers, for instance PDO with a degradation time around 6 months, might be a good fit for these applications, for example, by FFF of PDO filament.

Non-bioresorbable polymers, such as PEEK and acrylics, are two prominent polymers extensively used in dentistry. As alternatives to metallic and ceramic materials in clinical practice, PEEK dental implants ranging from surgical guides, aligners, fixed and removable denture, bridges, and others seem to provide patient comfort and satisfaction due to the lightweight, durability, wear resistance, and biocompatibility [26, 86]. Evonik has manufactured medical-grade PEEK for extensive dental applications. FFF is still the leading technology in 3D printing of PEEK. Lack of printing precision currently limits its wide dental application by 3D printing. With the fast advances in new additive manufacturing technology, personalised PEEK dental devices by 3D printing are expected to provide more convenience to dentists and comfort to patients. Acrylic polymers have been well adapted in 3D-printing dentistry, not only in clinics and dental labs, but are also commercially available for mass production. Invisalign® custom-made clear aligners are examples of 3D-printed commercial products by SLA, with Align Technology's proprietary technology including acrylic materials [87] to straighten teeth. ClearX, a clear SMP sheet developed by K Line Europe, has been 3D printed into 4D aligners with shape-shifting property within the oral temperature range that can be easily modified and reused [88].

3D-printing technology can help create novel dental implants with rough or porous surface finishing. However, seeming to be attractive in many medical applications, these rough or porous surfaces in dentistry may generate challenges many years later [89]; for example, discoloration and promotion of bacterial or fungal growth, leading to odours. Yet, 3D printing has not been widely used in fabrication of crowns or restorations due to a lack of material providing sufficient mechanical properties, accuracy, and fit to tooth preparations. Material innovation is key to further adapting 3D printing in dentistry. The ability to produce complex geometries and accurate dimensions, which cannot be produced by other techniques such as milling/machining, remains the unique advantage of 3D-printing technology in dentistry.

4.4.6 3D-Printed Orthopaedic Devices

Customised implants enabled by 3D printing are arguably the most ground-breaking aspect for orthopaedic surgery, for example, in joint replacements and bone reconstruction therapies. Without compromising the mechanical properties, PLA-based osteoconductive composites are an important and challenging problem for orthopaedic applications when the implants are used in load bearing applications. The inclusion of rigid fillers such as HA can increase the melt viscosity of the composites. Designing and printing the scaffolds with this composite composition would need to be taken into consideration to obtain scaffolds with desirable properties. Reliable fixation of a scaffold at the implantation site without additional binders which can be achieved by 3D printing of the composites, is extremely beneficial. 3D-printed scaffolds can be easily moulded to patient-specific bone defect sites; for example, as guided bone regeneration (GBR). In a rabbit calvaria model, 3D-printed PLA GBR demonstrated biocompatibility and integration with bone defect margin to enhance bone augmentation [90]. Photothermal-responsive shape memory bone scaffold was 3D printed from β-TCP/poly(lactic acid-*co*-trimethylene carbonate) and black phosphorous nanosheets and exposed to near-infrared radiation prior to implantation, enabling shape remoulding and precise fitting in irregular bone defects. Once implanted in the desired place, the temperature dropped to body temperature and the scaffold displayed mechanical properties similar to those of cancellous bone. This method successfully treated bone defects of irregular shapes [91].

There are a few end-use orthopaedic implants being 3D printed. Orthofix Medical has launched a series of 3D-printed spacer systems designed for use such as in Posterior Lumbar Interbody Fusion surgeries [92]. Onkos Surgical received 510 (k) clearance from the FDA for the modular 3D-printed BioGrip collars with nano HA treated surfaces [93]. 4Web Medical utilises engineering principles to produce 3D-printed spin implants that may actively participate in the healing process [94]. Cleared by the FDA for commercial sales in the US, Smith & Nephew now offers 3D-printed knee implants featuring porous surfaces that can bind more naturally to the bone without the need of cement [95]. In 2020, the first 3D-printed PCL-based scaffold, TruMatch Graft Cage in a highly customisable manner from Depuy Synthes, was implanted into a patient to restore long bone injuries for the first time [96].

4.5 Regulatory

Regulatory agencies play a significant role in the medical device manufacturing process. These bodies are governmental organisations with legal authority to select devices that can be sold on the market through a scientific support and risk analysis process. At a fundamental level, they assign accountability of risk to the maker of a device through the legal system.

In the US, prior to 1906, there was minimal oversight into the manufacturing and distribution of food and drug products. In 1938, the Federal Food, Drug, and Cosmetic Act (FD&C) was passed and ultimately led to the modern FDA. This legislation extended previous regulation to overseeing of medical products, drugs, cosmetics, and therapeutic medical devices. It also granted authority for inspections at facilities manufacturing these products. Device regulation further evolved in 1976, with the designation of device classes and the pathways for Premarket Approval (PMA) and premarket notifications (510 k). Additional amendments from this period introduced pathways for investigational devices through the Investigational Device Exemption (IDE) process. New post market requirements also included registration of manufacturing establishments, tracking of device adverse events, and gave the FDA power to ban devices deemed unsafe or not effective. After 1990, regulation was extended to improve post market device surveillance, allow device recalls, and provided flexibility for 'Humanitarian Device Exemption (HDE)' devices targeted towards rare diseases [97].

Due to the rate and scale of the modern global market, regulatory bodies worldwide have evolved throughout history to serve their region's interests. The FDA has been the most centralised organisation historically, with activities in countries that manufacture products marketed only in the US. The modern European Medicines Agency (EMA), headquartered in Amsterdam, has acted to harmonise regulatory activities across Europe [98]. Regulatory bodies also exist in other countries including Japan, China, the United Kingdom, Canada, India, Russia, Singapore, Australia, Brazil, and Switzerland [99]. Many international standard think tanks have aided manufacturers and regulatory bodies in developing harmonised language, test methods, and processes to improve efficiency; however, each organisation acts independently, and devices are not typically approved automatically across markets.

The designation of a medical device by a regulatory agency is highly dependent on its intended use and the risk associated with treatment. This requirement has not been altered with the proliferation of 3D-printing technology. Any printed device intended to treat a condition, regardless of manufacturing means, would still be subject to the parameters outlined under guidance. It follows that anyone designing or printing a device would then be responsible for its lifecycle once used [100].

As 3D-printing manufacturing extends to more devices, regulatory definitions need to be refined and clearly outlined. Risk-based strategies are often employed to identify a device's classification; however, quantitatively defining terms like 'low risk' or 'healthcare facility' become more complex when devices can be printed easily [101]. Consider an anatomical model that is 3D printed in a hospital's surgery suite. If the model were used as a surgery guide for a specific patient there would likely be higher risk involved than if the model were used in a general training scenario. The regulatory impact of each case might be highly divergent. Due to the small and extensible footprint of many 3D-printing technologies, the lines can become blurred between a medical service provider and a medical device manufacturer. Whilst producing patient matched parts is highly convenient and efficient, manufacturing these devices at the point-of-care could potentially cause the provider to fall under the designation of a medical device manufacturer. This adds an increased level of complexity, additional overseeing, and expense to the provider [102]. Rarely are small-scale medical providers staffed to maintain the quality system framework sufficient for regulatory body overseeing. Any design manufacturing parameters must be controlled so that all parts fall within the scope of the device classification and any validated specifications. This requirement extends to any computers, software, or hardware involved in the manufacturing chain. Furthermore, for each device that is produced, the burden of proof remains with the manufacturer to assure that it is cleared through the appropriate classification system [103].

Prior to 2012, there were limited 3D-printed devices cleared for use outside of the dental market. A highly visible case in 2013, involving a lifesaving, 3D-printed tracheal stent/splint, was cleared for emergency use by the FDA [65]. Shortly after, building interest and maturation of the technology led to the formation of the 'Additive Manufacturing Working Group.' This group further improved guidance related to AM from the FDA through interactive discussions and technical guidance [104]. In 2021, the FDA issued a further public request for a response from the industry. Close collaboration with industry members and regulatory bodies should further broaden the range of 3D-printed devices that are cleared on the market in the future.

The agility and extensibility unique to 3D printing played an unprecedented part during the COVID-19 pandemic. In the early stages of the pandemic, the availability of personal protective equipment and lifesaving equipment, such as ventilators, became scarce. To minimise loss of life, the FDA granted Emergency Use Authorization (EUA) for an array of medical devices. This cleared the use of existing devices to be used across their designation with modification. For example, certain devices such as anaesthesia gas machines and positive pressure breathing devices could be modified or used as ventilators in emergency settings. More than 50 EUA medical device clearances were granted during this period and included parts such as the Formlabs BiPAP adapter used to convert sleep apnea machines into ventilators [105, 106]. Additionally, a network of Non-Traditional Producers (NTPs) consisting of hospitals, hobbyists, 3D-printing manufacturers, governmental agencies, and universities used 3D printers to produce over 50 million parts in a 5-month period. These parts included face shield components, nasal swabs, ear savers and mask parts, and ventilator components. Parts were designed by experts and then submitted to the National Institute of Health's 3D Print Exchange. The reviewed models were given a 'Clinically Reviewed' status [107].

4.6 Future Perspectives

Rapid progress of nearly every aspect of 3D-printing technology in recent years has led to more sophisticated, patient or site-specific medical implants. This trend will continuously evolve in the direction of making 3D or 4D printing an easily adaptable technique and tool in personalised medical devices. Solutions might be found, or progress made, to resolve current limitations or challenges. Ideally, 3D-printed medical devices having sound mechanical properties and surface characteristics to mimic surrounding tissue and encourage cell attachment and proliferation are beneficial to biomimetic implants. To achieve this, not only would such a polymer or material be available, but also a 3D-printing technique would be well tailored amongst the materials, printers, and processes. In certain applications; for example, FFF printed devices' function and structure sometimes are constrained by the current printing resolution threshold. More fine detailed structure can be achieved by; for example, SLA, DLP, or PμSLA. Yet, the prevailing materials in the SLA process bearing photo-reactive moiety, selection of such a resin for medical implants requires careful consideration of its biocompatibility and regulatory compliance.

Bioresorbable polymers eliminate the need for subsequent secondary surgery to remove medical implants, which reduce medical care costs and improve patients' comfort. Nonetheless, safe elimination of the degradation by-products from the body and a predictable degradation time window are still the key fundamental requirements in development of new bioresorbable polymers to fit various biomedical needs via 3D printing.

Smart materials, such as stimuli responsive polymers, and rapid advancement in 3D printing enabled the possibility of a more fascinating 4D printing technology which would lead to more innovative solution and advanced technology towards more personalised treatment. 4D printing remains in its early stage and behind more prevailing 3D-printing techniques due to the limited availability of suitable printable resins. Combination of multiple and appropriate stimuli responsive materials can provide a wonderful opportunity in medical implants design, fabrication, and applications creativity. For example, smart medical devices with different functions (e.g., functions in gradient) and degradation rates to match complex tissue regeneration and functions requirement expect to be enabled by 3D/4D printing individually or through a printing process combination. Current materials, smart materials, and 3D/4D-printed structures are still mainly investigated at the macroscale, represented by such as mechanical properties and surface finish. Towards an understanding of materials for 3D/4D printing at the microscale, for instance, interaction with localised microenvironment or response to multiple physiological signals would lead to more precisely controlled, multifunctional, point-of-care, and biomimetic medical implants.

Last but not least, similar to other new technologies, high cost and lack of sufficient data to comply with regulatory requirements remain challenges for 3D printing which are vital to individualised medical implants. Satisfaction of patient-specific treatment and increased patient compliance propels the further advance of 3D-printing technology.

References

[1] www.globenewswire.com/news-release/2022/01/24/2371842/0/en/3D-Printed-Medical-Devices-Market-is-reaching-a-valuation-of-US-1-839-2-Million-by-2028-Comprehensive-Research-Report-by-FMI (accessed 27 January 2022).

[2] Jones, R., Haufe, P., Sells, E. et al. (2011). RepRap: the replicating rapid prototyper. *Robotica* **29**: 177.

[3] Hull, C.W. (1986). *Apparatus for production of three-dimensional objects by stereolithography*. US 4,575,330 A.

[4] Crump, S.S. (1992). *Apparatus and method for creating three-dimensional objects*. US 5,121,329 A.

[5] Noorani, R. (2017). *3D Printing Technology, Applications, and Selection*, 1st ed. Boca Raton, FL: CRC Press.

[6] Guvendiren, M., Molde, J., Soares, R.M.D. et al. (2016). Designing biomaterials for 3D printing. *ACS Biomaterials Science and Engineering* **2**: 1679.

[7] Mardis, N.J. (2018). Emerging technology and applications of 3D printing in the medical field. *Molecular Medicine* **115**: 368.

[8] Duan, B., Wang, M., Zhou, W.Y. et al. (2010). Three-dimensional nanocomposite scaffolds fabricated via selective laser sintering for bone tissue engineering. *Acta Biomaterialia* **6**: 4495.

[9] Weems, A.C., Perez-Madrigal, M.M., Arno, M.C. et al. (2020). 3D printing for the clinic: examining contemporary polymeric biomaterials and their clinical utility. *Biomacromolecules* **21**: 1037.

[10] Zhang, J.F., Jones, S., Wang, D. et al. (2019). Influence of thermal annealing on mechanical properties and *in vitro* degradation of poly(*p*-dioxanone). *Polymer Engineering Science* **59**: 1701.

[11] Gentsch, R., Nikoukar, M., Acreman, K. et al. (2022). *Biocompatible polymer powders for additive manufacturing*. EP3676309 B1.

[12] www.healthcare.evonik.com/en/medical-devices/bioresorbable-polymers/3d-printing (accessed 27 January 2022).

[13] Zhang, J.F., Wang, Y., Lam, M.L. et al. (2017). Cross-linked poly(lactic acid)/dextran nanofibrous scaffolds with tunable hydrophilicity promoting differentiation of embryoid bodies. *Materials Today Communications* **13**: 306.

[14] Salamanca, E., Tsao, T.C., Hseuh, H.W. et al. (2021). Fabrication of polylactic acid/β-tricalcium phosphate FDM 3D printing fiber to enhance osteoblastic-like cell performance. *Frontiers in Materials* **8**: 2296.

[15] Agueda, J.R.H.S., Chen, Q., Maalihan, R.D. et al. (2021). 3D printing of biomedically relevant polymer materials and biocompatibility. *MRS Communications* **11**: 197.

[16] Liu, F., Chen, Q., Liu, C. et al. (2018). Natural polymers for organ 3D bioprinting. *Polymers* **10**: 1278.

[17] Koprnicky, J., Safka, J., and Ackermann, M. (2018). Using of 3D printing technology in low cost prosthetics. *Materials Science Forum* **919**: 199.

[18] Boorla, R. and Prabeena, T. (2019). Fabrication of patient specific knee implant by fused deposition modeling. *Materials Today* **18**: 3638.

[19] (2021 10 June). PEEK for medical 3D printing expands patient-specific applications. *Plastic Today*.

[20] Vidakis, N., Petousis, M., Velidakis, E. et al. (2022). Medical-grade polyamide 12 nanocomposite materials for enhanced mechanical and antibacterial performance in 3D printing applications. *Polymers* **14**: 440.

[21] Xiao, J. and Gao, Y. (2017). The manufacture of 3D printing of medical grade TPU. *Progress in Additive Manufacturing* **2**: 117.

[22] Lin, Y., Umebayashi, M., Abdallah, M.N. et al. (2019). Combination of polyetherketoneketone scaffold and human mesenchymal stem cells from temporomandibular joint synovial fluid enhances bone regeneration. *Scientific Reports* **9**: 472.

[23] Hossain, M., Navaratne, R., and Peric, D. (2020). 3D printed elastomeric polyurethane: viscoelastic experimental characterizations and constitutive modelling with nonlinear viscosity functions. *International Journal of Non-Linear Mechanisms* **126**: 103546.

[24] Teran, M., Bishop, M., Kimura, T. et al. (2019). *Light based hearing systems, apparatus, and methods*. US 10,178,483.

[25] Dawwood, A. and Marti, B.M. (2015). 3D printing in dentistry. *British Dental Journal* **219**: 521.

[26] Halleem, A. and Javaid, M. (2019). Polyether ether ketone (PEEK) and its manufacturing of customised 3D printed dentistry parts using additive manufacturing. *Clinical Epidemiology in Global Health* **7**: 654.

[27] Verma, S., Sharma, N., Kango, S. et al. (2021). Developments of PEEK (polyetheretherketone) as a biomedical material: a focused review. *European Polymer Journal* **147**: 110295.

[28] www.medical.evonik.com/en/new-osteoconductive-vestakeep-fusion-peek-filament-for-3d-printed-implants-170049 (accessed 1 March 2022).

[29] Liu, Z., Zhang, P., Yan, M. et al. (2019). Additive manufacturing of specific ankle-foot orthoses for persons after stroke: a preliminary study based on gait analysis data. *Mathematical Biosciences Engineering* **16**: 8134.

[30] Verbelena, L., Dadbakhsh, S., van den Eynde, M. et al. (2017). Analysis of the material properties involved in laser sintering of thermoplastic polyurethane. *Additive Manufacturing* **15**: 12.

[31] Plummer, K., Vasquez, M., Majewski, C. et al. (2012). Study into the recyclability of a thermoplastic polyurethane powder for use in laser sintering. *Proceedings of the IMechE British Journal of Engineering and Manufacturing* **226**: 1127.

[32] Sood, A.K., Ohdar, R.K., and Mahapatra, S.S. (2010). Parametric appraisal of mechanical property of fused deposition modelling processed parts. *Materials and Design* **31**: 287.

[33] Falahati, M., Ahmadvand, P., Safaee, S. et al. (2020). Smart polymers and nanocomposites for 3D and 4D printing. *Materials Today* **40**: 215.

[34] Zhang, J.F., Xue, T., Jones, M.S. et al. (2021). *Multi-block shape memory bioresorbable polymers*. US 2021/0017329 A1.

[35] Bodaghi, M., Noroozi, R., Zolfagharian, A. et al. (2019). 4D printing self-morphing structures. *Materials* **12** (8): 1353.

[36] Zhang, J.F., Xue, T., Jones, M.S. et al. (2021). *Rigid resorbable materials with polymer and organic fillers*. US 2021/0015969 A1.

[37] Andreu, A., Su, P.C., Kim, J.H. et al. (2021). 4D printing materials for vat photopolymerization. *Additive Manufacturing* **44**: 102024.

[38] Seo, J.W., Shin, S.R., Park, Y.J. et al. (2020). Hydrogel production platform with dynamic movement using photo-crosslinkable/temperature reversible chitosan polymer and stereolithography 4D printing technology. *Tissue Engineering and Regenerative Medicine* **17**: 423.

[39] Tamay, D.G., Usal, T.D., Alagoz, A.S. et al. (2019). 3D and 4D printing of polymers for tissue engineering applications. *Frontiers in Bioengineering and Biotechnology* **7**: 164.

[40] Bartolo, A.D. and Melchels, F.P.W. (2020). Prolonged recovery of 3D printed, photo-cured polylactide shape memory polymer networks. *APL Bioengineering* **4**: 036105.

[41] Fer, G.L. and Becker, M.L. (2020). 4D printing of resorbable complex shape-memory poly(propylene fumarate) star scaffolds. *ACS Applied Materials and Interfaces* **12**: 22444.

[42] Merckle, D., Constant, E., Cartwright, Z. et al. (2021). Ring opening copolymerization of four-dimensional printed shape memory polyester photopolymers using digital light processing. *Macromolecules* **54**: 2681.

[43] Weems, A.C., Armo, M.C., Yu, W. et al. (2021). 4D polycarbonates via stereolithography as scaffolds for soft tissue repair. *Nature Communications* **12**: 3771.
[44] www.4dbiomaterials.co.uk (accessed 31 January 2022).
[45] Kazantseva, N. (2018). Main factors affecting the structure and properties of titanium and cobalt alloys manufactured by the 3D printing. *Journal of Physical Conference Series* **1115**: 042008.
[46] Azgomi, N., Tetteh, F., Duntu, S.H. et al. (2021). Effect of heat treatment on the microstructural evolution and properties of 3D-printed and conventionally produced medical-grade Ti6Al4V ELI alloy. *Metal and Materials Transactions A* **52**: 3382.
[47] Zocca, A., Elsayed, H., Bernardo, E. et al. (2015). 3D-printed silicate porous bioceramics using a non-sacrificial preceramic polymer binder. *Biofabrication* **7**: 025008.
[48] Clare, A., Chalker, P., Davies, S. et al. (2008). Selective laser sintering of barium titanate–polymer composite films. *Journal of Material Science* **43**: 3197.
[49] Gao, C., Yang, B., Hu, H. et al. (2013). Enhanced sintering ability of biphasic calcium phosphate by polymers used for bone scaffold fabrication. *Materials Science and Engineering C* **33**: 3802.
[50] Song, X., Li, W., Song, P. et al. (2015). Selective laser sintering of aliphatic-polycarbonate/hydroxyapatite composite scaffolds for medical applications. *International Journal of Advanced Manufacturing Technology* **8**: 15.
[51] Tan, K., Chua, C., Leong, K. et al. (2003). Scaffold development using selective laser sintering of polyetheretherketone–hydroxyapatite biocomposite blends. *Biomaterials* **24**: 3115.
[52] Lorrison, J., Dalgarno, K., and Wood, D. (2005). Processing of an apatite-mullite glass-ceramic and an hydroxyapatite/phosphate glass composite by selective laser sintering. *Journal of Materials Science: Materials in Medicine* **16**: 775.
[53] Khatri, B., Lappe, K., Habedank, M. et al. (2018). Fused deposition modeling of abs-barium titanate composites: a simple route towards tailored dielectric devices. *Polymers* **10**: 20734360.
[54] Lee, K.S., Kim, R.H., Yang, D.Y. et al. (2008). Advances in 3D nano/microfabrication using two-photon initiated polymerization. *Progress in Polymer Science* **33**: 631.
[55] Bertsch, A., Zissi, S., Jezequel, J. et al. (1997). Microstereophotolithography using a liquid crystal display as dynamic mask-generator. *Microsystem Technology* **3**: 42.
[56] Salimon, A.I., Senatov, F.S., Kalyaev, V. et al. (2020). Shape memory polymer blends and composites for 3D and 4D printing applications. In: *3D and 4D Printing of Polymer Nanocomposites Materials, Processes, Applications, and Challenges* (ed. K.K. Sadasivumi, K. Deshmukh, and M.A. Almaadeed). Amsterdam: Elsevier.
[57] Bodaghi, M., Damanpack, A.R., and Liao, W.H. (2016). Self-expanding/shrinking structures by 4D printing. *Smart Materials Structures* **20**: 1.
[58] Yeazel, T.R. and Becker, M.L. (2020). Advancing toward 3D printing of bioresorbable shape memory polymer stents. *Biomacromolecules* **21**: 3957.
[59] Ge, Q., Sakhaei, A.H., Lee, H. et al. (2016). Multimaterial 4D printing with tailorable shape memory polymers. *Scientific Reports* **6**: 31110.
[60] Wei, H., Zhang, Q., Yao, Y. et al. (2017). Direct-write fabrication of 4D active shape-changing structures based on a shape memory polymer and its nanocomposite. *ACS Applied Materials and Interfaces* **9**: 876.
[61] Kuang, X., Chen, K., Dunn, C.K. et al. (2018). 3D printing of highly stretchable, shape-memory, and self-healing elastomer toward novel 4D printing. *ACS Applied Materials and Interfaces* **10**: 7381.
[62] De Marco, C., Alcantara, C.C.J., Kim, S. et al. (2019). Indirect 3D and 4D printing of soft robotic microstructures. *Advanced Materials Technology* **4**: 1900332.
[63] Govindarajan, T. and Shandas, R. (2019). Microgrooves encourage endothelial cell adhesion and organization on shape-memory polymer surfaces. *ACS Applied Biological Materials* **2**: 1897.
[64] http://medicine.umich.edu/dept/otolaryngology/3d-airway-printed-splint (accessed 27 January 2022).
[65] Zopf, D.A., Nelson, M.E., and Ohye, R.G. (2013). Bioresorbable airway splint created with a three-dimensional printer. *New England Journal of Medicine* **368**: 2043.
[66] Les, A.S., Ohye, R.G., Filbrun, A.G. et al. (2019). 3D-printed, externally-implanted, bioresorbable airway splints for severe tracheobronchomalacia. *Laryngoscopy* **129**: 1763.
[67] Morrison, R.J. (2015). Mitigation of tracheobronchomalacia with 3D-printed personalized medical devices in pediatric patients. *Science Translational Medicine* **7**: 285ra64.

[68] Sood, V., Green, G.E., Les, A. et al. (2021). Advanced therapies for severe tracheobronchomalacia: a review of the use of 3D-printed, patient-specific, externally implanted, bioresorbable airway splints. *Seminar on Thoracic Cardiovascular Surgery Pediatric Cardiac Surgery Annuls* **24**: 37.

[69] Huang, Y., Wu, W., Liu, H. et al. (2021). 3D printing of functional nerve guide conduits. *Burn Trauma* **9**: tka011.

[70] Miao, S., Zhu, W., Castro, N.J. et al. (2016). 4D printing smart biomedical scaffolds with novel soybean oil epoxidized acrylate. *Scientific Reports* **6**: 27226.

[71] Advincula, R.C., Dizon, J.R., Caldona, E.B. et al. (2021). On the progress of 3D-printed hydrogels for tissue engineering. *MRS Communications* **11**: 539.

[72] Chung, J.J., Jim, J., Kim, S.H. et al. (2020). Toward biomimetic scaffolds for tissue engineering: 3D printing techniques in regenerative medicine. *Frontiers in Bioengineering and Biotechnology* **8**: 586406.

[73] Senatov, F.S., Niazz, K.V., Zadorozhnyy, M.Y. et al. (2016). Mechanical properties and shape memory effect of 3D-printed pla-based porous scaffolds. *Journal of Mechanical Behavior and Biomedical Materials* **57**: 139.

[74] Inverbizzi, M., Turri, S., Levi, M. et al. (2018). 4D printed thermally activated self-healing and shape memory polycaprolactone-based polymers. *European Polymer Journal* **101**: 169.

[75] Zhang, B., Skelly, J.D., Maalouf, J.R. et al. (2019). Multifunctional scaffolds for facile implantation, spontaneous fixation, and accelerated long bone regeneration in rodents. *Science Translational Medicine* **11**: 7411.

[76] Wang, Y.J., Jeng, U.S., and Hsu, S.H. (2018). Biodegradable water-based polyurethane shape memory elastomers for bone tissue engineering. *ACS Biomaterials Science and Engineering*. **4**: 1397.

[77] Zhou, G., Jiang, H., Yin, Z. et al. (2018). *In vitro* regeneration of patient-specific ear-shaped cartilage and its first clinical application for auricular reconstruction *EBioMedicine*. **28**: 287.

[78] www.sciencedaily.com/releases/2006/04/060404084412.htm (accessed 9 February 2022).

[79] www.poietis.com (accessed 7 February 2022).

[80] www.bellaseno.com (accessed 8 February 2022).

[81] www.organovo.com (accessed 9 February 2022).

[82] Dawood, A., Marti, B.M., Sauret-Jackson, V. et al. (2019). 3D printing in dentistry. *British Dental Journal* **11**: 520.

[83] Jiang, N., Yang, Y., Zhang, L. et al. (2021). 3D-printed polycaprolactone reinforced hydrogel as an artificial TMJ disc. *Journal of Dental Research* **11**: 1302.

[84] Rasperini, G., Pilipchuk, S.P., Flanagan, C.L. et al. (2015). 3D-printed bioresorbable scaffold for periodontal repair. *Journal of Dental Research*. **94**: 153S.

[85] Chen, R.S., Hsu, S.H., Chang, H.H. et al. (2021). Challenge tooth regeneration in adult dogs with dental pulp stem cells on 3D-printed hydroxyapatite/polylactic acid scaffolds *Cells*. **10**: 3277.

[86] Prechtel, A., Stawarczyk, B., Hickel, R. et al. (2020). Fracture load of 3D printed PEEK inlays compared with milled ones, direct resin composite fillings, and sound teeth. *Clinical Oral Investigations* **24**: 3457.

[87] Chen, Y., Li, C., Kaza, S. et al. (2017). *Dental materials using thermoset polymers*. US 20170007362A1 01: 12.

[88] Elshazly, T.M., Keilig, L., Alkabani, Y. et al. (2022). Potential application of 4D technology in fabrication of orthodontic aligners. *Frontiers in Materials* **8**: 794536.

[89] Quirynen, M., Abarca, M., van Assche, N. et al. (2007). Impact of supportive periodontal therapy and implant surface roughness on implant outcome in patients with a history of periodontitis. *Journal of Clinical Periodontology*. **34**: 805.

[90] Yun, J., Lee, J., Ha, C. et al. (2021). The effect of 3-D printed polylactic acid scaffold with and without hyaluronic acid on bone regeneration. *Journal of Periodontology*. **1**.

[91] Wang, C., Yue, H., Liu, J. et al. (2020). Advanced reconfigurable scaffolds fabricated by 4D printing for treating critical-size bone defects of irregular shapes. *Biofabrication* **18**: 045025.

[92] www.orthofix.com (accessed 8 February 2022).

[93] www.onkossurgical.com (accessed 8 February 2022).

[94] www.4webmedical.com (accessed 8 February 2022).

[95] http://Fiercebiotech.com/medtech/smith-nephew-kicke-135m-for-cementless-knee-implant-maker-engage-surgical (accessed 8 February 2022).
[96] http://orthoworld.com/first-case-with-depuy-synthes-trumatch-graft-cage (accessed 10 February 2022).
[97] www.fda.gov/medical-devices/overview-device-regulation/history-medical-device-regulation-oversight-united-states (accessed 13 February 2022).
[98] Norman, G. (2016). Drugs and devices: comparison of European and US approval processes. *JACC: Basic Transactions in Science* **1**: 399.
[99] Jain, A.B., Mollet, A., and Szucs, T.D. (2017). Regulatory watch: structural and procedural characteristics of international regulatory authorities. *Nature Reviews of Drug Discovery* **16**: 594.
[100] Prima, M.D., Coburn, J., Hwang, D. et al. (2016). Additively manufactured medical products – the FDA perspective. *3D Printing in Medicine* **2**: 1.
[101] http://resources.asme.org/poc3dp-events (accessed 13 February 2022).
[102] Christensen, A. and Rybicki, F.J. (2017). Maintaining safety and efficacy for 3D printing in medicine. *3D Printing in Medicine* **3**: 1.
[103] http://www.carbon3d.com/resources/whitepaper/additive-manufacturing-a-primer-on-regulatory-affairs (accessed 13 February 2022).
[104] http://www.fda.gov/medical-devices/3d-printing-medical-devices/3d-printing-medical-devices-point-care-discussion-paper (accessed 13 February 2022).
[105] Moshkovits, I. and Shepshelovich, D. (2022). Emergency use authorizations of covid-19–related medical products. *JAMA of International Medicine* **182**: 228.
[106] https://formlabs.com/company/press/fda-authorization-adapters (accessed 13 February 2022).
[107] Manero, A., Smith, P., Koontz, A. et al. (2020). Leveraging 3D printing capacity in times of crisis: recommendations for covid-19 distributed manufacturing for medical equipment rapid response. *International Journal of Environmental. Research in Public Health* **17**: 4634.

5

3D Printed Implants for Long-Acting Drug Delivery

Aikaterini Dedeloudi[1], Sune Andersen[2], Peyton Hopson[3] and Dimitrios A. Lamprou[1]

[1] *School of Pharmacy, Queen's University Belfast, Belfast, UK*
[2] *Research & Development Department, Janssen, Beerse, Belgium*
[3] *Advanced Engineering and Technology Department, Johnson & Johnson, Jacksonville, Florida, US*

5.1 Introduction

3D printing (3DP), as an emerging technology, has contributed to the development of innovative medicinal products, targeting towards the sustainability and durability of the administered therapy. It can be applied effectively in developing implants for tissue engineering and drug delivery systems (DDSs) with long-acting profiles, since it provides a therapeutic prediction before medical intervention. The use of Computer-Aided Design (CAD) and real-time monitoring techniques can optimise the material selection, the ability in modifying geometry and dimensions of printed scaffolds, and subsequently the efficacy and the quality of the product, facilitating the development of adjustable and customised multifunctional therapeutic systems.

The development of long-acting 3D-printed implantable DDSs concern novel therapeutic patient-centred approaches targeting towards an improved patient compliance. Contrary to conventional therapies administered to the gastrointestinal (GI) tract, implantable devices avoid first-pass metabolism, GI degradation, and compliance issues. Moreover, this alternative method of therapy provides exceptional potential in local and controlled drug release and in personalising drug dose administration, aiming at a targeted therapeutic approach and maintenance of a steady state of the disease status and progress [1].

3D Printing of Pharmaceutical and Drug Delivery Devices: Progress from Bench to Bedside, First Edition.
Edited by Dimitrios A. Lamprou, Dennis Douroumis and Sheng Qi.

Implantable devices have a considerable perspective in preventing, relieving, and slowing the progression of chronic diseases such as cardiovascular, pulmonary, and inflammatory diseases, and cancer. Formulations of diverse implantable delivery systems, characterised by long-acting release profiles, can be fundamental in targeting chronic diseases, consequently providing better survival prospects in patients [2, 3]. Crucial parameters that play a significant role in developing a long-acting release profile are the physico-chemical properties of materials used (e.g., solubility, swellability, erosion, thermal properties, mechanical stiffness, roughness), manufacturing processes, and clinical profiles of patients [4].

With 3DP technology, tailor-made scaffolds can be designed with specified drug release profiles through formulation development for both the drug and the implant. Moreover, 3DP can be implemented in creating diverse-based dosage forms by using a variety of combined technologies and materials [5–7]. In detail, implantable and injectable scaffolds describe the two major categories of 3D-printed DDSs, which can be further classified into passive and active, and into degradable or non-degradable, according to the materials used [1]. As 3DP is a novel approach in manufacturing innovative DDSs, some of the current challenges involve high cost of manufacturing and long production times [8]. Safety issues concerning the therapeutic threshold of the implantable DDSs have also to be evaluated, since surgical incorporation is involved; thus, toxic effects may be provoked and cause malfunction. Consequently, it is important that all critical process parameters (e.g., physiological, biopharmaceutical, technological) related to critical quality attributes of the manufacturing process and formulation for implantable devices are considered, with regard to safety and efficacy [9].

In this chapter, an elaboration of the feasibility of 3DP in developing implantable devices for long-acting drug delivery is examined.

Definition Clarification:

The term 'scaffolds' concern all the types of devices or DDSs that are tissue-incorporated (e.g., implantable or injectable). However, the terms 'implants' and 'implantable' devices are used interchangeably.

5.2 Types of 3D-Printed Scaffolds

As 3D-printed DDSs are characterised by complexity and multifunctionality, they can be classified into two wide categories: implantable and injectable scaffolds (Figure 5.1). Implantable scaffolds are be further classified into passive and active implants and as biodegradable and non-biodegradable, according to the formulation materials used (Figure 5.2). The main aspect that differentiates passive and active systems is the release rate of the active pharmaceutical ingredient (API). In passive implants, the release rate cannot be modulated; however, active implants are stimulated by external factors [10].

5.2.1 Implantable Scaffolds

5.2.1.1 Passive Implants

The manufacturing process of passive implantable systems is technically simple, as they are solid formulations consisting of biocompatible materials. The main mechanism of drug transport in these systems is based on passive diffusion and the drug kinetic release depends on several parameters, such as the physico-chemical properties and the concentration of drug and excipients, and also the mechanical and surface properties of the final formulation [10].

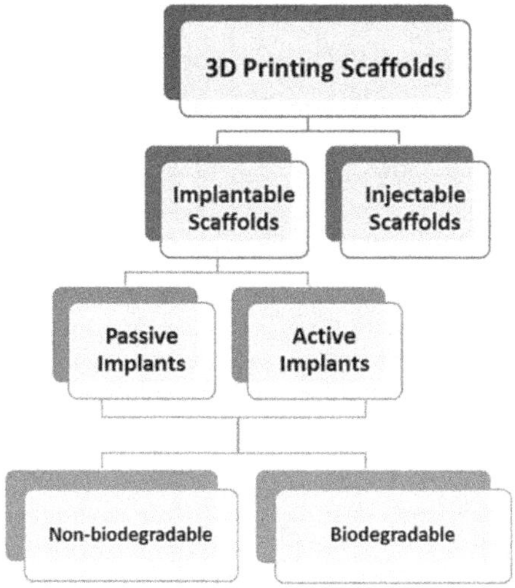

Figure 5.1 *Classification of 3D-printed scaffolds.*

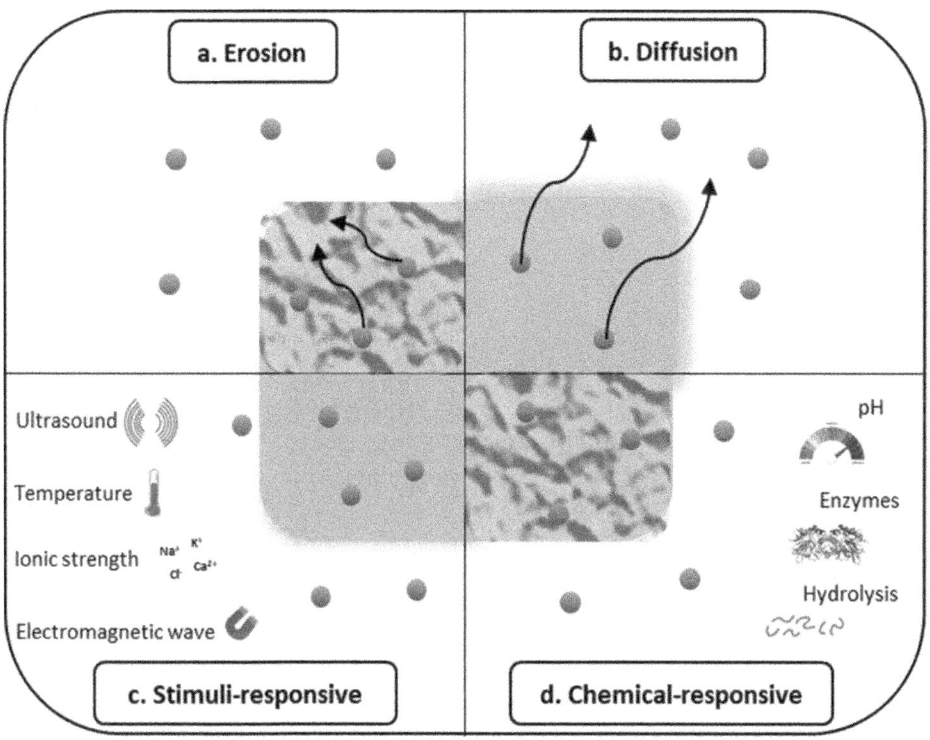

Figure 5.2 *Mechanisms of drug release in passive (a,b) and active implants (c,d).*

5.2.1.1.1 Non-Biodegradable Passive Implants

Polymers, ceramics, and metals are the three major categories of materials that are used in the formulation of 3D-printed non-biodegradable implants. Polymers concern the acrylic resins which are thermoplastic, including vinyl monomers such as styrene and vinyl acetate, and thermoset, also including a high level of methyl methacrylate in their structure [1, 10, 11]. Non-biodegradable polymers are used in monolithic or reservoir type implants [1]. Metal printing materials that are thermo-conductive, such as titanium and stainless steel [12], are mainly used in tissue engineering, whilst ceramic substrates such as carbides, nitrides, and metal-oxides are less used [13].

Applications of polymer-based non-biodegradable implants are used either in drug delivery or in tissue reinforcement systems; however, metal and ceramic-based systems concern scaffolds for tissue engineering. Even though non-biodegradable implantable devices are considered as long-acting DDSs, they have limitations in their use, as they need to be removed when the drug concentration needs to be eliminated [13, 14].

5.2.1.1.2 Biodegradable Passive Implants

Biodegradable implants are characterised as polymeric-based implants that are considered to be subjected to progressive elimination when they are inserted into tissues. Biodegradable polymers which are regularly used are poly(lactic acid) (PLA) [6], poly(caprolactone) (PCL) [6, 7], and poly(lactic-co-glycolic acid) (PLGA) [15]. These polymers are considered efficient when utilised for long-acting implants, providing an extensive release profile of the API. Moreover, biodegradable implantable devices are characterised as safe, biocompatible, and easily applicable, as they are either absorbed or excreted by the body, consequently extraction after implantation is not needed. However, their complex manufacturing, the limited variation of polymers available, the ambiguous regulatory aspects of their use, and the variability in clinical release profile of the drug, define significant limitations that lessen their broad application [16].

5.2.1.2 Active Implants

Active implants are characterised by an on-demand drug release mechanism, activated by magnetic fields, polymeric valves, and manual operators' electrical and thermal energy [4]. Moreover, AM is an upcoming technology which is applied in formulating stimulated active scaffolds based on smart biomaterials [17]. These systems are used for treating diseases from a chronotherapeutic perspective, such as cancer, diabetes, or osteoporosis. Nevertheless, these types of implants may present drawbacks, concerning their manufacturing complexity and unsafe operation system (e.g., electrical-based implants), the continuous need of a power supply, and the invasive surgery for tissue removal of those that are non-biodegradable [18].

Active implants are classified into two categories: single and multiple reservoir systems. The pump structure of the former consists of a drug reservoir; however, the latter system consists of more than one drug compartment [4, 18].

5.2.2 Injectable Scaffolds

The design of long-acting injectable scaffolds is based on formulating systems that resemble the physiological properties of the tissue, in which they are incorporated [19]. Injectable scaffolds are considered as *in situ* forming systems, exposed into phase transition from liquid to gel or solid state, when they are inserted into the tissue [20]. Materials that present suitable engineering properties and a non-toxic, bioavailable, and biodegradable profile, are used in manufacturing injectable DDSs. As far as the administration route is concerned, materials used should present a considerably high shear viscometrical flow and a long-term stability formation. In this way, an optimum 'through the needle' administration and a progressive biodegradation of the product are ensured [19]. Furthermore, these systems are mainly used for local-based therapies and the drug release mechanism depends on diffusion phenomena [20].

Stimuli responsive polymers (e.g., pH, temperature, infrared), that are characterised by *in situ* gelling properties, enhance the material flow and consequently the drug release [21]. Therefore, injectable scaffolds represent a substantial type of extensive drug release delivery systems, as they provide a better patient compliance profile without invasive surgical procedures and with low medical and manufacturing costs [22]. Moreover, injectable scaffolds are suitable formulations for administering high molecular weight (MW) active substances, such as bioactive proteins, growth factors [19], and nanoparticles [23]. However, they present stability issues and lyophilisation procedures have to be considered during their manufacturing, not only for preserving stability of bioactive incorporated ingredients but also for excipients prone to chemical, electrochemical, mechanical, and thermal stability [24].

Injectable scaffolds are classified into two categories: organogels and hydrogels [19]. Organogels are mainly formed by low MW organogelators (LMOG) (e.g., sterol-, phospholipid-, fatty and amino acid-based organogels), polymeric organogelators (POG) (e.g., polyesters, poly(alkylene), poly(ethylene glycol)), and cellulose-based polymers. Polymeric organogels tend to form simulated 3D polymer-linked environments that entrap the drug and define its release rate from the system. Injectable scaffolds based on organogels are characterised as depot systems efficient in transferring hydrophobic or lipophobic drugs for a variety of chronic diseases [21]. In contrast, injectable scaffolds, that are based on hydrogels, consist of hydrophilic polymers such as poly(amides), poly(anhydrites), polyesters, and polysaccharides [22]. They form complex crosslinked systems with various entrapment efficiencies, depending on the MW of the polymeric chains and their hydration potential. The used polymers create inner pores that permit the progressive release of the drug, consequently, its bioavailability. These systems can be applied in cell-based therapies due to their multifunctionality and adaptability in different aqueous environments [23].

5.2.3 Innovative 3D-Printed Scaffolds

An upcoming perspective on printing 3D scaffolds concerns the use of innovative polymers and applied technologies, which contribute to drug release profile improvement, biocompatibility issues, and ease in scale-up production. 4D printing technology is concerned with an innovative technology for tissue engineering and therapeutic devices manufacturing. It is defined as a time-related technology, using materials with shape memory properties, which can be modified during time progression. Materials that possess time-dependent transformation properties are composite hydrogels, liquid crystal elastomers, and smart materials [25, 26].

Shape Memory Polymers (SMP) are multifunctional stimuli responsive smart biomaterials that can be easily adjusted to different biological environments by altering their physico-chemical properties. Their activation is generated by physical stimulations (e.g., temperature, pH, light, water, etc.), providing Shape Memory Effects (SME), depending on time lapse [25]. Bioprinting methods based on smart materials can be useful in incorporating sensitive substances (e.g., proteins, cells, genes), thus creating novel pharmaceutical systems that enable a stable long-acting release of the drug and a targeted therapy [27, 28]. Bio-fabricated scaffolds can be either used for replacing tissues and/or drug release, and are characterised as pharmacologically active, clinically compatible, and stable [29, 30]. Moreover, using 4D printing technology in manufacturing diverse scaffolds, which are based on metamaterials, can provide an optimum therapeutic approach in controlling drug release and therapeutic levels in blood [31].

When designing novel therapeutic scaffolds based on innovative biomaterials, regulatory aspects have to be examined. Where a more complex and multifunctional system is concerned, more profound and extensive research aspects for safety and efficacy are considered.

5.3 Critical Parameters in Designing 3D-Printed Implantable Scaffolds

Scaffold architecture features play a significant role in determining drug release rates. Different combinations of materials, physico-chemical characteristics of active substances, and applied manufacturing techniques are critical in defining biological interactions and subsequently therapeutic response (Figure 5.3). Designing parameters, which affect the *in vivo* performance of scaffolds, are related to their structural, mechanical, and biological properties [32].

In the following section, structural, mechanical, and biological attributes concerning 3D-printed implantable scaffolds are examined.

5.3.1 Structural Characteristics

Passing from 2D to 3D dimensions, shape, geometry, surface, and porosity define attributes, which affect stability and biological parameters of printed systems. Material combinations that form multifunctional structures with convoluted and intricate topology features need to be evaluated. Therefore, implantable scaffolds concerning pharmaceutical systems with proper mechanical stability, encapsulating, and transferring a variety of therapeutic and functional materials (e.g., active substances, genes, cells), allows their delivery into target tissue, determining their release rate and stability. Moreover, 3D-structured systems

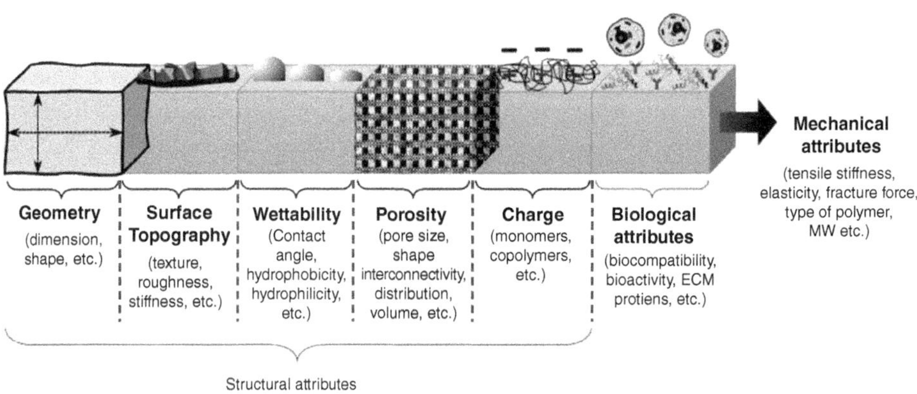

Figure 5.3 Critical attributes in designing 3D-printed implantable scaffolds.

are described by their ability to absorb and to deliver natural nutrients, and to excrete generated metabolic waste.

5.3.1.1 *Geometry of Implants*

Dimensions and geometrical characteristics of implants are significant parameters in defining and evaluating adhesive and compatibility potential, drug release rate, and cell and tissue growth. Examining the diversity of scaffold dimensions and shapes, geometry seems to correlate with a variety of functional processes that differ from the macroscopic to microscopic scale. Studies have shown that cell and tissue regeneration are observed macroscopically and, rather than the vascularisation process, mass and drug flow are detected in a microscopic view [32].

Moreover, 3DP technology helps in designing and manufacturing scaffolds with desirable geometrical and mechanical features. Farmer et al. have examined the relation of geometrical features of 3D-printed mesh implants on their mechanical properties. In this research, results showed that material and mesh geometry (e.g., layer height, circle or rectangular shape of mesh) affected mechanical properties (e.g., tensile stiffness, elasticity, fracture force, elongation at break) and drug release rate. Additionally, the use of materials with high tensile strength can cause tissue deterioration in contrast to low tensile strength [33]. Bidan et al. studied the impact of different geometries of scaffolds in tissue growth rate. When analysing the growth law in convex and non-convex shapes, it is obvious that proliferation of cells is directly related to curvature adjustments. 'Non-convexity' determining tissue growth rate in the pore size and 'circularity' determining the number of pores adapting in the scaffold are two major parameters defining tissue deposition rate. In detail, branch-like convex shapes (e.g., cross-shaped) can be easily attached to tissue, due to their geometrical complexity. The more curvature of the geometry of the scaffold, the faster the tissue growth rate. As non-convex curvature produces circular characteristics, due to degradation scaffold procedures, cell attachment and subsequently tissue proliferation decelerates. In this study, results showed that from cross- to circular-shaped scaffolds, a diminishing rate on tissue proliferation rate is observed [34].

5.3.1.2 *Porosity Properties and Pore Features*

Porosity features are crucial parameters in developing implantable long-acting DDSs. The type of material and the manufacturing process can determine size, shape, volume, distribution, and interconnectivity of pores created in scaffolds. Therefore, these characteristics can define accurately mechanical strength, adhesion, and tissue-implant compatibility, as well as mass transfer rate and drug release rate. The size of created pores and channels is responsible for the creation of a dynamic system with mass flow properties that are renewable and sustainable [32, 35].

In a study by Gupte et al., examining the cell life and tissue proliferation in nanofibrous microporous scaffolds, it was shown that pore size and porosity of implants are directly associated with vascularisation and bone formation. A variety of pore size distributions, small (125–250 μm), medium (250–425 μm), and large (425–600 μm), was examined. Bone volume and thickness increment appeared in scaffolds with greater pore sizes. Moreover, in small-sized pore scaffolds, a chondrogenesis and cartilage morphology resulted, opposed to larger-sized pore scaffolds, which permitted vascularsation and consequently bone formation [36]. Consequently, porosity parameters have a great impact in tissue regeneration and drug release rates, which involve clinical performance.

5.3.1.3 Surface Properties

Surface of implantable DDSs is the first part interacting with the tissue. Topographic characteristics regarding the texture, roughness, and stiffness of implants can define their abilities of adhesion hydration, swellability, erosion, and drug release rate. Moreover, surface topology is crucial for modulating cell–implant interactions, evaluating micro-environmental biocompatibility and therapeutic outcome; thus, determining the pharmacodynamic and pharmacokinetic potential of the system [35].

Interphase interactions between the surface of implants and the tissue environment depend on physico-chemical properties of the surface. Hydrophilicity and hydrophobicity, stiffness and roughness, charge of functional groups, and contact angle, are crucial parameters which characterise the interaction potential of implant with tissue. The hydrophilicity of surfaces is estimated with the contact angle [32].

In a study by Shen et al., hydrophobic and cationic surfaces are efficient in preventing microorganism development. Covalent bonding surface additions (e.g., hydroxylated titanium alloy) contribute to surface stabilisation with specialised characteristics (e.g., antimicrobial potential) that are efficient in providing long-lasting functionality and applicability. Moreover, external triggers (e.g., hydrothermal process, augment hydrophilicity, roughness, and porosity features) contribute to an increase in hydrophobicity achieved with silane addition [37].

5.3.2 Mechanical Properties

Polymeric properties related to mechanical features, such as type of polymer, MW, crosslinking modifications, polymer density, iso-elasticity between materials, toughness, strut thickness, and angle can affect cell adhesion and penetration to a certain degree. Mechanical properties of scaffolds can determine interactions with cells. Cells can detect and distinguish surface characteristics, which are related to mechanical properties of the scaffold; therefore, adhesion, proliferation, and differentiation of cells are feasible when there is implant–cell compatibility [38]. Consequently, the choice of materials is directly related to the manufacturing method selected that reflects on the attributes of the final scaffold. Hence, it is essential that material selection is examined under ranking procedures, which concern process parameters (e.g., temperature, pressure, speed, etc.), shape of the scaffold (hollow, circular, non-circular, etc.), physical attributes (e.g., thickness, surface roughness, etc.), and economics and costs [39].

5.3.3 Biological and Physiological Parameters

A variety of physiological parameters play a significant role in implant tissue incorporation: cellular adhesion, degradation and absorption rates, and biocompatibility characteristics regarding the potential of scaffold implantation.

5.3.3.1 Cellular Adhesion

Examining the micro-environmental conditions of the tissue after scaffold implantation, a variety of cell and protein interactions is observed. Physico-chemical characteristics of scaffolds, concerning surface and mechanical aspects, are the most crucial in determining scaffold incorporation potential, cell and immune system response, drug release rate, and pharmacological and pharmacokinetic profiles. Integrins are proteins that are mediator factors in facilitating cell–implant interactions. Integrins, as signalling molecules, are primarily attached in implants and activate cell clustering and adhesion. Henceforth, cells transmit signals for cell proliferation and extracellular matrix (ECM) deposition and differentiation. Therefore, by modifying scaffold structure and surface topography, different mechanisms and processes of cell activation are created [40].

5.3.3.2 *Absorption and Degradation Rates*

The absorption profile of an active substance is a crucial parameter in formulating long-acting implants. Solubility and permeability are criteria that characterise the Biopharmaceutics Classification System (BCS) class of an active substance and determine its release rate and physical interactions with the material used. Hence, release rate is required to be slower than absorption rate, in order to achieve an extend drug release profile [41]. Long-acting implantable scaffolds aim at drug bioavailability enhancement since they avoid the first-pass metabolism, by releasing the drug directly in the systematic circulation. Therefore, implantable scaffolds concern local drug delivery and targeted DDSs and are beneficial candidates for increasing pharmacological potency of low-dose drugs [2].

Attributes that determine and influence the degradation rate of biodegradable scaffolds involve properties of utilised materials. Physical characteristics such as MW, chemical structure and copolymer ratio, glass transition temperatures (T_g), polymer structural morphology (crystalline, amorphous), and chain orientation influence the potential of degradation process. Along with implant elimination, tissue regeneration occurs; thus, degradation rate of implants should be followed by cell proliferation and differentiation [2].

5.3.3.3 *Biocompatibility Aspects*

Implantable devices that are designed for tissue regeneration should be defined by biocompatible features. Considering that the utilised materials can trigger immune system and consequently inflammatory responses, it is essential that they possess biocompatible characteristics. Hence, modification of materials (e.g., PEGylation, surface coatings, micro-scale roughness) can circumvent immune system identification, enhancing tissue growth and scaffold stability maintenance [42].

5.4 Critical Parameters in Selecting Materials for 3D-Printed Scaffolds

There is a range of materials used in the manufacturing of 3D-printed scaffolds as DDSs, which vary according to their therapeutic functionality and the anatomical region where they are designed to be inserted. Particularly, the examination of the materials' physicochemical properties, addressed for long-acting scaffolds, is a crucial step, which has to be initially performed. Characteristics of biomaterials are related to surface features such as charge, stiffness, degradation, and wettability, with their efficiency in promoting functional interactions between cells and DDSs, characterised by compatibility, adaptability, and degradability properties [1]. Consistently, biomaterials have the potential to determine drug release rate in tissues, in supporting cell proliferation and differentiation, and in progressive physical elimination. Moreover, in order to ensure and ameliorate adhesive properties of the system and release rate of the drug, biomaterials and accordingly scaffolds can be modified [44]. Therefore, their surface can be amended either physically (e.g., particle size, shape) [44], chemically (ligand addition) [45], or by applying coatings [7].

The most common materials used concern the formulation of the polymer-, metal-, and ceramic-based scaffolds. Metallic and ceramic materials are applied in bioprinting scaffolds for tissue restoration [46]. On the other hand, polymeric-based scaffolds are used mainly in DDSs [47]. As far as polymeric materials are concerned, these are classified according to their source into natural and synthetic and according to their physico-chemical properties into biodegradable and non-biodegradable [1]. Natural polymers are sensible in their modification, most of them possessing poor mechanical and stability properties when used individually, and their physico-chemical properties are not easily defined due to their complexity [48]. However, they are capable of chemical structural modifications, which enhance their mechanical properties [49]. Contrarily, synthetic polymers are vastly applied in 3DP due to their mechanical stiffness and their well-defined physical and chemical characteristics. Synthetic polymers are broadly used due to their stability and predictable functionality [47]. Moreover, with reference to biodegradability properties, these may concern either natural or synthetic polymers. Non-biodegradable differ from biodegradable polymeric materials, as they present biodegradable and bioresorbable limitations, and surgical invasion is required for their removal [50] (Table 5.1).

Table 5.1 *Categories of biomaterials used in 3D-printed implantable scaffolds.*

Biomaterials		Material-Based properties	References
Natural polymers	Cellulose (bacterial and plant derived), chitosan, silk fibroin, collagen, hyaluronic acid	• Bioadhesive • Biocompatible • Biodegradable • Renewable • Abundant availability • Biomimetic • Promote cell viability and tissue ingrowth • Non-toxic • Low immunogenicity • Pore formation	[5, 57–70]
Synthetic polymers	Poly(lactic acid) (PLA), poly(lactic-co-glycolic acid) (PLGA), poly(glycolic acid) (PGA), poly(caprolactone) (PCL)	• Thermoplastic • Biodegradable • Biocompatible • Bioresorbable • High mechanical strength • Controllable degradation • Reproducible • Lack of biological signals	[6, 7, 26, 52, 66, 70–73]
	poly(siloxanes), poly(ethylene-vinyl acetate) (PEVA)	• Non-biodegradable • Non-immunogenic • Reproducible • Tailored mechanical properties and shapes	[71, 74–77]
	poly(urethanes)	• Versatile (biodegradable or non-biodegradable) • Stimuli responsive • Biostable • Elastomeric • Thermoplastic (TPU) • Structure modifications	[72–75]

Table 5.1 (Continued)

Biomaterials		Material-Based properties	References
Ceramics	Inorganic metal compounds and/or calcium salts, phosphate salts	• Bioresorbable	[16, 46, 76–78]
Metals	Titanium, cobalt-chromium, stainless steel, tantalum, gold, magnesium, gallium, Iron alloys	• High mechanical properties	[79, 80]

5.4.1 Materials Used in 3D-Printed Long-Acting Scaffolds

5.4.1.1 Natural Polymers

Natural-based biopolymers are widely used in implant fabrication technologies, due to their biodegradability and tissue incorporation bio-suitability. Cellulose, chitosan, silk fibroin, collagen, hyaluronic acid, and other natural co-polymers and proteins are claimed as prospective materials for 3D-printed implants.

Bacterial and plant derived cellulose are vastly used in hydrogel-based 3DP scaffold formulations, either in drug delivery or in tissue enhancement, as they possess considerable swelling properties and significant mechanical integrity. They can be easily modified with crosslinking adjustments, which contribute to mechanical advancement properties of scaffolds [5, 51–55]. Chitosan is characterised by high mechanical strength attributes; thus, it is used in tissue engineering 3D-printed scaffolds. It is also ordinarily applied in modified release DDSs as a long-term degradation polymer [56–58]. Silk fibroin is another natural polymer which possesses high mechanical stability and is feasible in crosslinking modifications. Moreover, it is defined by high elasticity and low viscosity properties that permit the fabrication of low-temperature 3D-printed systems. Therefore, silk fibroin is efficient in injectable DDSs, hydrogels, bioadhesive systems, and tissue regeneration scaffolds [48 52, 56–60]. Collagen is mainly utilised in tissue regeneration of 3D-printed scaffolds and in DDSs for the delivery of biological factors, as it is characterised by low mechanical and fast degradation properties. However, collagen is sensitive to manipulation and is potentially used in low-temperature 3D-printed systems [48, 58, 59, 61]. Hyaluronic acid can be applied in 3DP-implantable scaffolds for tissue engineering. It is mostly used in combination with other materials, as it is characterised by low viscoelasticity and hydrophilicity. Hence, it promotes cell proliferation, migration, differentiation, and inflammation and wound healing processes [61–64].

5.4.1.2 Synthetic Polymers

Synthetic polymers that are used in implantable scaffold formulations are classified into three categories: biodegradable, non-biodegradable, and versatile.

Thermoplastic and biodegradable poly(esters) (e.g., poly(lactic acid) (PLA), poly(glycolic acid) (PGA), poly(lactic-co-glycolic acid) (PLGA), and poly(caprolactone) (PCL)) are defined by significant mechanical properties, such as high mechanical strength, thermoplasticity, and versatility in surface modifications. Moreover, they are characterised by low biological activity and biocompatibility, bioresorbability, and controllable degradation mechanisms; hence, they are broadly applied in 3D-printed implantable DDSs or tissue engineering systems [6, 7, 15, 45, 62, 64–67].

Non-biodegradable polymers, such as poly(siloxanes) and poly(ethylene-vinyl acetate) (PEVA), are synthetic polymers that differ from biodegradable in drug release mechanisms. Biodegradable polymers permit drug release, both through diffusion of drug and/or degradation of the matrix. Applications of non-biodegradable polymers are implemented mostly in scaffolds concerning tissue regeneration, due to their adjustable mechanical properties and shapes, chemical and thermal stability, non-immunogenic and antimicrobial properties, and compatibility with natural tissues [65, 68–71].

Versatile synthetic polymers (e.g., poly(urethanes)) are assessed as either biodegradable or non-biodegradable, according to their synthetic process, thus can appear in a rubber- or polyol-based structure. They are considered biostable with notable thermoplastic and elastomeric properties. Regarding their diverse functionality, such as their facility in structure modifications and their stimuli responsiveness, they are effectively incorporated into DDSs and tissue engineering 3D-printed scaffolds [72–75].

5.4.1.3 Ceramics and Metals

Ceramic 3D-printed implantable systems mainly consist of inorganic metal compounds and/or calcium salts and phosphate salts (e.g., silicon dioxide (SiO_2), calcium sulfate (CS), α- and β-tricalium phosphate (β-TCP, α-TCP), hydroxyapatite (HA)). Pursuant to their bioresorbable, osteoinductive and brittle properties are vastly proposed and applied in 3D-printed bone implants [16, 46, 76–78]. As far as metallic-based 3D-printed scaffolds are concerned, a variety of metals and metal alloys are used (e.g., titanium, cobalt-chromium, stainless steel, tantalum, gold, magnesium, gallium, iron alloys). High corrosion resistance and high mechanical strength properties, surface roughness, and tissue-implant interlock potential are significant parameters in designing implantable scaffolds. Metallic biomaterials are mostly applied in 3D-printed implants for tissue engineering and regeneration [80, 81].

5.4.1.4 Composites

Composite material combinations describe multiparametric systems characterised by variability. The functionality of each material affects the properties (physico-chemical, biological, mechanical) of the system and define the quality of the final product. Along with material combination properties, the formulation process parameters are considered as crucial and properly defined by screening processes, comparative analyses, and evaluations, in order to specify different formulation performances and incompatibilities. Parameters regarding the function, cost, and environmental characteristics of the materials and machines used are part of a sequential algorithm concerning decision-making for the design criteria [39].

5.5 Manufacturing Techniques for Implantable Scaffolds

Critical material attributes (CMA) and critical process parameters (CPP) are considered crucial factors in designing and developing novel pharmaceutical implantable scaffolds [1]. Polymers that are used in long-acting scaffold formulations are considered to be suitable with the applied method and show biocompatible properties [81]. Various techniques have been developed for the manufacturing of polymer-based implantable DDSs, with the most common techniques being hot-melt extrusion (HME), compression, injection moulding, solvent casting, and 3DP [1]. An overview of the advantages and disadvantages of the techniques is given in Table 5.2.

Table 5.2 *Advantages and disadvantages of 3DP-coupled manufacturing methods for 3D-printed implantable scaffolds.*

Manufacturing method	Advantages	Disadvantages	References
Hot-melt extrusion	• Variety of thermoplastic materials/polymers available • Continuous manufacturing process • Scalable • Solvent free method • Suitable for lipophilic drugs	• Unsuitable for thermosensitive materials • High-temperature application	[84–89]
Compression	• Suitable for thermosensitive or solvent sensitive materials	• Limited application for hydrophilic materials • Unpredictable surface characteristics: unsuitable for controlled drug release rate profiles	[44, 47]
Injection moulding	• Variety in use of thermoplastic materials • Suitable for incorporating low soluble drugs into solid disperse systems • Ease in creating multi-shaped solid formulations • Modifying DDSs	• High-cost equipment • Slow formulation procedure for tailor-made implants, due to mould modification requirements	[95]
Solvent-casting	• Cost-effective • Coating of implants: long-acting drug release profile. • Relatively lower temperatures for solvent removal compared to HME.	• Not scalable • Need of a volatile solvent (stability issues)	[2, 7, 8, 87]
3D printing	• Personalised implantable scaffolds • Variety of material combinations • Variety in types of 3DP technique (e.g., FDM, SLA, SLS, etc.) • Ease in combination with different manufacturing techniques (e.g., HME, injection moulding, solvent-casting, etc.)	• Challenging in scale-up process. • Capacity • Cost	[3, 6, 44, 81]

5.5.1 Hot-Melt Extrusion

Hot-melt extrusion (HME) is well-known as a pharmaceutical manufacturing technology, especially for amorphous solid dispersion [82]. The principle of the method is based on melting and mixing a variety of thermoplastic excipients, such as aliphatic polyesters and active substances [82, 83]. This manufacturing technique involves simultaneous melting and continuous mixing of the formulation materials. The mixing process is controlled through the designs of the screw and combinations of various screw elements and temperature zones, i.e., the process includes zones with various mixing elements and temperatures. Finally, the melt is extruded through an orifice as a filament, which for tableting is milled to obtain granules, but for 3DP is deposited to form the desired geometry (Figure 5.4). HME is capable of continuous manufacturing and a fully scalable process that combines a broad variety of dosage forms produced by less unit operations. Moreover, it is a solvent free method and has, for example, been described as suitable in enhancing the solubility of lipophilic drugs [84, 85]. However, there are some limitations in its use, as it is considered an unsuitable technique for thermosensitive drugs, where there is a limited number of thermal stable polymers and a great energy density needs to be applied during high process temperature thermal manufacturing [86, 87].

A combined technological method of HME-3DP can be applied in modifying release system implants [88, 89]. An HME-FDM combination has shown that HME defines the crystallisation of polymers, which affects the mass and content uniformity of filament (e.g., thermal stability, melt viscosity, etc.), consequently the rigidity of filament, which determines the product's mechanical properties (e.g., tensile strength, yield stress, etc.). On the other hand, extruded filament can be printed in different shapes and geometries with FDM; thus, creating customised products. HME-FDM coupling can be used in a single-step manufacturing procedure, aiming at a more cost-effective and efficient process for printing tailor-made implants [91, 92].

5.5.2 Compression

The compression manufacturing technique is considered an appropriate method in formulating solid implantable scaffolds which consist of materials that are thermosensitive or solvent sensitive. It is a three-step, simple procedure, which includes the API, excipients, and lubricant grinding, blending, and compression. Initially, API and excipients are ground, and then subsequently mixed along with the added lubricant, and finally compressed

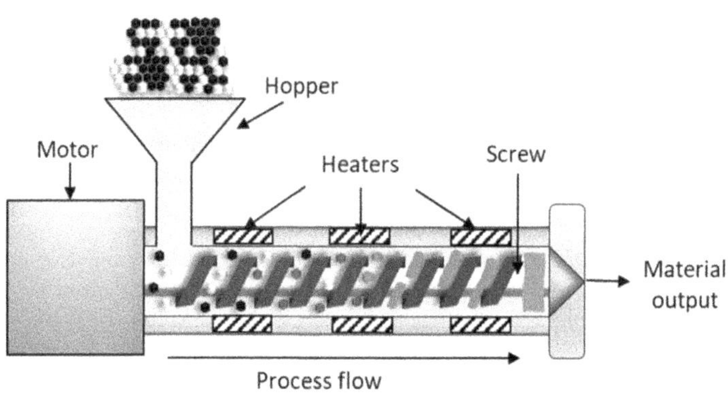

Figure 5.4 *Illustration of a typical hot-melt extruder.*

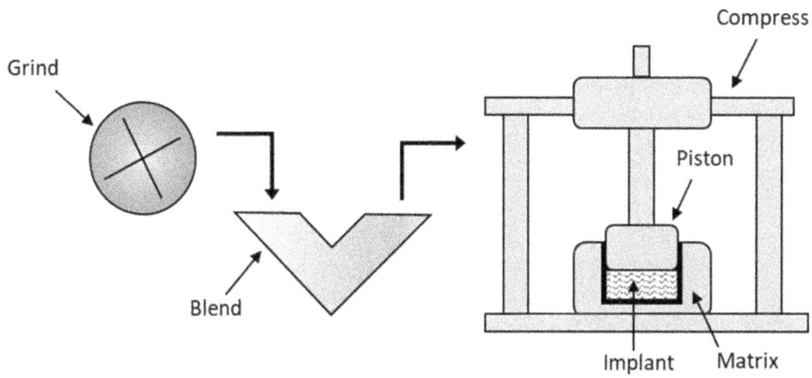

Figure 5.5 Illustration of a typical compression machine.

directly into a dimension-defined matrix (Figure 5.5). Materials have to be selected by good flowability and compressibility properties [93].

However, there is a limitation in using hydrophilic materials for prolonged release implant formulations, as an initial burst effect can accelerate the release rate of the drug. The compression method does not provide a predictable release profile due to the diversity of surface features and topology of the implant, thus it is not used extensively as a manufacturing method for implantable scaffolds [44, 47].

5.5.3 Injection Moulding

Injection moulding technology can be applied in formulating implants that consist of thermoplastic polymers [1]. This technique includes heat, injection moulding, and solidify of the polymers. In more detail, the injection moulding machine consists of two parts, the injection unit and the moulding unit. Materials are inserted into the injection unit, where they undergo melting and subsequently injection procedure (Figure 5.6). During the manufacturing process, temperature, pressure, and speed affect the quality of the final product. Moreover, it is mainly combined with a twin-screw HME, as a first-step manufacturing procedure, in order to improve the dispersion of materials [93, 94].

Injection moulding permits the formulation of low soluble drugs into solid dispersal systems, thus enhancing their solubility properties. This method can be used for modifying drug release systems, as it enables creating multi-shaped solid formulations. However,

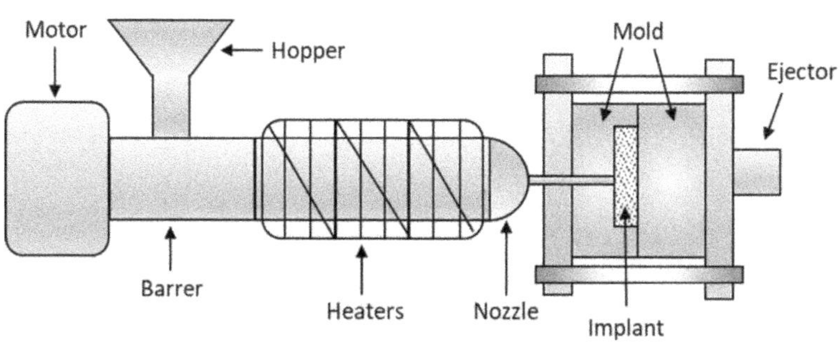

Figure 5.6 Illustration of a typical injection moulding machine.

injection moulding requires high-cost equipment, and is a slow formulation procedure for personal-based implants, compared to the fused filament fabrication (FFF) technique, due to mould-modification requirements [95].

5.5.4 Solvent Casting

The principal of the solvent casting technique concerns the initial dissolution of polymers and active substances in the presence of a volatile solvent. After the stage of material dissolution, the solution is casted into an implant scaffold followed by solvent evaporation (Figure 5.7) [8]. Solvent casting is also used in coating long-acting implantable DDSs. Coated implants are considered more efficient as they are defined by fine-tuning properties, providing a more controlled and precise drug release profile [7].

Solvent casting is a simple and cost-effective technique, in which lower HME evaporating temperatures can be applied. Nevertheless, constraints in production scale-up are common, due to mixing speed, heating, drying, and parameters that affect implant quality attributes, such as the uniformity of content, thickness, and surface, etc. [87]. Moreover, the significant use of volatile solvents, especially the use of organic solvents, can provoke stability issues in the final product, as their removal may be difficult and infeasible [96].

5.5.5 3D Printing

Additive manufacturing (AM), or 3DP, is considered as an innovative technology in developing personalised implantable DDSs. 3DP enables the combination of biomaterials with different physico-chemical properties in designing and formulating simple or diverse structures that are characterised by their variable pharmacological and pharmacokinetic potential [6, 81]. Furthermore, 3DP can be combined with different techniques, such as HME, injection moulding, and solvent casting with the prospect of developing implantable systems with long-acting drug release profiles [44]. 3D-printed implants are gaining ground in medicinal field and treatments, for example in the areas of contraception, bone healing, cancer, and chronic diseases [3].

5.5.6 Scale-Up in 3D-Printing Process for the Manufacturing of Scaffolds

The need of scaling up in AM is essential, due to the increasing demand for tailor-made drug delivery devices; however, there are many considerations that set limitations in several stages concerning their production development process and market disposition.

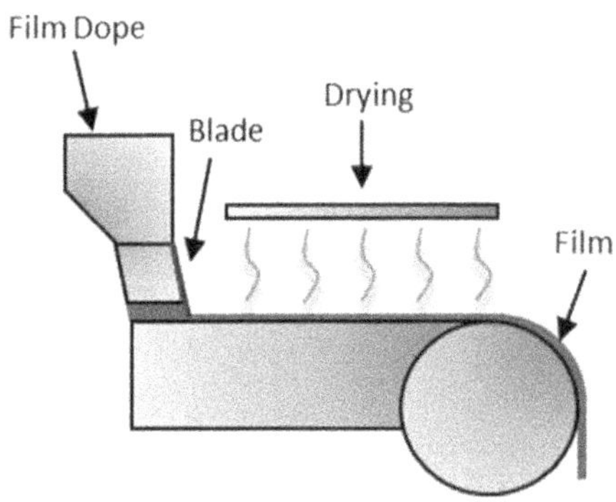

Figure 5.7 Illustration of a typical solvent casting machine.

Engineering equipment and applied technology that reflect in product quality should be developed and assured by precise standards and criteria. Moreover, guidelines that refer to regulatory policy should be established in accordance with scale-up steps, in order to ensure a quality assurance framework. Therefore, quality and material attributes, continuous manufacturing process, and ongoing regulatory oversights concerning industrial production scale, are essential to be examined in real time [79, 81].

5.6 Drug Release Mechanism of Long-Acting 3D-Printing Polymeric Implantable Systems

Mechanisms which determine the release kinetic profile of drugs from long-acting implantable systems are based on a variety of parameters and variables that concern properties of utilised materials and their combinations used, drug incorporation protocols, manufacturing methods of scaffolds, geometry, surface chemistry, and topology of scaffolds (Table 5.3). Hence, degradation, erosion, or swelling effects of polymeric scaffolds define the dissolution, desorption, or diffusion mechanisms of drug release [99, 100]. Considering the

Table 5.3 *Advantages and disadvantages of drug release mechanisms of long-acting implantable DDSs.*

Type of long-acting implant	Drug release mechanism	Advantages	Disadvantages	References
Monolithic type	• Passive diffusion • Erosion	• Hydrophilic or hydrophobic systems • Ease in manufacturing process	• Burst effect issues	[102]
Reservoir type	• Passive diffusion	• Polymer properties define the release rate of the drug • API form (crystalline vs. amorphous) • Controlled delivery systems • Target-specific therapies	• High-cost manufacturing process	[103]
Swelling-controlled	• Swelling	• Polymeric hydrogel-based systems with a high-MW drug load capacity • High absorption of water • Degradable-based systems • Stimuli responsive systems • Variability in polymers used	• Stability issues (need for lyophilisation)	[5, 47]
Drug-eluting coated	• Passive diffusion	• Enzymatic or chemical degradation protection	• Non-biodegradable systems	[7]

poly-parametric effects that affect drug release mechanisms, kinetic release models are examined under the prism of different analytical models, such as statistical methods (e.g., ANOVA), model independent methods (e.g., MDT, DE%, f_1, f_2 index), and model dependent methods (e.g., Zero order, Korsmeyer–Peppas, Hixon–Crowell). These approaches concern examples for the mathematical evaluation and subsequently the characterisation of drug release mechanism from DDSs [99, 101].

Diffusion-controlled systems, monolithic and reservoir type, are the most common mechanism release models in implantable polymeric devices. In monolithic type systems, the API is mixed with excipients in a matrix and its release rate is determined by the rate of diffusion rather than the rate of dissolution of the active substance. However, in reservoir type systems, the molecules of API are in contact with the polymeric membrane, diffuse through the cavities and pores of the polymer, and the release rate of active substance is determined by the permeability properties of the drug and the thickness of the membrane [4].

5.7 Outlining Regulatory Framework for 3D-Printed Implantable Scaffolds

Reviewing the regulatory aspects of 3DP-implantable devices, it is obvious that the area is characterised by gaps and ambiguities, due to multiparametric manufacturing procedures and complexity in defining the conditions of personalising drug administration. Parameters regarding novel engineering processes, software, and innovative materials, set challenges to regulatory agencies in enacting legislations and normative frameworks that assure safety and effectiveness of AM devices [104, 105].

Regulation frameworks of 3D-printed implantable systems are not only circumscribed on final product features but also on applied technologies used during the manufacturing process. In more detail, designing software (CAD), printing equipment, materials (polymers, cells, genes), and product development processes, are parameters that should be provisioned through a regulatory perspective. Furthermore, regulation aspects which concern the point of manufacturing, either centralised or point-of-care (hospitals, clinics, health centres) have to be considered, as 3DP can either be used for custom or industrial manufacturing procedures [106–108]. Industrial manufacturing of 3D-printed implants would facilitate the wider distribution of personalised-based medicinal products with a perspective in adjustable properties (anatomical size adjustments) and multi-purpose applications (medical diagnosis, disease treating or prevention).

Quality issues through AM procedure are crucial to be evaluated and determined [104]. The application of process analytical technology (PAT) methodology is essential in examining the consecutive stages of implant manufacturing as well as attributes of the final product. Moreover, implementing and observing the product quality consistency process (feedback control) facilitates defining and controlling the manufacturing process regarding the final product [97, 109]. Implantable delivery systems have to be defined also by quality criteria that are in compliance with Good Manufacturing Practice (GMP), in adherence to International Council for Harmonisation (ICH) safety and efficacy guidelines, and to follow Quality Management System (QMS) that comply with International Organisation for Standardization (ISO) standards (Figure 5.8) [106, 108].

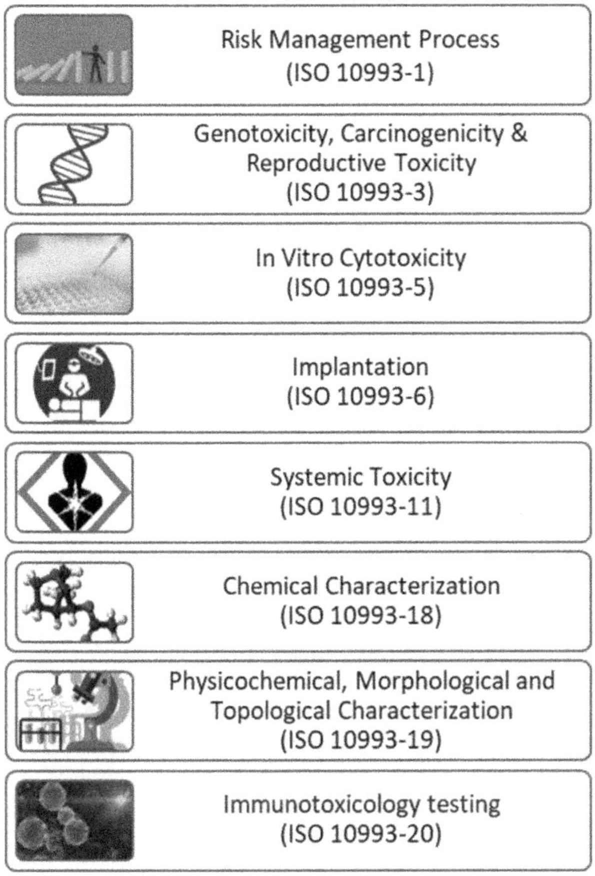

Figure 5.8 *Biological evaluation of medical devices, according to ISO quality standards.*

As far as the market disposition process of AM implants is concerned, it cannot always be described by simple and direct steps. Steps that are essential concern prime product categorisation (centralised or point-of-care), prime deliberation with sponsors (throughout product development and production), clinical trials, within the stages and final implant production quality evaluation (before market distribution), quality analysis and control of product characteristics that are already in market distribution, compliance with quality standards and criteria (e.g., GMP, ISO), and assessment and reconsideration conductions about the amelioration of the commercial product. When considering all these stages, it is crucial that each step of product disposition in the market is authorised and accredited by regulatory observation [106].

5.7.1 Commercial Implantable Scaffolds

Commercial implantable drug-loaded products comprise formulations that are not based on 3DP technology. Reviewing the literature, approved pharmaceutical implants involve birth control, eye, and cancer treatment (Table 5.4), which concern matrix and reservoir-based systems (Table 5.5) [111].

Table 5.4 *Commercial implantable DDSs.*

Eye	Retisert® (Fluocinolone), Ozurdex® (Dexametasone), Vitrasert® (Ganciclovir)
Cancer	Gliadel® (Prolifeprosan and Carmustine), Vantas®/Supprelin LA® (Histrelin), Viadur® (Leuprolide), Zoladex® (Goserelin)
Birth control	Mirena®/Norplant (Levonorgestrel), Nexplanon® (Etonogestrel), NuvaRing™ (Etonogestrel and ethinyl estradiol), Paragard® (Copper)

Table 5.5 *Drug release systems of commercial implantable DDSs.*

Matrix implants	Gliadel® (intracranial implant), Ozurdex® (ocular implant), Zoladex® (subcutaneous implant)
Reservoir implants	Iluvien® (ocular implant), Jadelle® (Levonorgestrel, subcutaneous implant)

5.8 Conclusions

Overviewing DDSs categories, oral drug formulations are considered the most prevalent in pharmaceutical therapies; however, implantable systems are constantly gaining ground. The main objective of implant formulation addresses safety and efficacy parameters and is directly associated with the manufacturing process. Drug-loaded implantable scaffolds contribute to the amelioration and enhancement of drug stability and bioavailability issues, diminish side effects, and reduce provoked toxicity. However, possible implant compatibility issues and implant rejection can occur, due to inapt tissue integration and material–cell reactions. Furthermore, according to the type of implant, passive or active, the release rate of drug can be predicted. Drug release in passive implants is determined by the choice of material and drug, the surface properties, and the porosity of the implant; however, the drug release rate in active implants varies according to the activation mechanism used.

Manufacturing technique selection is related to material properties (e.g., mechanical, physico-chemical, etc.) and process scalability parameters (e.g., cost-effective, industrial scaling up, etc.). Moreover, fabrication techniques of implantable scaffolds are broad and mostly coupled with 3DP techniques. 3DP techniques are emerging as a technology, which enables the initial rational design of personalised fabricated diverse DDSs with multifunctional therapeutic potential. Thus, it is a patient-centred technology targeting optimised pharmacological and pharmacokinetic drug profiles.

References

[1] Stewart, S.A., Domínguez-Robles, J., Donnelly, R.F., and Larrañeta, E. (2018). Implantable polymeric drug delivery devices: classification, manufacture, materials, and clinical applications. *Polymers (Basel)* **10** (12): 1379.

[2] Abdelkader, H., Fathalla, Z., Seyfoddin, A. et al. (2021). Polymeric long-acting drug delivery systems (LADDS) for treatment of chronic diseases: inserts, patches, wafers, and implants. *Advanced Drug Delivery Reviews* **177**: 113957. Available from: https://doi.org/10.1016/j.addr.2021.113957.

[3] Corduas, F., Mancuso, E., and Lamprou, D.A. (2020). Long-acting implantable devices for the prevention and personalised treatment of infectious, inflammatory and chronic diseases. *Journal of Drug Delivery Science and Technology* **60** (June): 101952. Available from: https://doi.org/10.1016/j.jddst.2020.101952.

[4] Quarterman, J.C., Geary, S.M., and Salem, A.K. (2021). Evolution of drug-eluting biomedical implants for sustained drug delivery. *European Journal of Pharmaceutics and Biopharmaceutics* **159** (September 2020): 21–35. Available from: https://doi.org/10.1016/j.ejpb.2020.12.005.

[5] Hu, X., Yang, Z., Kang, S. et al. (2020). Cellulose hydrogel skeleton by extrusion 3D printing of solution. *Nanotechnology Reviews* **9** (1): 345–353.

[6] Stewart, S.A., Domínguez-Robles, J., McIlorum, V.J. et al. (2020). Development of a biodegradable subcutaneous implant for prolonged drug delivery using 3D printing. *Pharmaceutics* **12** (2): 105.

[7] Stewart, S.A., Domínguez-Robles, J., McIlorum, V.J. et al. (2020). Poly(caprolactone)-based coatings on 3D-printed biodegradable implants: a novel strategy to prolong delivery of hydrophilic drugs. *Molecular Pharmaceutics* **17** (9): 3487–3500.

[8] Łyszczarz, E., Brniak, W., Szafraniec-Szczęsny, J. et al. (2021). The impact of the preparation method on the properties of orodispersible films with aripiprazole: electrospinning vs. casting and 3D printing methods. *Pharmaceutics* **13** (8): 122.

[9] Weems, A.C., Pérez-Madrigal, M.M., Arno, M.C., and Dove, A.P. (2020). 3D printing for the clinic: examining contemporary polymeric biomaterials and their clinical utility. *Biomacromolecules* **21**: 1037–1059.

[10] Kumar, A. and Pillai, J. (2018). Chapter 13: implantable drug delivery systems: an overview. In: *Nanostructures for the Engineering of Cells, Tissues and Organs* (ed. Grumezescu, A.M.), 473–512. William Andrew Publishing. Available from: http://dx.doi.org/10.1016/B978-0-12-813665-2.00013-2.

[11] Nunes-pereira, J., Ribeiro, S., Ribeiro, C. et al. (2015). Poly (vinylidene fluoride) and copolymers as porous membranes for tissue engineering applications. *Polymer Testing* **44**: 234–241. Available from: https://doi.org/10.1016/j.polymertesting.2015.05.001.

[12] Zakhary, K.E. and Thakker, J.S. (2017). Emerging biomaterials in trauma. *Oral Maxillofacial Surgery Clinics of North America* **29** (1): 51–62. Available from: http://dx.doi.org/10.1016/j.coms.2016.08.010.

[13] Kirby, B., Kenkel, J.M., Zhang, A.Y. et al. (2021). Three-dimensional (3D) synthetic printing for the manufacture of non-biodegradable models, tools and implants used in surgery: a review of current methods. *Journal of Medical Engineering & Technology* **45** (1): 14–21. Available from: https://doi.org/10.1080/03091902.2020.1838643.

[14] Koutsamanis, I., Paudel, A., Alva Zúñiga, C.P. et al. (2021). Novel polyester-based thermoplastic elastomers for 3D-printed long-acting drug delivery applications. *Journal of Controlled Release* **335** (May): 290–305.

[15] Naseri, E., Butler, H., Macnevin, W. et al. (2020). Low-temperature solvent-based 3D printing of PLGA: a parametric printability study printability study. *Drug Development and Industrial Pharmacy* **46** (2): 173–178. Available from: https://doi.org/10.1080/03639045.2019.1711389.

[16] Jammalamadaka, U. and Tappa, K. (2018). Recent advances in biomaterials for 3D printing and tissue engineering. *Journal of Functional Biomaterials* **9** (1): 22.

[17] Polley, C., Distler, T., Detsch, R. et al. (2020). 3D printing of piezoelectric barium titanate-hydroxyapatite scaffiolds with interconnected porosity for bone tissue engineering. *Materials (Basel)* **13** (7): 1773.

[18] Lee, S.H., Kim, B.H., Park, C.G. et al. (2018). Implantable small device enabled with magnetic actuation for on-demand and pulsatile drug delivery. *Journal of Controlled Release* **286** (April): 224–230. Available from: https://doi.org/10.1016/j.jconrel.2018.07.037.

[19] Anitua, E., Pino, A., Troya, M. et al. (2018). A novel personalized 3D injectable protein scaffold for regenerative medicine. *Journal of Materials Science: Materials in Medicine* **32** (1): 18–25. Available from: http://dx.doi.org/10.1007/s10856-017-6012-6.

[20] Li, Z., Mu, H., Larsen, S.W. et al. (2021). An *in vitro* gel-based system for characterizing and predicting the long-term performance of PLGA *in situ* forming implants. *International Journal of Pharmaceutics* **609**: 121183.

[21] Esposito, C.L., Tardif, V., Sarrazin, M. et al. (2020). Preparation and characterization of 12-HSA-based organogels as injectable implants for the controlled delivery of hydrophilic and lipophilic therapeutic agents. *Materials Science and Engineering C* **114:** 110999. Available from: https://doi.org/10.1016/j.msec.2020.110999.

[22] Manoukian, O.S., Arul, M.R., Sardashti, N. et al. (2018). Biodegradable polymeric injectable implants for long-term delivery of contraceptive drugs. *Journal of Applied Polymer Science* **135** (14): 46068.

[23] Ying, G., Jiang, N., Parra-cantu, C. et al. (2020). Bioprinted injectable hierarchically porous gelatin methacryloyl hydrogel constructs with shape-memory properties. *Advanced Functional Materials* **30** (46): 2003740.

[24] Ho, D., Leguyader, C., Srinivasan, S. et al. (2021). Fully synthetic injectable depots with high drug content and tunable pharmacokinetics for long-acting drug delivery. *Journal of Controlled Release* **329**: 257–269. Available from: https://doi.org/10.1016/j.jconrel.2020.11.030.

[25] Gao, B., Yang, Q., Zhao, X. et al. (2016). 4D bioprinting for biomedical applications. *Trends in Biotechnology* **34** (9): 746–756. Available from: http://dx.doi.org/10.1016/j.tibtech.2016.03.004.

[26] Kuang, X., Roach, D.J., Wu, J. et al. (2019). Advances in 4D printing: materials and applications. *Advanced Functional Materials* **29** (2): 1805290.

[27] Wang, B., Di, P.J., Browe, D.C. et al. (2021). Affinity-bound growth factor within sulfated interpenetrating network bioinks for bioprinting cartilaginous tissues. *Acta Biomaterialia* **128**: 130–142.

[28] Sun, Y., You, Y., Jiang, W. et al. (2020). Generating ready-to-implant anisotropic menisci by 3D-bioprinting protein-releasing cell-laden hydrogel-polymer composite scaffold. *Applied Materials Today* **18**: 100469. Available from: https://doi.org/10.1016/j.apmt.2019.100469.

[29] Costa, J.B., Silva-correia, J., Oliveira, J.M., and Reis, R.L. (2017). Fast setting silk fibroin bioink for bioprinting of patient-specific memory-shape implants. *Advanced Healthcare Materials* **6** (32): 1701021.

[30] Nulty, J., Freeman, F.E., Browe, D.C. et al. (2021). 3D bioprinting of prevascularised implants for the repair of critically-sized bone defects. *Acta Biomaterialia* **126**: 154–169. Available from: https://doi.org/10.1016/j.actbio.2021.03.003.

[31] Li, J., Yang, F., Long, Y. et al. (2021). Bulk ferroelectric metamaterial with enhanced piezoelectric and biomimetic mechanical properties from additive manufacturing. *ACS Nano* **15** (9): 14903–14914.

[32] Fernandes, C., Suares, D., Dhawan, V., and Prabhu, P. (2018). Chapter 15: Nanostructured polymer scaffolds for tissue engineering technology. In: *Fundamentals of Nanoparticles* (eds. A. Ahmed Barhoum and A.S.H. Makhlouf), 451–483. William Andrew Publishing. William AndereW. Available from: http://dx.doi.org/10.1016/B978-0-323-51255-8/00015-X.

[33] Farmer, Z.L., Utomo, E., Domínguez-Robles, J. et al. (2021). 3D printed estradiol-eluting urogynecological mesh implants: influence of material and mesh geometry on their mechanical properties. *International Journal of Pharmaceutics* **593**: 120145. doi: 10.1016/j.ijpharm.2020.120145.

[34] Bidan, C.M., Kommareddy, K.P., Rumpler, M. et al. (2013). Geometry as a factor for tissue growth: towards shape optimization of tissue engineering scaffolds. *Advanced Healthcare Materials* **2** (1): 186–194.

[35] Chen, X., Fan, H., Deng, X. et al. (2018). Scaffold structural microenvironmental cues to guide tissue regeneration in bone tissue applications. *Nanomaterials* **8** (11): 1–15.

[36] Gupte, M.J., Swanson, W.B., Hu, J. et al. (2018). Pore size directs bone marrow stromal cell fate and tissue regeneration in nanofibrous macroporous scaffolds by mediating vascularization. *Acta Biomaterialia* **82**: 1–11. Available from: https://doi.org/10.1016/j.actbio.2018.10.016.

[37] Shen, J., Gao, P., Han, S. et al. (2020). A tailored positively-charged hydrophobic surface reduces the risk of implant associated infections. *Acta Biomaterialia* **114**: 421–430. Available from: https://doi.org/10.1016/j.actbio.2020.07.040.

[38] Chen, S., Guo, Y., Liu, R. et al. (2018). Colloids and surfaces: Biointerfaces tuning surface properties of bone biomaterials to manipulate osteoblastic cell adhesion and the signaling pathways for the enhancement of early osseointegration. *Colloids Surfaces B Biointerfaces* **164**: 58–69. Available from: https://doi.org/10.1016/j.colsurfb.2018.01.022.

[39] Uz Zaman, U.K. Rivette, M., Siadat, A., and Mousavi, S.M. (2018). Integrated product-process design: material and manufacturing process selection for additive manufacturing using multi-criteria decision making. *Robotics and Computer-Integrated Manufacturing* **51**: 169–180. Available from: https://doi.org/10.1016/j.rcim.2017.12.005.

[40] Robotti, F., Bottan, S., Fraschetti, F. et al. (2018). A micron-scale surface topography design reducing cell adhesion to implanted materials. *Scientific Reports* **8** (1): 10887. Available from: http://dx.doi.org/10.1038/s41598-018-29167-2.

[41] Kaduri, M., Poley, M., Adir, O., and Krinsky, N. (2018). Biocompatibility, biodegradation and excretion of polylactic acid (PLA) in medical implants and theranostic systems. *Chemical Engineering Journal* **340**: 9–14. Available from: https://doi.org/10.1016/j.cej.2018.01.010.

[42] Wang, Y., Sun, L., Mei, Z. et al. (2020). 3D printed biodegradable implants as an individualized drug delivery system for local chemotherapy of osteosarcoma. *Materials & Design* **186**: 108336. Available from: https://doi.org/10.1016/j.matdes.2019.108336.

[43] Liu, J., Tagami, T., and Ozeki, T. (2020). Fabrication of 3D-printed fish-gelatin-based polymer hydrogel patches for local delivery of PEGylated liposomal doxorubicin. *Marine Drugs* **18** (6): 325.

[44] John, A.A., Subramanian, A.P., Vellayappan, M.V. et al. (2015). Review: physico-chemical modification as a versatile strategy for the biocompatibility enhancement of biomaterials. *RSC Advances* **5** (40): 39232–39244.

[45] Frone, A.N., Batalu, D., Chiulan, I. et al. (2020). Morpho-structural, thermal and mechanical properties of PLA/PHB/cellulose biodegradable nanocomposites obtained by compression molding, extrusion, and 3D printing. *Nanomaterials* **10** (51): 51.

[46] Zhang, Z., He, F., Wang, B. et al. (2021). Biodegradable PGA/PBAT blends for 3D printing: material performance and periodic minimal surface structures. *Polymers (Basel)* **13** (21): 3757.

[47] Vijayavenkataraman, S., Kuan, L.Y., and Lu, W.F. (2020). 3D-printed ceramic triply periodic minimal surface structures for design of functionally graded bone implants. *Materials & Designs* **191**: 108602. Available from: https://doi.org/10.1016/j.matdes.2020.108602.

[48] Yelles MC, H.-B., Tran Tan, V., Danede, F. et al. (2017). PLGA implants: how Poloxamer/PEO addition slows down or accelerates polymer degradation and drug release. *Journal of Controlled Release* **253**: 19–29. Available from: http://dx.doi.org/10.1016/j.jconrel.2017.03.009.

[49] Lee, H., Yang, G.H., Kim, M. et al. (2018). Fabrication of micro/nanoporous collagen/dECM/silk-fibroin biocomposite scaffolds using a low temperature 3D printing process for bone tissue regeneration. *Materials Science and Engineering C* **84**: 140–147. Available from: https://doi.org/10.1016/j.msec.2017.11.013.

[50] Wu, Q., Therriault, D., and Heuzey, M.C. (2018). Processing and properties of chitosan inks for 3D printing of hydrogel microstructures. *ACS Biomaterials Science & Engineering* **4** (7): 2643–2652.

[51] Maturavongsadit, P., Paravyan, G., Kovarova, M. et al. (2021). A new engineering process of biodegradable polymeric solid implants for ultra-long-acting drug delivery. *International Journal of Pharmaceutics X* **3**: 100068. Available from: https://doi.org/10.1016/j.ijpx.2020.100068.

[52] Gutierrez, E., Burdiles, P.A., Quero, F. et al. (2019). 3D printing of antimicrobial alginate/bacterial-cellulose composite hydrogels by incorporating copper nanostructures. *ACS Biomaterials Science & Engineering* **5** (11): 6290–6299.

[53] Huang, L., Du, X., Fan, S. et al. (2019). Bacterial cellulose nanofibers promote stress and fidelity of 3D-printed silk based hydrogel scaffold with hierarchical pores. *Carbohydrate Polymers* **221**: 146–156.

[54] Oladapo, B.I., Oshin, E.A., and Olawumi, A.M. (2020). Nano-Structures & Nano-Objects Nanostructural computation of 4D printing carboxymethylcellulose (CMC) composite. *Nano-Structures & Nano-Objects [Internet]* **21**: 100423. Available from: https://doi.org/10.1016/j.nanoso.2020.100423.

[55] Sultan, S. and Mathew, A.P. (2018). 3D printed scaffolds with gradient porosity based on a cellulose nanocrystal hydrogel. *Nanoscale* **10** (9): 4421–4431.

[56] Podgorbunskikh, E.M., Bychkov, A.L., Ryabchikova, E.I., and Lomovsky, O.I. (2020). The effect of thermomechanical pretreatment on the structure and properties of lignin-rich plant biomass. *Molecules* **25** (4): 995.

[57] Rojas-martínez, L.E., Flores-hernandez, C.G., López-marín, L.M., and Martinez-hernandez, A.L. (2020). 3D printing of PLA composites scaffolds reinforced with keratin and chitosan: effect of geometry and structure. *European Polymer Journal* **141**: 110088. Available from: https://doi.org/10.1016/j.eurpolymj.2020.110088.

[58] Zhu, L., Chen, S., Liu, K. et al. (2020). 3D poly (L-lactide)/chitosan micro/nano fibrous scaffolds functionalized with quercetin-polydopamine for enhanced osteogenic and anti-inflammatory activities. *Chemical Engineering Journal* **391**: 123524.
[59] Sun, Y., Yang, C., Zhu, X. et al. (2019). 3D printing collagen/chitosan scaffold ameliorated axon regeneration and neurological recovery after spinal cord injury. *Journal of Biomedical Materials Research Part A* **107A** (9): 1898–1908.
[60] Li, X.H., Zhu, X., Liu, X.Y. et al. (2021). The corticospinal tract structure of collagen/silk fibroin scaffold implants using 3D printing promotes functional recovery after complete spinal cord transection in rats. *Journal of Materials Science: Materials in Medicine* **32**: 1–13. Available from: http://dx.doi.org/10.1007/s10856-021-06500-2.
[61] Cebe, P., Hu, X., Kaplan, D.L. et al. (2013). Beating the heat-fast scanning melts silk beta sheet crystals. *Scientific Reports* **3** (1): 1–7.
[62] Lee, C.-F., Hsu, Y.-H., Lin, Y.-C. et al. (2021). 3D printing of collagen/oligomeric proanthocyanidin/oxidized hyaluronic acid composite scaffolds for articular cartilage repair. *Polymers* **13** (18): 3123.
[63] Hsieh, Y.H., Shen, B.Y., Wang, Y.H. et al. (2018). Healing of osteochondral defects implanted with biomimetic scaffolds of poly(ε-caprolactone)/hydroxyapatite and glycidyl-methacrylate-modified hyaluronic acid in a minipig. *International Journal of Molecular Sciences* **19** (4): L 1125.
[64] Shokri, A., Ramezani, K., Jamalpour, M.R. et al. (2021). *In vivo* efficacy of 3D-printed elastin–gelatin–hyaluronic acid scaffolds for regeneration of nasal septal cartilage defects. *Journal of Biomedical Materials Research Part B: Applied Biomaterials.* **110** (3): 614–624.
[65] Später, T., Mariyanats, A.O., Syachina, M.A. et al. (2020). *In vitro* and *in vivo* analysis of adhesive, anti-inflammatory, and proangiogenic properties of novel 3D printed hyaluronic acid glycidyl methacrylate hydrogel scaffolds for tissue engineering. *ACS Biomaterials Science & Engineering* **6** (10): 5744–5757.
[66] Yeo, T., Ko, Y.G., Kim, E.J. et al. (2021). Promoting bone regeneration by 3D-printed poly(glycolic acid)/hydroxyapatite composite scaffolds. *Journal of Industrial and Engineering Chemistry* **94**: 343–351. Available from: https://doi.org/10.1016/j.jiec.2020.11.004.
[67] Tappa, K., Jammalamadaka, U., Weisman, J.A. et al. (2019). 3D printing custom bioactive and absorbable surgical screws, pins, and bone plates for localized drug delivery. *Journal of Functional Biomaterials* **10** (2): 17.
[68] Wang, W., Zhang, B., Li, M. et al. (2021). 3D printing of PLA/n-HA composite scaffolds with customized mechanical properties and biological functions for bone tissue engineering. *Composites Part BL Engineering* **224**: 109192. Available from: https://doi.org/10.1016/j.compositesb.2021.109192.
[69] Kumar, S., Hota, S.K., Pais, R. et al. (2021). Implications of advance biomaterials in development of new contraceptive devices. *Journal of Materials Nanoscience* **8** (1): 23–34. Available from: https://pubs.thesciencein.org/journal/index.php/jmns/article/view/249.
[70] Smeets, R., Precht, C., Hahn, M. et al. (2017). Biocompatibility and osseointegration of titanium implants with a silver-doped polysiloxane coating: an *in vivo* pig model. *International Journal of Oral & Maxillofacial Implants.* **32** (6): 1338–1344.
[71] Liravi, F. and Toyserkani, E. (2018). A hybrid additive manufacturing method for the fabrication of silicone bio-structures: 3D printing optimization and surface characterization. *Materials & Design* **138**: 46–61. Available from: https://doi.org/10.1016/j.matdes.2017.10.051.
[72] Barthes, J., Courtial, E., Brenet, E. et al. (2019). Using 3-D printing and bioprinting technologies for personalized implants. In: *Advances in Biomechanics and Tissue Regeneration*, 269–286. Academic Press. Available from: http://dx.doi.org/10.1016/B978-0-12-816390-0.00014-5.
[73] Kim, S. and Liu, S. (2018). Smart and biostable polyurethanes for long-term implants. *ASC Biomaterials Science & Engineering* **4** (5): 1479–1490.
[74] Salimi, S., Wu, Y., Barreiros, M.I.E. et al. (2020). Polymer chemistry a 3D printed drug delivery implant formed from a dynamic supramolecular polyurethane. *Polymer Chemistry* **11** (20): 3453–3464.
[75] Domínguez-Robles, J., Mancinelli, C., Mancuso, E. et al. (2020). 3D printing of drug-loaded thermoplastic polyurethane meshes: a potential material for soft tissue reinforcement in vaginal surgery. *Pharmaceutics* **12** (1): 63.

[76] Fouladian, P., Kohlhagen, J., Arafat, M. et al. (2020). Three-dimensional printed 5-fluorouracil eluting polyurethane stents for the treatment of oesophageal cancers. *Biomaterials Science* **8** (23): 6625–6636.
[77] Żebrowski, R., Walczak, M., Korga, A. et al. (2019). Effect of shot peening on the mechanical properties and cytotoxicity behaviour of titanium implants produced by 3D printing technology. *Journal of Healthcare Engineering* **2019**: 8169538.
[78] Safonov, A., Maltsev, E., Chugunov, S. et al. (2020). Design and fabrication of complex-shaped ceramic bone implants via 3D printing based on laser stereolithography. *Applied Sciences* **10** (20): 7138.
[79] Chen, S., Shi, Y., Zhang, X., and Ma, J. (2020). Evaluation of BMP-2 and VEGF loaded 3D printed hydroxyapatite composite scaffolds with enhanced osteogenic capacity *in vitro* and *in vivo*. *Materials Science and Engineering C* **112**: 110893. Available from: https://doi.org/10.1016/j.msec.2020.110893.
[80] Kim, S.C., Jo, W.L., Kim, Y.S. et al. (2019). Titanium powder coating using metal 3D printing: a novel coating technology for cobalt–chromium alloy implants. *Tissue Engineering and Regenerative Medicine* **16** (1): 11–18. Available from: https://doi.org/10.1007/s13770-018-0168-0.
[81] Velásquez-García, L.F. and Kornbluth, Y. (2021). Biomedical applications of metal 3D printing. *Annual Review of Biomedical Engineering* **23**: 307–338.
[82] Konta, A.A., García-Piña, M., and Serrano, D.R. (2017). Personalised 3D printed medicines: which techniques and polymers are more successful? *Bioengineering* **4** (4): 79
[83] Tran, P.H.L., Lee, B.J., and Tran, T.T.D. (2021). Recent studies on the processes and formulation impacts in the development of solid dispersions by hot-melt extrusion. *European Journal of Pharmaceutics and Biopharmaceutics* **164**: 13–19. Available from: https://doi.org/10.1016/j.ejpb.2021.04.009.
[84] Repka, M.A., Bandari, S., Kallakunta, V.R. et al. (2018). Melt extrusion with poorly soluble drugs: an integrated review. *International Journal of Pharmaceutics* **535** (1–2): 68–85. Available from: http://dx.doi.org/10.1016/j.ijpharm.2017.10.056.
[85] Bagde, A., Patel, N., Patel, K. et al. (2021). Sustained release dosage form of noscapine HCl using hot melt extrusion (HME) technique: formulation and pharmacokinetics. *Drug Delivery Translational Research* **11** (3): 1156–1165.
[86] Fan, R., Chuan, D., Hou, H. et al. (2021). Development and evaluation of a novel biodegradable implants with excellent inflammatory response suppression effect by hot-melt extrusion. *European Journal of Pharmaceutical Sciences* **166**: 105981. Available from: https://doi.org/10.1016/j.ejps.2021.105981.
[87] Ren, Y., Mei, L., Zhou, L., and Guo, G. (2019). Recent perspectives in hot melt extrusion-based polymeric formulations for drug delivery: applications and innovations. *AAPS PharmSciTech* **20** (3): 92.
[88] Shadambikar, G., Kipping, T., Di-gallo, N. et al. (2020).Vacuum compression molding as a screening tool to investigate carrier suitability for hot-melt extrusion formulations. *Pharmaceutics* **12** (11): 1019.
[89] Lee, J., Song, C., Noh, I. et al. (2020). Hot-melt 3d extrusion for the fabrication of customizable modified-release solid dosage forms. *Pharmaceutics* **12** (8): 738.
[90] Melocchi, A., Inverardi, N., Uboldi, M. et al. (2019). Retentive device for intravesical drug delivery based on water-induced shape memory response of poly(vinyl alcohol): design concept and 4D printing feasibility. *International Journal of Pharmaceutics* **559**: 299–311. Available from: https://doi.org/10.1016/j.ijpharm.2019.01.045.
[91] Kempin, W., Franz, C., Koster, L. et al. (2017). Assessment of different polymers and drug loads for fused deposition modeling of drug loaded implants. *European Journal of Pharmaceutics and Biopharmaceutics* **115**: 84–93. Available from: http://dx.doi.org/10.1016/j.ejpb.2017.02.014.
[92] Ponsar, H., Wiedey, R., and Quodbach, J. (2020). Hot-melt extrusion process fluctuations and their impact on critical quality attributes of filaments and 3D-printed dosage forms. *Pharmaceutics* **12** (6): 511.
[93] Ervasti, T., Niinikoski, H., Mäki-Lohiluoma, E. et al. (2020). The comparison of two challenging low dose APIs in a continuous direct compression process. *Pharmaceutics* **12** (3): 279.

[94] Sohail, M., Zafar, S., Yousef, B. et al. (2021). A review of emerging technologies enabling improved solid oral dosage form manufacturing and processing. *Advanced Drug Delivery Reviews* **178**: 113840. Available from: https://doi.org/10.1016/j.addr.2021.113840.

[95] Fuenmayor, E., Forde, M., Gately, N.. et al. (2019). Comparison of fused-filament fabrication to direct compression and injection molding in the manufacture of oral tablets. *International Journal of Pharmaceutics* **558**: 328–340. Available from: https://doi.org/10.1016/j.ijpharm.2019.01.013.

[96] Fuenmayor, E., O'Donnell, C., Gately, N. et al. (2019). Mass-customization of oral tablets via the combination of 3D printing and injection molding. *International Journal of Pharmaceutics* **569**: 118611. Available from: https://doi.org/10.1016/j.ijpharm.2019.118611.

[97] Abdelhakim, H.E., Williams, G.R., Craig, D.Q.M. et al. (2020). Human mouthfeel panel investigating the acceptability of electrospun and solvent cast orodispersible films. *International Journal of Pharmaceutics* **585**: 119532. Available from: https://doi.org/10.1016/j.ijpharm.2020.119532.

[98] Basit, A.W. and Gaisford, S. (ed.) (2018). *3D Printing of Pharmaceuticals*. Springer.

[99] Yu, Y., Moncal, K.K., Li, J. et al. (2016). Three-dimensional bioprinting using self-assembling scalable scaffold-free 'tissue strands' as a new bioink. *Scientific Reports* **6** (1): 1–11.

[100] Korte, C. and Quodbach, J. (2018). 3D-printed network structures as controlled-release drug delivery systems: dose adjustment, API release analysis and prediction. *AAPS PharmSciTech* **19**: 3333–3342.

[101] Caloria, I.R., Bragaa, G., Jesusb, P. et al. (2020). Polymer scaffolds as drug delivery systems. *European Polymer Journal* **129**: 109621. Available from: https://doi.org/10.1016/j.eurpolymj.2020.109621.

[102] Irizarry, R., Skomski, D., Chen, A. et al. (2018). Theoretical modeling and mechanism of drug release from long-acting parenteral implants by microstructural image characterization. *Industrial & Engineering Chemistry Research* **57** (45): 15329–15345.

[103] Shapira-Furman, T., Serra, R., Gorelick, N. et al. (2019). Biodegradable wafers releasing Temozolomide and Carmustine for the treatment of brain cancer. *Journal of Controlled Release* **295**: 93–101. Available from: https://doi.org/10.1016/j.jconrel.2018.12.048.

[104] Johnson, L.M., Krovi, S.A., Li, L. et al. (2019). Characterization of a reservoir-style implant for sustained release of tenofovir alafenamide (TAF) for HIV pre-exposure prophylaxis (PrEP). *Pharmaceutics* **11** (7): 315.

[105] Adamo, J.E., Grayson, W.L., Hatcher, H. et al. (2018). Regulatory interfaces surrounding the growing field of additive manufacturing of medical devices and biologic products. *Journal of Clinical and Translational Science* **2** (5): 301–304.

[106] Gioumouxouzis, C.I., Karavasili, C., and Fatouros, D.G. (2019). Recent advances in pharmaceutical dosage forms and devices using additive manufacturing technologies. *Drug Discovery Today* **24** (2): 636–643. Available from: https://doi.org/10.1016/j.drudis.2018.11.019.

[107] International Coalition of Medicines Regulatory Authorities (ICMRA) (2019). ICMRA innovation project 3D bio-printing case study: summary of discussions and considerations for the new informal innovation. *ICMRA Innovation Project*. 1–23. Available from: http://www.icmra.info/drupal/sites/default/files/2019-10/ICMRA_Innovation_WS3_3D_Bio-printing.pdf (accessed 12 February 2022).

[108] Al-jawadi, S., Capasso, P., and Sharma, M. (2018). The road to market implantable drug delivery systems: a review on US FDA ' s regulatory framework and quality control requirements. *Pharmaceutical Development and Technology* **23** (10): 953–963.

[109] Stanco, D., Urbán, P., Tirendi, S. et al. (2020). 3D bioprinting for orthopaedic applications: current advances, challenges and regulatory considerations. *Bioprinting* **20**: e00103.

[110] Maniruzzaman, M. and Nokhodchi, A. (2017). Continuous manufacturing via hot-melt extrusion and scale up: regulatory matters. *Drug Discovery Today* **22** (2): 340–351. Available from: http://dx.doi.org/10.1016/j.drudis.2016.11.007.

[111] Ragelle, H., Rahimian, S., Guzzi, E.A. et al. (2021). Additive manufacturing in drug delivery: innovative drug product design and opportunities for industrial application. *Advanced Drug Delivery Reviews* **178**: 113990.

6

Wound Dressings by 3D Printing

Joshua Boateng

School of Science, Faculty of Engineering and Science, University of Greenwich, Medway Campus, Kent, UK

6.1 Wound Healing Process

Wound healing is a biologically complex but well-organised process, involving various distinct but overlapping biochemical, physiological, and molecular phases (or stages). These phases are categorised as haemostasis, inflammation, proliferation (granulation tissue formation), and re-epithelialisation/remodelling, as depicted in Figure 6.1 [1]. The transition between wound healing phases depends on the differentiation and maturation of different types of essential cells such as keratinocytes, fibroblast, mast cells, and macrophages [2]. These are mediated by various biological molecules such as growth cytokines, growth factors, and enzymes, some of which are stimulatory, whilst others can be destructive, such as the matrix metalloproteinases. Any major changes in any of the above components could result in stalling of one or more phases, and subsequently in the development of chronic (hard-to-heal) wounds. The wound healing stages are summarised briefly below but they need to be well understood to develop formulations (dressings) and novel technologies. A full critical evaluation of the wound healing phases is beyond the scope of this chapter and the reader is therefore referred to more comprehensive texts [3–5].

3D Printing of Pharmaceutical and Drug Delivery Devices: Progress from Bench to Bedside, First Edition.
Edited by Dimitrios A. Lamprou, Dennis Douroumis and Sheng Qi.

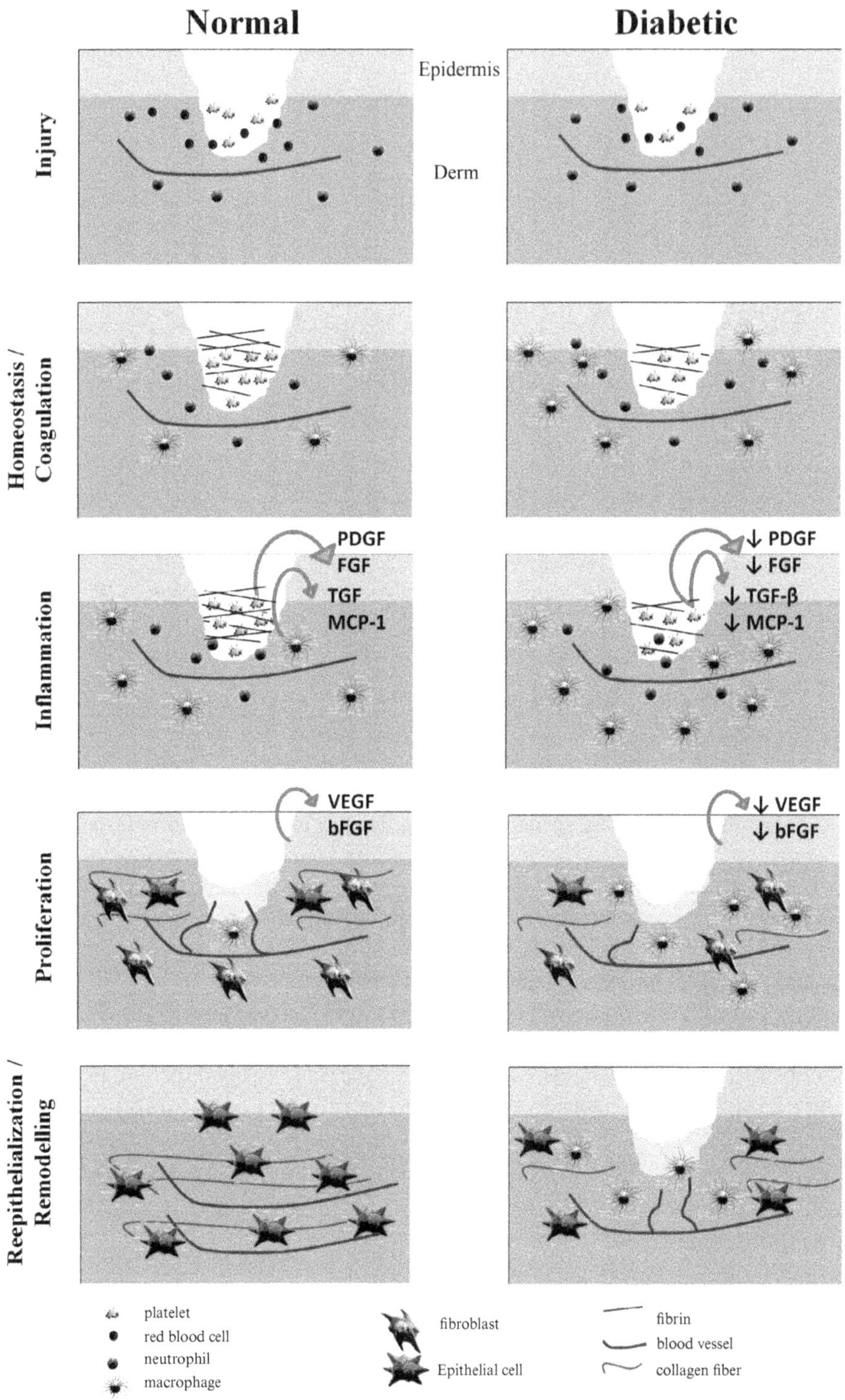

Figure 6.1 *Differences in healing processes of normal and diabetic wounds. (Source: Adapted from [6].) [Platelet derived growth factor (PDGF), transforming growth factor beta (TGF-β), monocytes chemoattractant protein 1 (MCP-1), epidermal growth factor (EGF), vascular endothelial growth factors (VEGF), and basic fibroblast growth factor (bFGF)].*

6.1.1 Haemostasis/Coagulation

Immediately after injury, bleeding occurs to flush out bacteria and/or antigens from the wound site and triggers haemostasis which is initiated by formation of a fibrin plug (coagulation) consisting of platelets and provides an instant protection from the external environment and primary coverage of the wound [1]. Aggregated platelets secrete cytokines (including leukotriene B4, platelet factor IV) and several growth factors (such as PDGF, TGF-β, MCP-1, IGF-1, EGF) that recruit neutrophils and monocytes to the wound area, which are extremely important for the first phase of new tissue formation process [7].

6.1.2 Inflammation

The inflammatory phase is initiated simultaneously with haemostasis, sometimes from a few minutes after injury up to 24 hours and continues for about 3 days [3]. Once haemostasis has been achieved, the blood vessels dilate (vasodilation) to allow pro-inflammatory substances (growth factors) such as PDGF, TGF-β, MCP-1, IGF-1, and EGF, which attract the migration of leukocytes, macrophages, and lymphocytes into the wound site [8].

6.1.3 Proliferation

The proliferative phase occurs just after 72 hours and lasts for 2–4 weeks after injury. The phase is characterised by fibroblast migration, collagen synthesis, angiogenesis, and granulation tissue formation [7]. The migration of fibroblasts activated by TGF-β and PDGF produces the matrix proteins fibronectin, hyaluronan, and later collagen and proteoglycans; subsequently rebuilding the extracellular matrix (ECM) to determine the synthesis and remodelling of the tissue [9]. The synthesis of collagen plays a central role in wound repair because of fibroblasts migration.

6.1.4 Re-epithelisation/Remodelling

The last phase of wound healing is re-epithelisation or remodelling, also known as maturation. It occurs simultaneously with the development of granulation tissue and continues for several months up to 2 years after wound closure [3]. Regeneration (remodelling) of the tissue is characterised by continuous collagen synthesis and its breakdown as ECM.

6.1.5 Wound Classification

The Wound Healing Society defines a wound as the 'breakdown or disruption of normal anatomical structure of skin and a disturbance to its function to the immune system leading to serious morbidity and mortality [3].' Wounds can be broadly classified as acute and chronic.

Acute wounds are superficial wounds arising from mechanical injuries such as surgical incisions, scratches, and tears and typically affecting the epidermis and superficial dermis. These wounds heal within the expected timeframe of 8–12 weeks and progress through the orderly sequential wound healing phases (i.e., haemostasis, inflammation, proliferation, maturation, and remodelling [4]).

Chronic wounds, on the other hand, experience disruption in one of more of the above wound healing phases [1] and result from tissue destruction of skin layers including epidermis, dermis, and subcutaneous fat tissue. This causes prolonged healing times beyond 12 weeks, which may remain for several months. Chronic wounds are usually associated with underlying conditions such as diabetes, chronic renal failure, and hepatic failure which affects the balance between wound bioburden and the patient's immune system [10, 11]. Furthermore, the occurrence of polymicrobial and mixed infections, coupled with poor initial management and treatment of the wound could contribute to poor healing of chronic wounds [12, 13]. Chronic wounds include diabetic foot ulcers, leg ulcers, pressure ulcers, venous ulcers, arterial ulcers, and trauma, and their incidence has been exacerbated by several factors such as obesity, ageing population, alcoholism, smoking, stress, poor nutrition, or prolonged pressure [2].

6.1.6 Wound Dressings

Wound dressings play a key role in the management of both acute and chronic wounds. Generally, dressings should possess certain ideal characteristics, as summarised in Figure 6.2 [3, 4, 10, 14–17]). The choice of wound dressings depends on several factors that affect wound healing, such as heavy exudate, microbial load, biofilms, severity of pain, and infection.

However, no single dressing fulfils all the requirements of healing, due to complexities of chronic wounds. Generally, dressings can be classified into three main categories [3, 4] and these are summarised briefly in Table 6.1.

6.1.7 3D Printing

Various additive manufacturing (AM) techniques (3D printing) have recently been employed for the production of different scaffolds, including films possessing controlled micro-architecture and geometry [18]. These techniques present huge potential as an alternative for film dressing formation [19, 20]. The approach involves the printing of polymers and bioactive substances to fabricate biocompatible film scaffolds that replace traditional solvent cast approaches for producing films.

The controlled microstructures of these scaffolds can significantly enhance dressing performance by aiding material exchange, adhesion, migration of cells, and ultimately tissue regeneration in the wounds [20, 21]. In addition, the use of 3D printing and bioprinting methods provides more flexibility and repeatability when compared with other scaffold preparation methods [22]. This is because it is possible to design functional 3D structures

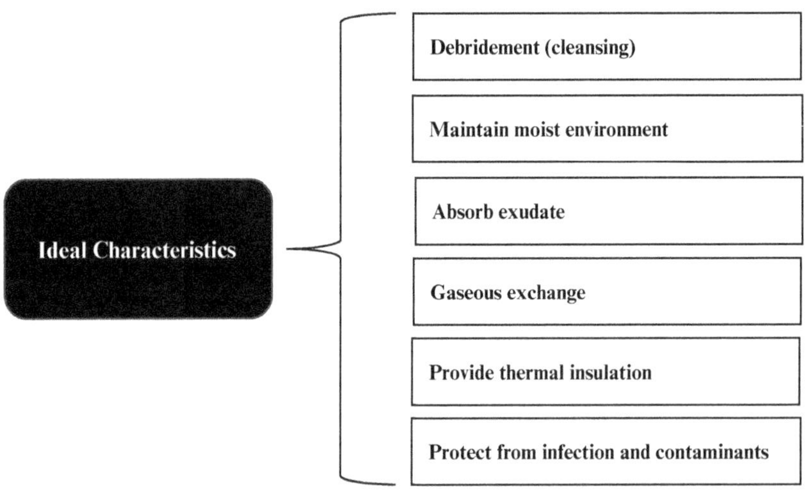

Figure 6.2 Summary of the ideal characteristics of a wound dressing.

Table 6.1 *Summary of the classification of different dressings, showing evolution from traditional to advanced dressings.*

	Traditional dressings	Modern moist dressings	Advanced dressings
Key features	• Used as coverings and for absorbing blood and exudate and for cleansing • No active part in wound healing	• Maintain a moist environment to facilitate proper cell migration and proliferation. • No active part in wound healing	• Generally made from biomaterials such as collagen, hyaluronic acid, chitosan, alginates, and elastin • High biocompatibility and biodegradability and ability to take active part in the wound healing process • Can be incorporated with biological agents such as growth factors, nucleic acids, and stem cells for delivery to the wound site [4]. • Impregnation of naturally derived agents such as aloe vera, honey, silver, iodine that make them medicated drug delivery dressings • The polymer-based systems can maintain moist environment
Advantages	• Good as secondary dressing covering for moist or advanced dressings • They are very readily available, cheap, and easy to use. For example, gauze dressings made of woven or nonwoven fibres of cotton, rayon polyester, or a combination of both, are used in the packing of open surgical and cavity wounds • Sterile gauze pads are also used as packing of open wounds to absorb exudates and fluids	• Create a moist environment to facilitate effective wound healing • Provides adequate protection for the wound and control of exudate levels • Film dressings allow investigation of the wound without removal • Hydrogels are good for providing autolytic debriding action (softening) of necrotic tissue due to high moisture content • Generally cost-effective	• Biologically active; therefore, have therapeutic/pharmacological effects • Very effective in treating chronic wounds

(Continued)

Table 6.1 *(Continued)*

	Traditional dressings	Modern moist dressings	Advanced dressings
Limitations	• Passive action with no active part in wound healing process • Can cause dryness and trauma upon removal • Poor moisture handling ability, therefore leaks when saturated with moisture • The use of these dressings is limited when the outer surface becomes moistened by heavy exudate and external fluids thus resulting in maceration to the wound sites. • Removal of these dressings can promote cross-contamination of the wounds by dispersion of bacteria into the air • Creams and gels have poor residence time on the wound surface and cause messiness	• Modern dressings are passive, target only one phase of wound healing • They depend on the body's natural ability to heal itself, which is compromised in chronic wounds, therefore can be ineffective • Hydrogels and films have poor moisture handling capacity and therefore not suitable for highly exuding wounds	• Expensive, as experienced health personnel are required for application • In the case of skin grafts, another wound has to be created in order to harvest fresh skin to replace lost tissue. Such wounds can be very painful and prone to also becoming infected
Examples	Cotton, gauze, creams, gels	Hydrogels, hydrocolloids, alginates, films, foams,	Tissue engineered substitutes, collagen, and hyaluronic acid-based dressings and scaffolds, drug loaded biological, and modern moist dressings

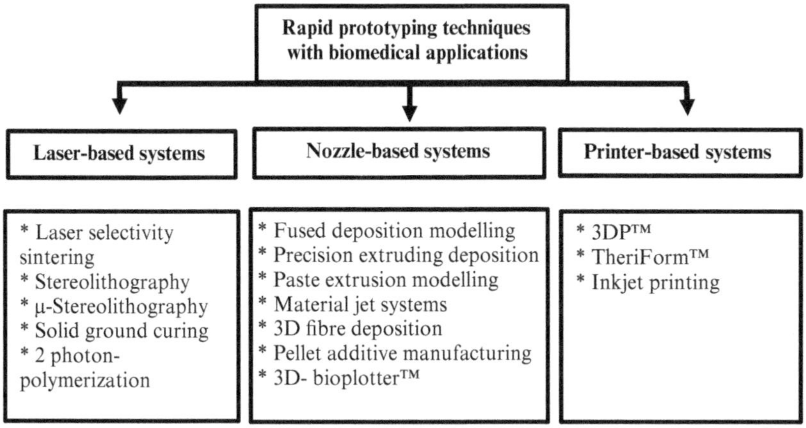

Figure 6.3 *Classification of RP techniques with biomedical applications. (Source: Adapted from [27] with permission.)*

possessing pre-determined size and porosities with the help of computer-aided design (CAD) software followed by printing with an automated 3D printer [23–25]. Furthermore, they compare with other methods of fabricating scaffolds (films) employment of 3D bioprinting.

Rapid prototyping (RP), also sometimes referred to as AM or solid freeform techniques, refers to a class of technologies that can automatically construct physical models from CAD data. By using the RP system, almost any shape can be produced [18, 26]. The main RP techniques that can produce scaffolds for biomedical applications, such as wound dressings, have been summarised in Figure 6.3.

6.1.8 3D-Printed Dressings

3D-printed dressings are of increasing current interest in academic research and industry R&D settings, especially due to the possibility of personalised production for specific wounds and patients, depending on factors such as size, shape, and location of the wound. The possibility of scanning the contours of a specific wound to produce scaffolds or dressings, having the exact same dimensions as that of the target wound, is an interesting prospect, which will most likely influence the future of wound management and tissue regeneration.

Furthermore, advancements in biomaterial polymer science, bioengineering, and tissue culture technologies have opened a new field of 3D bioprinting, where 3D scaffolds mimicking the natural dermis and epidermis can be laden with relevant skin cells to produce 'artificial skin' that significantly helps to trigger better skin regeneration in hard-to-heal chronic wounds [28].

Though most of the polymers used in modern and advanced biomaterial-based dressings are effective, their use in 3D printing of dressings is challenging, as most are not printable in their native form. Therefore, they require some form of modification to improve their gel viscosities to enable them to be printed without losing their structure and collapse and represents one of the current knowledge gaps exercising various researchers and practitioners. In addition, several researchers have investigated the use of new inks/bioinks from both natural and synthetic sources [29]. The rest of this chapter will critically review and evaluate the published literature around 3D-printed and 3D-bioprinted dressings and/or scaffolds in the form of case studies, as outlined below.

6.2 Case Studies

Yang et al. [30] reported on 3D-printed antibacterial dressing comprising *N*-halamine/TiO_2 bound to gelatin methacrylate combined with xanthan gum in a composite system, as illustrated in Figure 6.4. The 3D-printed dressings exhibited excellent swelling and moisture handling properties as well as good antibacterial activity, with 100% elimination of *Escherichia coli* and *Staphylococcus aureus* within 60 minutes and prevented bacterial biofilm formation. The 3D-printed dressings were biocompatible and demonstrated accelerated wound healing effect in mice and showed good potential for use in wound healing.

Shafiee et al. [31], designed biomimetic dressing comprising medical grade polycaprolactone via 3D printing and demonstrated microporous architecture and mechanical properties that helped to reduce skin scarring during wound healing. The 3D-printed scaffolds were further seeded with human gingival tissue multipotent mesenchymal stem/stromal cells and compared with full thickness excisional wound in rats over a six-week period. The former showed a decrease in wound contracture followed by significant improvements in skin regeneration (including decreased scarring) through increased granulation and re-epithelialisation when compared with the control groups.

Altan et al. [32] proposed a composite chitosan-xanthan gum patch via a 3D-printing approach, with the aim to achieve cell-mimicking, adhesion, proliferation, and differential characteristics for wound healing applications. The study involved preliminary investigations of viscosity of xanthan gels at different concentrations to determine ease of printing, which was achieved at higher xanthan concentrations. Furthermore, higher xanthan content resulted in better mechanical, hydrophilic, and slow degradation characteristics. The optimised patches showed improvement in cell proliferation rate, adhesion and viability, with potential for use as dressing for wound healing.

Hafezi et al. [33] reported on a 3D-printed chitosan dressing crosslinked with genipin to be applied to chronic wounds. The gels required plasticisation with polyethylene glycol prior to printing to achieve flexible and mechanically stable film dressing matrices. The 3D-printed

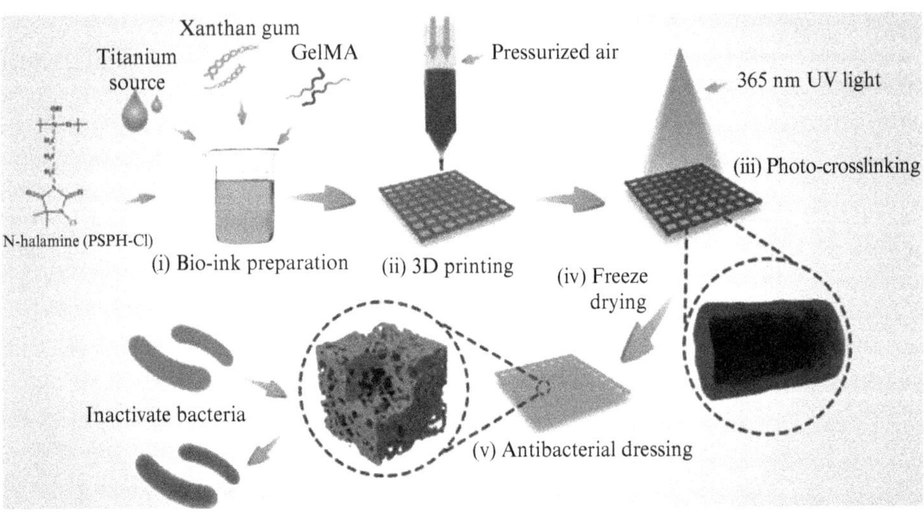

Figure 6.4 *Schematic illustration of the fabrication of antibacterial 3D-printed dressings. (Source: tarasov_vl/Adobe Stock Photos.)*

films were characterised for their functional physico-chemical properties, including tensile, fluid handling, mucoadhesion, drug dissolution, morphology, and cell viability. Films comprising equal proportions of chitosan and plasticiser were deemed optimum, demonstrating amorphous nature, molecular interactions between the three main components, and appropriate adhesion and moisture handling ability. Finally, the 3D-printed scaffolds demonstrated high biocompatibility with human skin fibroblast cell lines, showing greater than 90% of cell viability after 48 hours, which suggests their non-toxic nature and ability to sustain cell proliferation, which is essential for effective wound healing. In a follow-up study [34], the same group employed extrusion-based 3D bioprinting to print cell-laden chitosan-genipin bioinks to obtain composite scaffolds containing skin cells (keratinocytes and fibroblasts). A summary of their bioprinting setup is provided in the schematic shown in Figure 6.5, with crosslinked alginate used as the base layer before bioprinting of the cell laden chitosan–genipin constructs. Nuclear magnetic resonance spectroscopy was used to confirm the chitosan–genipin crosslinking.

An MTT assay showed high cell viability values above 85% for keratinocytes and fibroblast cells within the bioinks, both before and after bioprinting, showing the gentle processing conditions which maintained the cell viability as well as confirming the biocompatibility of the scaffold matrices containing the cells. In addition, cell viability was further confirmed using live–dead assay to differentiate the living cells from non-viable ones using confocal microscopy. The confocal microscopy imaging showed a high cell viability for the bioprinted constructs at each time point, as shown in Figure 6.6, and confirmed that the number of cells increased during the testing period from day 0 to day 7. Finally, scanning electron microscopy showed a microporous structure that is expected to support cell growth and allow supply of nutrients to the cells, while the keratinocytes and fibroblasts showed a polygonal and spindle shape, respectively.

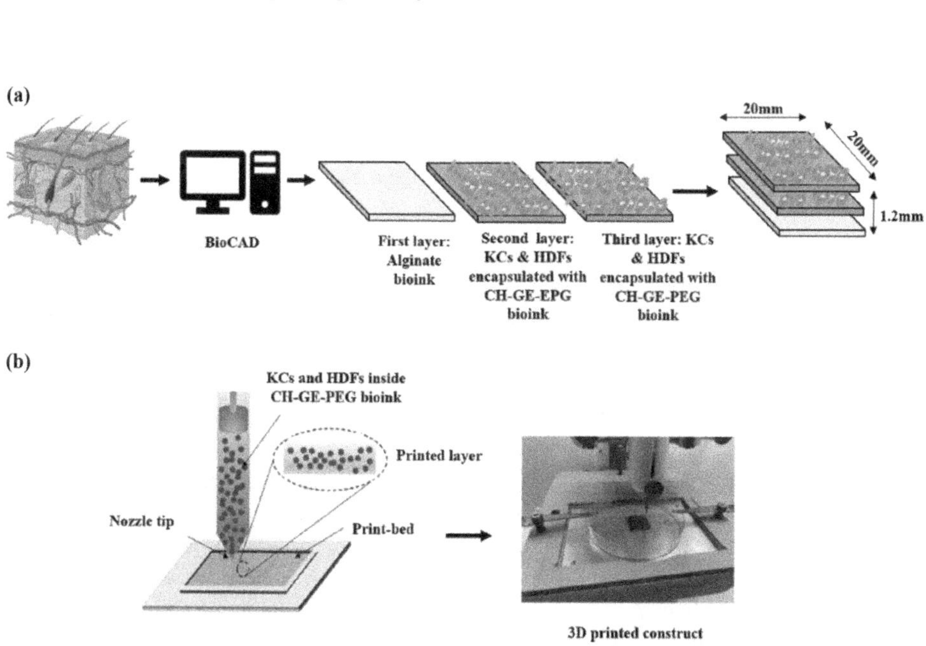

Figure 6.5 (a) Schematic illustration of the design of each layer of skin 3D-printed constructs. (b) 3D bioprinting of alginate layer and KCs and HDFs encapsulated with CH-GE-PEG layers. (Source: [34]/MDPI/CC BY 4.0.)

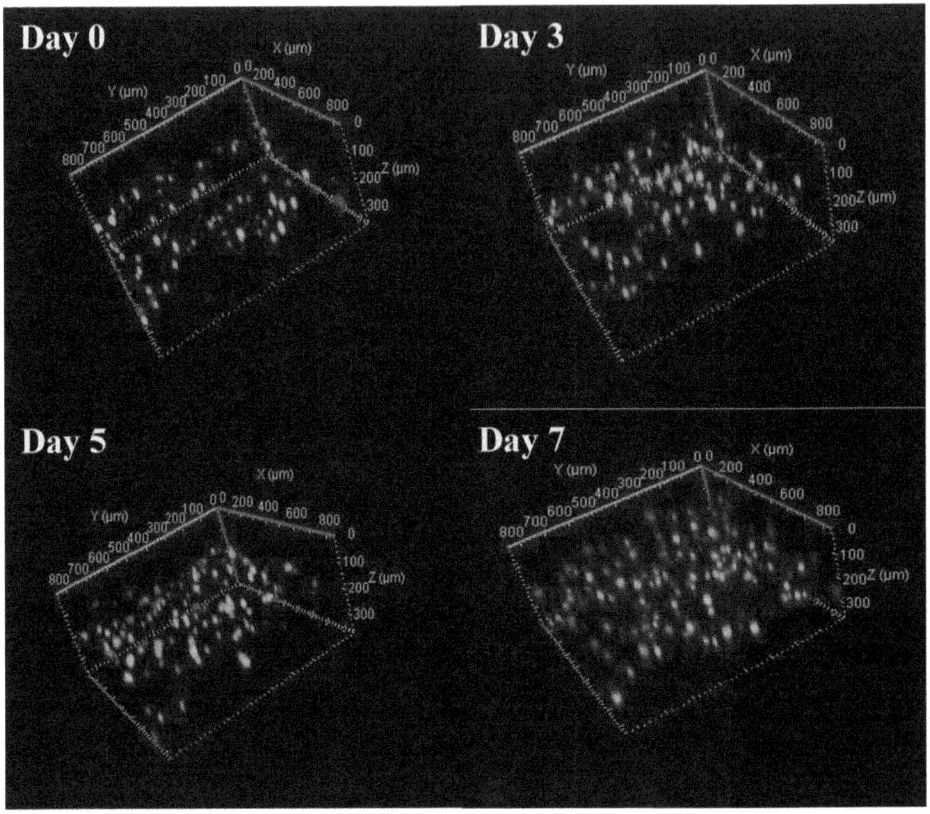

Figure 6.6 *Confocal images of bioprinted CH-GE-PEG cell-laden constructs throughout 7 days showing live (light spots) and total number of the cells (dark spots). All images shown in this graph are representative of three independent experiments with similar results. (Source: Reproduced from [34] under the Creative Commons Attribution (CC BY) licence.)*

Long et al. [35] developed polysaccharide-based biodegradable 3D-printed composite hydrogel dressing scaffolds (Figure 6.7) comprising pectin and chitosan encapsulating lidocaine as a model drug with immediate release characteristics. The scaffolds were designed in cubic mesh format using Solid works with grid size of 2.50 mm^{-2} and line width of 0.50 mm and printed using an extrusion-based printer. The resulting scaffolds were characterised for fluid handling (swelling and water absorption), morphology, porosity, bioadhesion, chemical interactions, and *in vitro* drug release characteristics. The 3D-printed scaffolds exhibited dimensional integrity and ability to self-adhere to the skin with high swelling ratio and water absorption for maintaining a moist wound healing environment. Furthermore, loading of lidocaine did not affect functional performance of the scaffolds and were able to release the drug over six hours, which was explained by the Korsmeyer–Peppas model.

Muwaffak et al. [36] reported on the development and characterisation of patient-specific 3D-scanned and 3D-printed polycaprolactone-based wound dressings (Figure 6.8) with antimicrobial properties. The authors followed various formulation approaches to finally obtain the 3D-printed dressings. First, filaments containing different antimicrobial metals

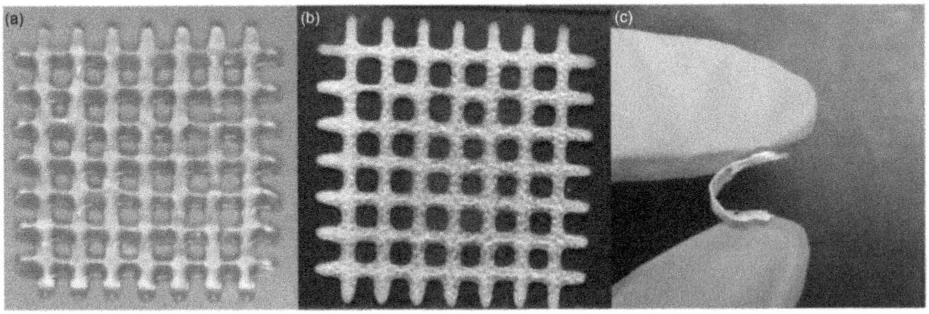

Figure 6.7 *3D-printed chitosan-pectin (CS-PEC) hydrogel scaffold: a) freshly printed; b) lyophilised; and c) flexibility of a lyophilised scaffold. (Source: Reproduced from [35] with permission from Elsevier.)*

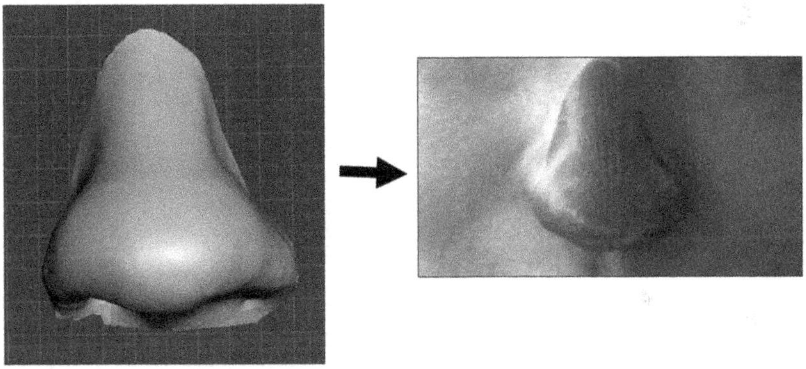

Figure 6.8 *3D scan model of a nose (left) and the printed wound dressing of this model with Cu-PCL (right). (Source: Reproduced from [36] with permission from Elsevier.)*

including silver, zinc, and copper were prepared using hot-melt extrusion. Subsequently, they used 3D scanning to generate 3D models of nose and ears as a means of producing dressings with customised shapes for specific patients. The extruded filaments containing different amounts of the metals were used to produce 3D-printed dressings and the metal release tested by means of inductively coupled plasma atomic emission spectroscopy and demonstrated fast initial release in the first 24 hours followed by a more sustained release up to 72 hours. Further *in vitro* antibacterial activity showed that the 3D-printed dressings loaded with silver and copper exhibited more potent bactericidal activity against selected bacteria than zinc loaded equivalent.

In a recent study, Teoh et al. [37] reported on the development of personalised 3D-printed hydrogel-based dressings (Figure 6.9) that could be loaded with multiple drugs with tuneable dosages based on varied architectures. The hydrogel dressings were based on chitosan methacrylate, using lidocaine and levofloxacin as model analgesic (pain) and antibiotic (infection) to achieve medicated scaffolds able to take an active part in the wound healing process. The researchers were able to customise the drug loading amounts and with varied release rates by loading them at different locations within the dressings and also changing the thickness of blank (drug-free) layers surrounding the drug-loaded layers. The study

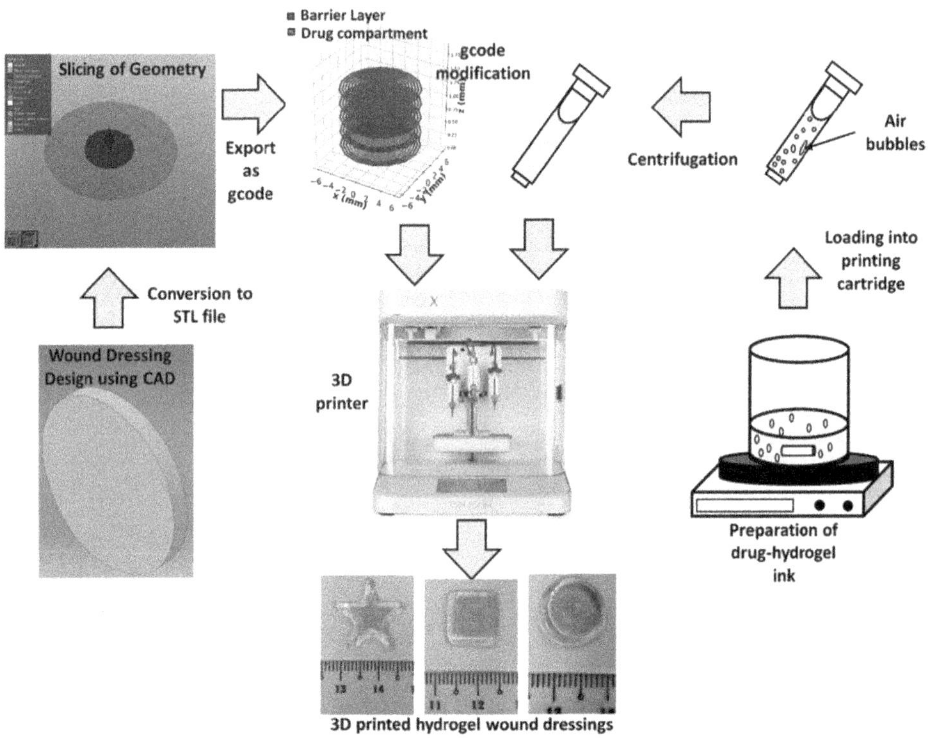

Figure 6.9 *Process scheme used for the fabrication of 3D-printed hydrogel wound dressings in this study. For images of 3D-printed hydrogel wound dressings, indocyanine green was only used to distinguish between the drug compartment and the barrier layer. (Source: Reproduced from [37] with permission from Elsevier.)*

clearly showed the potential of 3D printing to develop dressings able to meet specific and unique patient needs for achieving personalised wound therapeutics.

There are several other reported studies on 3D-printed dressings and some of these are summarised in Table 6.2.

6.3 Summary/Conclusions

Novel methods such as 3D printing and 3D bioprinting have the capability to be a platform for producing polymeric films for the individual patient. This will possibly resolve the issue of the pharmaceutical industry and pharmacies to meet the future demand of customised (personalised) medicine [44].

Table 6.2 *Summary of 3D-printed wound dressings reported in the literature.*

Type of dressing/bioink material(s)	Loaded actives	3D-printing method	Functional characterisation	Reference
Sodium Alginate-Polyethyleneglycol	*Satureja cuneifolia* extract	Extrusion 3D printer	FTIR, SEM, cell viability, tensile, swelling, antibacterial assay, drug release, cell attachment and proliferation	[38]
Poly(Dimethylsiloxane)	Silver nanoparticles	Extrusion 3D-bioprinter	SEM-EDS, STEM, contact angle, *in vivo* wound healing, cyto-toxicity, non-fouling assay, *in vitro* antibacterial assay	[39]
Carboxymethylcellulose	Platelet-rich plasma (autologous growth factors)	Extrusion 3D printer	Swelling, FTIR-ATR, hardness, cohesiveness, adhesion, growth factor release, chorioallantoic membrane (CAM) assay, cell proliferation and migration (*in vitro*), *in vivo* wound healing assessment	[40]
Calcium Chloride Cross-linked Sodium Alginate	–	Extrusion-based 3D bioprinter	FTIR, swelling, degradation percentage, wettability, zeta potential, cell viability assay, live-dead assay by fluorescence microscopy, real-time PCR	[41]
Mesoporous Bioglass Modified With Pluronic Acid and Sodium Alginate	Exosomes	Extrusion based cryogenic 3D printer	XRD, FTIR, MTT assay, cell adhesion, live/dead cell staining, *in vivo* animal wound healing assay, blood perfusion, Western blot	[42]
Gelatin Methacryloyl	*N*-halamine modified ceria nanoparticles	Inkjet printer	Swelling, filament collapse test, antimicrobial test, *in vitro* cytotoxicity, hemocompatibility, whole blood clotting, cutaneous wound healing in mice	[43]

Abbreviations: Fourier-Transform Infrared Spectroscopy (FTIR); Scanning electron microscopy (SEM), MTT, XRD, PCR, semi-crystalline elastomeric (SEM)-EDS, STEM.

References

[1] Falanga, V. (2005). Wound healing and its impairment in the diabetic foot. *Lancet* **366** (9498): 1736–1743.

[2] Guo, S. and Dipietro, L.A. (2010). Factors affecting wound healing. *Journal of Dental Research* **89** (3): 219–229.

[3] Boateng, J., Matthews, K., Stevens, H., and Eccleston, G. (2008). Wound healing dressings and drug delivery systems: a review. *Journal of Pharmaceutical Sciences* **97** (8): 2892–2923.

[4] Boateng, J. and Catanzano, O. (2015). Advanced therapeutic dressings for effective wound healing: a review. *Journal of Pharmaceutical Sciences* **104** (11): 3653–3680.

[5] Ather, S. and Harding, K.G. (2018). *Wound Management and Dressings*, **2019** (ed. S.J. Tate), 1–22. The Textile Institute Book Series.

[6] Moura, L.I.F., Dias, A.M.A., Carvalho, E., and De Sousa, H.C. (2013). Recent advances on the development of wound dressings for diabetic foot ulcer treatment: a review. *Acta Biomaterialia* **9** (7): 7093–7114.

[7] Enoch, S. and Price, P. (2004). Cellular, molecular and biochemical differences in the pathophysiology of healing between acute wounds, chronic wounds and wounds in the aged. *World Wide Wounds*. ISSN: 1369-2607 (Revision 1).

[8] Demidova-Rice, T.N., Hamblin, M.R., and Herman, I.M. (2012). Acute and impaired wound healing: pathophysiology and current methods for drug delivery. Part 1: Normal and chronic wounds: biology, causes, and approaches to care. *Advances in Skin & Wound Care* **25** (7): 304–314.

[9] Grinnell, F. (1994). Fibroblasts, myofibroblasts, and wound contraction. *Journal of Cell Biology* **124** (4): 401–404.

[10] Sarheed, O., Ahmed, A., Shouqair, D., and Boateng, J. (2016). *Antimicrobial Dressings for Improving Wound Healing: New Insights into Ancient Challenges* (ed. V. Alexandrescu), 377–402. InTech. https://doi.org/10.5772/63961.

[11] Nawazz, Z. and Bentley, G. (2011). Surgical incisions and principles of wound healing. *Surgery* **29** (2): 59–62.

[12] Roudbary, M., Razi, F., Mohammadi, S.R. et al. (2014). Distribution of mycotic and bacterial infections among the patients with diabetic foot ulcers referred to diabetes and metabolic disorder clinic, Tehran, Iran 2011–2012. *European Journal of Experimental Biology* **4** (4): 20–25.

[13] Dowd, S., Delton Hanson, J., Rees, E. et al. (2011). Survey of fungi and yeast in polymicrobial infections in chronic wounds. *Journal of Wound Care* **20** (1): 40–47.

[14] Hilton, J.R., Williams, D.T., Beuker, B. et al. (2004). Wound dressings in diabetic foot disease. *Clinical Infectious Diseases: An Official Publication of the Infectious Diseases Society of America* **39** (Suppl 2): 100–103.

[15] Dhivya, S., Padma, V.V., and Santhini, E. (2015). Wound dressings: a review. *BioMedicine* **5** (4): 24–28.

[16] Sood, A., Granick, M.S., and Tomaselli, N.L. (2014). Wound dressings and comparative effectiveness data. *Advances in Wound Care* **3** (8): 511–529.

[17] Chang, K.W., Alsagoff, S., Ong, K.T., and Sim, P.H. (1998). Pressure ulcers: randomised controlled trial comparing hydrocolloid and saline gauze dressings. *The Medical Journal of Malaysia* **53** (4): 428–431.

[18] Hafezi, F., Kucukgul, C., Ozler, S.B., and Koc, B. (2015). *Bioprinting: Application of Additive Manufacturing in Medicine* (ed. A. Bandyopadhyay and S. Bose). s.l: CRC Press Florida, USA.

[19] Leong, K.F., Cheah, C.M., and Chua, C.K. (2003). Solid freeform fabrication of three-dimensional scaffolds for engineering replacement tissues and organs. *Biomaterials* **24** (13): 2363–2378.

[20] Liu, J., Chi, J., Wang, K. et al. (2016). Full-thickness wound healing using 3D bioprinted gelatin-alginate scaffolds in mice: a histopathological study. *International Journal of Clinical and Experimental Pathology* **9** (11): 11197–11205.

[21] Dean, D.M., Napolitano, A.P., Youssef, J., and Morgan, J.R. (2007). Rods, tori, and honeycombs: the directed self-assembly of microtissues with prescribed microscale geometries. *FASEB Journal* **21** (14): 4005–4012.

[22] Tong, C., Hao, H., Xia, L. et al. (2016). Hypoxia pretreatment of bone marrow-derived mesenchymal stem cells seeded in a collagen-chitosan sponge scaffold promotes skin wound healing in diabetic rats with hindlimb ischemia. *Wound Repair Regeneration* **24** (1): 45–56.

[23] Murphy, S.V. and Atala, A. (2014). 3D bioprinting of tissues and organs. *Nature Biotechnology* **32** (8): 773–785.
[24] Blaeser, A., Campos, D.D.F., Weber, M. et al. (2013). Biofabrication under fluorocarbon: a novel freeform fabrication technique to generate high aspect ratio tissue-engineered constructs. *BioResearch Open Access* **2** (5): 374–384.
[25] Zhang, Y. (2013). Study on the preparation of genipin crosslinked chitosan microspheres of resveratrol and *in vitro* release. *Journal of Polymer Research* **20** (175): 1–10.
[26] Srisairman, V. and Mathan Kumar, C.D. (2015). Rapid prototyping: additive/solid free form manufacturing in automobile engineering. *International Journal of Engineering Development and Research* **3** (4): 2321–9939.
[27] Billiet, T., Vandenhaute, M., Schelfhout, J. et al. (2012). A review of trends and limitations in hydrogel-rapid prototyping for tissue engineering. *Biomaterials* **33** (26): 6020–6041.
[28] Las Heras, K., Igartua, M., and Maria Hernandez, R. (2022). Cell-based dressings: a journey through chronic wound management. *Biomaterials Advances* **135**: 212738.
[29] Chinga-Carrasco, G., Ehman, N.V., and Area, M.C. (2019). Bagasse: a major agro-industrial residue as potential resource for nanocellulose inks for 3D printing of wound dressing devices. *Additive Manufacturing* **28**: 267–274.
[30] Yang, Z., Ren, X., and Liu, Y. (2021). Multifunctional 3D printed porous GelMA/xanthan gum based dressing with biofilm control and wound healing activity. *Materials Science & Engineering C* **131** (2021): 112493.
[31] Shafiee, A., Cavalcanti, A.S., Saidy, N.T. et al. (2021). Convergence of 3D printed biomimetic wound dressings and adult stem cell therapy. *Biomaterials* **268**: 120558.
[32] Altan, E., Turker, N., Hindy, O.A. et al. (2022). Investigation of 3D-printed chitosan-xanthan gum patches. *International Journal of Biological Macromolecules* **213**: 259–267.
[33] Hafezi, F., Scoutaris, N., Douroumis, D., and Boateng, J. (2019). 3D printed chitosan dressing crosslinked with genipin for potential healing of chronic wounds. *International Journal of Pharmaceutics* **560**: 406–415.
[34] Hafezi, F., Shorter, S., Tabriz, A.G. et al. (2020). Bioprinting and preliminary testing of highly reproducible novel bioink for potential skin regeneration. *Pharmaceutics* **12** (6): 550.
[35] Long, J., Etxabide Etxeberria, A., Nand, A.V. et al. (2019). A 3D printed chitosan-pectin hydrogel wound dressing for lidocaine hydrochloride delivery. *Materials Science and Engineering: C* **104** (2019): 109873.
[36] Muwaffak, Z., Goyanes, A., Clark, V. et al. (2017). Patient-specific 3D scanned and 3D printed antimicrobial polycaprolactone wound dressings. *International Journal of Pharmaceutics* **527** (1–2): 161–170.
[37] Heng Teoh, J., Muay Tay, S., Fuh, J., and Wang, C.-H. (2022). Fabricating scalable, personalized wound dressings with customizable drug loadings via 3D printing. *Journal of Controlled Release* **341**: 80–94.
[38] Ilhan, E., Cesur, S., Guler, E. et al. (2020). Development of *Satureja cuneifolia*-loaded sodium alginate/polyethylene glycol scaffolds produced by 3D-printing technology as a diabetic wound dressing material. *International Journal of Biological Macromolecules* **161**: 1040–1054.
[39] Shi, G., Wang, Y., Derakhshanfar, S. et al. (2019). Biomimicry of oil infused layer on 3D printed poly(dimethylsiloxane): non-fouling, antibacterial and promoting infected wound healing. *Materials Science and Engineering: C* **100**: 915–927.
[40] Diaz-Gomez, L., Gonzalez-Prada, I., Millan, R. et al. (2022). 3D printed carboxymethyl cellulose scaffolds for autologous growth factors delivery in wound healing. *Carbohydrate Polymers* **260**: 117768.
[41] Khoshnood, N., Zamanian, A., and Abbasi, M. (2021). The potential impact of polyethylenimine on biological behavior of 3D-printed alginate scaffolds. *International Journal of Biological Macromolecules* **178**: 19–28.
[42] Yiqiang, H., Bin, W., Xiong, Y. et al. (2021). Cryogenic 3D printed hydrogel scaffolds loading exosomes accelerate diabetic wound healing. *Chemical Engineering Journal* **426**: 130634.
[43] Yang, Z., Ren, X., and Liu, Y. (2021). *N*-halamine modified ceria nanoparticles: antibacterial response and accelerated wound healing application via a 3D printed scaffold. *Composites Part B: Engineering* **227**: 109390.
[44] Preis, M., Breitkreutz, J., and Sandler, N. (2015). Perspective: concepts of printing technologies for oral film formulations. *International Journal of Pharmaceutics* **494** (2): 578–584.

7

3D Printing of Hydrogels

Poornima Ramburrun and Yahya E. Choonara

Wits Advanced Drug Delivery Platform Research Unit, School of Therapeutic Sciences, Faculty of Health Sciences, Department of Pharmacy and Pharmacology, University of the Witwatersrand, Johannesburg, South Africa

7.1 Introduction

3D printing is defined as the spatial arrangement of biomaterial, living cells, and bioactive agents, by assembling them using a computer-aided layer-by-layer deposition technique for fabrication of 2D or 3D constructs for tissue engineering and drug delivery applications [1]. A gel is defined as a semi-solid jelly-like material, consisting of a liquid phase, with textural and mechanical properties that range from soft and weak to hard and tough. Featuring similar properties, a hydrogel is defined as a 3D crosslinked polymer network, which can absorb and retain large amounts of water. Hydrogels maintain a solid-like state, and can be either biodegradable or non-biodegradable, but do not dissolve in water; the liquid phase is absorbed through swelling of the polymer chains. This behaviour of hydrogels imparts excellent biomimetic properties to native mammalian tissues. Printing inks prepared from a solution of hydrogel precursors are rheologically modified to deposit continuous and smooth layers of ink during printing. To create a 3D-printed hydrogel construct, these layers are collected and stacked, or solidified in a layer-by-layer arrangement [2]. This method of fabrication has attracted the use of 3D-printed hydrogels for various pharmaceutical and biomedical applications due to its high precision and efficiency, thus paving the way to materialise the notion of on-demand personalised medicines [3].

3D Printing of Pharmaceutical and Drug Delivery Devices: Progress from Bench to Bedside, First Edition. Edited by Dimitrios A. Lamprou, Dennis Douroumis and Sheng Qi.

This chapter provides an overview of the biomedical applications of 3D-printed hydrogels, the types of hydrogel materials used in 3D printing, and the hydrogel printing techniques. The concept of printability and critical printing parameters that influence the 3D printing of hydrogels is discussed, as well as the regulatory and manufacturing considerations and limitations to drive this emerging technology to the pharmaceutical and clinical settings.

7.2 Applications of 3D-Printed Hydrogels

Hydrogel-based 3D printing is extensively explored in tissue engineering, regenerative medicine, drug delivery, cancer research, *in vitro* disease modelling, contact lenses, soft robotics, and wearable flexible electronics [4]. Hydrogels are used as 3D-printed scaffolds due to the ease of their design, synthesis, functionalisation, and versatility. Ultimately, their hydrophilic structures, flexibility, porosity, high water content, and swelling behaviour provides an environment that simulates native tissues and the extracellular matrix (ECM), including the potential for diffusion of biological molecules, nutrients, and gases [5]. Due to their functional groups, hydrogels can be designed to respond to external stimuli such as moisture, light, temperature, ions, pH, electrical signals, magnetic fields, and biochemical cues for use in various biomedical applications. Such stimuli responsive hydrogels are also known as bioresponsive or smart hydrogel materials. The investigation of 3D-printed hydrogels in biomedical and pharmaceutical applications is most popular in the research fields of tissue engineering, wound healing, and drug delivery.

7.2.1 Tissue Engineering

Tissue engineering utilises the concept of biology and biomaterials science for the objective of regenerating, restoring, or replacing damaged or defective tissues. The high biocompatibility and biomimetic properties of hydrogels makes them excellent materials for initiating, guiding, promoting, and supporting the growth of cells in the innate repair process of damaged tissues. These properties also make hydrogels suited for encapsulating a high cell density whilst maintaining cell viability – a current requisite for effective tissue regeneration alongside the inclusion of growth factors. With the availability of printing techniques to precisely control the fabrication 3D constructs using cell-laden hydrogels, the biomimicry of artificial organs can be enhanced, as the 3D hydrogel constructs can direct cell growth and morphology post printing [3].

3D-printed hydrogel scaffolds for tissue engineering applications are expected to fulfil several essential design criteria for optimal functioning, promotion of new tissue growth, and induction of minimal to no inflammatory or immune reaction from the host tissue. These design criteria are derived from material selection and the inherent physical characteristics (such as mechanical properties, biodegradability, porosity, and swelling behaviour), and biological properties (such as biocompatibility, bioadhesion, and bioactivity) of the hydrogel material [5]. Since 3D-printed hydrogels are investigated and designed for different tissue types (e.g., skin, bone, and neural tissues), the resulting functions and features of the hydrogel material may vary considerably as it is defined by its intended site of application or implantation, and the *in vivo* environmental exposure.

7.2.2 Wound Healing

Hydrogel materials are considered ideal for wound dressings as they fulfil two fundamental requirements for an environment supportive of wound healing, particularly burn wounds; the ability to control fluid loss from the body and the ability to maintain adequate hydration and moisture levels at the wound site, via the swelling and absorption properties of hydrogels [6]. 3D-printed hydrogel scaffolds for wound healing offers the following advantages for optimal healing: 1) customised formulations and incorporation of actives based on the individual's needs; 2) tailored-to-fit dimensional properties of the hydrogel dressing in terms of size, thickness, porosity, and surface area; 3) ease of drug loading for therapeutic or antibacterial effects; 4) availability of a wide range of natural and synthetic materials for hydrogel preparation of tuneable properties; 5) effective oxygen penetration imparted by the porosity and mesh-like hydrogel network structure; and 6) ability of lyophilised hydrogel matrices to absorb wound exudates and maintain a moist environment to promote cell proliferation and migration [7]. 3D-printed bioadhesive hydrogel wound dressings, with a gradually increasing rate of biodegradation, offer the added benefit of simple application and removal of the hydrogel between dressing changes to minimise secondary tissue damage, accidental removal of newly generated tissue, and scar formation [8]. These properties suggest that 3D-printed and lyophilised hydrogel scaffolds are suitable candidates for wound dressings and present the prospect for developing future personalised wound management therapies with high reliability and cost-effectiveness [7, 8].

7.2.3 Drug Delivery

The oral tablet is the most common pharmaceutical dosage form designed to deliver drugs to the human body. Tablets are made by the conventional large-scale method of powder milling and powder compression into solid dose units, a fast and economical method for pharmaceutical industries. However, these large-scale production methods are not suitable for the global demand of customised doses and personalised medicine; hence, the production of fixed tablet designs, shapes, and predetermined doses of the drug [9]. 3D-printing technology presents a strategy for developing patient-centred medicines in different types of dosage form designs for oral and external administration via reliably tailored drug combinations and release profiles based on the metabolic responses and pathophysiology of the patient [9, 10]. The versatility, biocompatibility, biodegradability, and biologically sensitive nature of hydrogel systems offer a suitable method for encapsulation and protection of drug compounds [6]. 3D-printed hydrogel-based tablets via extrusion-printing can yield on-demand dosage forms of different geometries, textures (i.e., hard swallowable tablets and soft chewable tablets), drug loading, and release profiles in a matter of minutes to accommodate customised oral drug delivery needs in paediatric to geriatric patient groups [11].

7.3 Types of Hydrogel Materials for 3D Printing

Hydrogels can be classified in several ways according to the origin of the polymer material, its ionic charge, mechanism of crosslinking, and its method of preparation (Figure 7.1) [4, 12]. Hydrogels can be prepared from natural polymers, synthetic polymers, or natural-synthetic hybrids. The blending of natural polymers with synthetic polymers yields hybrid hydrogels which constitute properties of the parent polymers. Understanding these

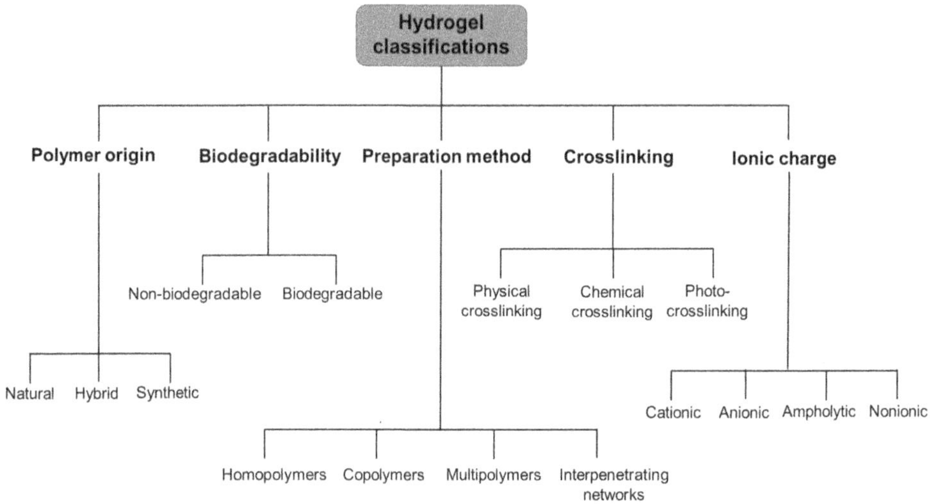

Figure 7.1 Schematic of hydrogel classifications.

classifications is important for realising its effects on hydrogel preparation and the possibility of utilising hydrogels for various biomedical applications based on their intrinsic physical and chemical properties [5]. Hydrophobic polymer networks produce hydrogels with limited water-absorption capacities compared to hydrophilic polymer networks that can absorb water many times their own weight. Hydrophilic functional groups in the hydrogel network, such as –COOH, –OH, $–NH_2$, –CONH–, $–CONH_2$, and $–SO_3H$, allow for the absorption of water; however, hydrogels can absorb liquids other than water, such as biological fluids [13].

7.3.1 Natural Polymers

Natural polymers are biological materials derived from either animal, plant, microbial, or marine sources. These materials can be further classified according to their primary chemical composition, polysaccharide-based materials, and protein-based materials, which can form gels when dispersed in water or aqueous liquid. Natural polymers are considered suitable for biomedical and tissue engineering applications due to their biocompatibility, biodegradability, and biologically recognisable moieties which support cellular activities [14]. These properties result from their inherent structural resemblance to native extracellular matrix (ECM); however, natural polymers may induce inflammatory or immune responses when introduced into the human body [12]. Despite their biomimetic appeal and improved support for cellular interactions, natural materials lack sufficient mechanical properties for biomedical applications [15]. Hydrogels made from natural polymers often require modification to adjust their textural and mechanical properties to match the environment at the application site. Due to the natural source of origin, natural polymer hydrogels may exhibit inconsistent properties between different batches of synthesised hydrogel, thus making uniformity of properties a challenge [16].

7.3.2 Synthetic Polymers

Synthetic polymers for the synthesis of 3D-printable hydrogels comprise mainly polyester, polyethylene, poloxamer, and acrylate polymers, which have low immunogenicity and do not induce any inflammatory or immune responses when introduced into the body. Synthetic polymers have the advantage to be custom designed to yield specific chemical, physical, and mechanical properties, biodegradability and functionality of high reproducibility by manipulating the polymerisation parameters, and composition of individual monomer units [14, 15]. Unlike natural polymer hydrogels, most synthetic polymer hydrogels are hydrophobic and lack inherent bioactive properties and cellular interactions [14, 17].

7.3.3 Natural-Synthetic Hybrid Polymers

To overcome the limitations of natural and synthetic hydrogel materials, natural polymers can be combined with synthetic polymers to create natural-synthetic or semi-synthetic hybrid hydrogels, where each material compensates a weakness in the other to offer improved chemical, physical, or biological properties (e.g., improved mechanical strength with bioactive properties). Hybrid hydrogels can be synthesised by photopolymerisation of graft polymers (natural polymer grafted onto a synthetic polymer, or synthetic polymer grafted onto a natural polymer) to create a crosslinked hydrogel network. The other approach involves crosslinking a blend of natural and synthetic polymer solutions to synthesise multifunctional hydrogels of adjustable properties such as crosslinking ability, bioadhesion, and degradation properties [13].

7.3.4 Ionically Charged Polymers

Hydrogels may be classified into four groups based on the presence or absence of an electrical charge on the polymer chains: 1) cationic (positively charged); 2) anionic (negatively charged); 3) ampholytic (contains both anionic and cationic groups); and 4) non-ionic (no charge, neutral). The swelling of ionic hydrogels (cationic, anionic, and ampholytic) is directed by the pH of the aqueous medium in which they are immersed, as pH determines the degree of dissociation of the ionic polymer chains. Cationic polymer chains dissociate at low pH values, therefore cationic hydrogels display enhanced swelling in acidic media. Anionic polymer chains dissociate at higher pH values, therefore anionic hydrogels display enhanced swelling in neutral to alkaline media. Since ampholytic hydrogels contain both acidic and basic groups in their structural repeating units, the charges are balanced at the isoelectric point of the polymer; where a change in the pH can change the overall ionic charge of the hydrogel [13]. The type of charge influences the selection of multivalent metal ions for physical crosslinking of the polymer chains, via electrostatic interactions, to form a 3D hydrogel network or a defined 3D-printed structure. Absent of charge, ionic hydrogels swell in aqueous medium due to water–polymer interactions [13].

7.3.5 Crosslinked Polymers

Hydrogels can be categorised according to the nature of crosslinking. The polymer chains in a hydrogel system can be crosslinked with each other: 1) chemically via covalent bonds between functional groups; or 2) physically via polymer chain entanglements or physical interactions such as ionic interactions, hydrogen bonds, or hydrophobic interactions [12, 14]. Covalent crosslinks form permanent junctions, whereas physical crosslinks form transient junctions which may be reversed upon the input of energy into the hydrogel system [14]. The mechanism by which polymers crosslink to form hydrogels is essential in the 3D-printing process for selecting a suitable printing technique that accommodates crosslinking of a gel solution either pre-printing, post-printing, or during printing to produce a defined hydrogel structure.

7.3.6 Method of Hydrogel Preparation

Synthetic polymers are classified according to their method of preparation, as they are man-made and the synthesis parameters during chemical polymerisation can be controlled. According to the method preparation, hydrogel materials can be classified into: 1) homopolymers; 2) copolymers; 3) multipolymers; and 4) interpenetrating networks (IPNs) [13].

Homopolymer hydrogels consist of a crosslinked polymer network of one type of hydrophilic monomer unit [4]. The structural framework of homopolymer hydrogels is defined by the polymerisation technique, nature of the monomer, and the crosslinker used [13]. Copolymer hydrogels consist of two different monomer units crosslinked to form a polymeric network, of which one monomer is hydrophilic to impart swelling properties to the hydrogel. Copolymeric hydrogels can be prepared by chemical crosslinking or physical crosslinking of the two monomer units. Chemical crosslinking occurs via polymerisation of both monomers using an initiator, multifunctional crosslinker, and a solvent, whereas physical crosslinking occurs by polyelectrolyte complexation, ionic interactions, polymer chain aggregations, or hydrogen bonding [13]. Multipolymer hydrogels consist of three or more monomer types crosslinked or polymerised to form a hydrogel network. By polymerising multiple co-monomer units, the properties of mechanical strength, biodegradability, and hydrogel chemical structure can be tailored [13].

Hydrogels forming IPNs consist of two interlaced polymer networks, without any polymer–polymer covalent linkages between the networks. Interpenetrating network hydrogels are formed by swelling of the first polymer network in the monomers of the second polymer, followed by crosslinking of the second polymer network so that an intermeshing network is formed [4]. Interpenetrating network hydrogels offer the advantages of high-order stability, improved hydrogel matrix stiffness and toughness, modifiable physical properties, and efficient drug-loading capacities compared to conventional hydrogels [13].

7.4 3D Printing Techniques for Hydrogels

3D-printing methods for biomedical applications can be classified according to their principal mechanism of operating (Figure 7.2); laser-based printing using the process of photopolymerisation, extrusion-based printing of polymers through a nozzle or orifice, and inkjet-based printing of material and binder using a printer [1]. Recent processes for hydrogel-based 3D printing utilises four types of techniques: 1) stereolithography; 2) laser-assisted bioprinting; 3) extrusion-based printing; and 4) inkjet-based printing [1].

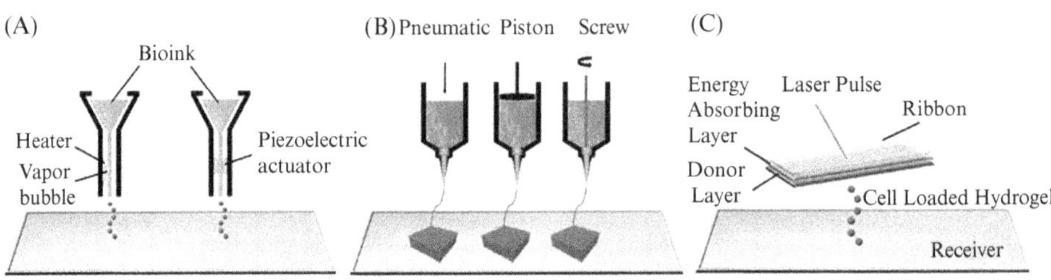

Figure 7.2 The three main operations of 3D printing for hydrogels: (A) inkjet-based printing; (B) extrusion-based printing; and (C) laser-based printing. (Source: reproduced with permission from reference [18] Copyright Creative Commons CC BY 4.0.)

7.4.1 Laser-Based 3D Printing

Laser-based systems for printing hydrogels (mainly stereolithography, two-photon polymerisation, and laser-induced forward transfer) involve the deposition of light energy in predetermined patterns, focused into a bed of liquid photo-crosslinkable polymeric resin to produce solid crosslinked hydrogel networks [19, 20]. A laser source, a reservoir of liquid photocurable resin, and a photo-initiator are required for this printing process. As long as they are miscible with the resin, the incorporation of drug or excipients do not require photopolymerisable functional groups as they become entrapped into the crosslinked matrix [20].

7.4.1.1 Stereolithography

In stereolithography, objects are solidified by scanning the surface of the photocurable material with UV light to produce a polymerised matrix through single photon absorption at the surface of the liquid material. 3D hydrogel constructs are achieved through solidification of the resin in a layer-by-layer process [6, 19]. Due to its poor penetrating ability, the UV light is absorbed within the first few micrometres by the liquid photocurable resin, thus achieving only microscale resolution with the stereolithography technique [21]. Therefore, μ-stereolithography was introduced to counter the limitations of conventional stereolithography from a resolution point-of-view [19]. Stereolithography may be used for 3D printing of oral drug-loaded hydrogel dosage forms (such as tablets) for controlled release. In such applications, the type and concentration of photo-initiator is essential to minimise toxicity when ingested. Pharmacologically non-toxic photo-initiators, such as riboflavin, is effective for fabricating dextran-methacrylate crosslinked hydrogels and stereolithography 3D-printed polyethylene glycol diacrylate crosslinked hydrogels [20, 22]. Although stereolithography could offer high printing accuracy, the photocuring process causes cell damage and limits its use for printing applications using multi-cellular bioinks [3].

7.4.1.2 Two-Photon Polymerisation

Two-photon polymerisation is an emerging 3D-printing technology that uses near-infrared femtosecond laser pulses to initiate polymerisation for the fabrication of high-resolution 3D constructs at both the micro- and nanoscales [6]. A high-resolution printability is critical for 3D hydrogels to simulate the native 3D microenvironment of tissues and to facilitate controlled drug release with optimal temporal and spatial distribution *in vitro* and *in vivo* [21]. Two-photon polymerisation is an advanced 3D-printing technique able to fabricate 3D hydrogels with a precise 3D configuration. Therefore, two-photon polymerisation technology has become an intensive area of research for the design and microfabrication of 3D hydrogels for applications in tissue engineering and drug delivery [21].

7.4.1.3 Laser-Induced Forward Transfer

The laser-induced forward transfer technique is a laser-assisted bioprinting method which consists of three main parts: 1) a pulsed laser source focused onto a thin layer of laser-absorbing material; 2) a donor substrate (ribbon); and 3) a receiving substrate for ejecting bioink droplets [4, 23]. The ribbon is a multilayer component comprised of a thin layer of transparent glass, quartz, or a laser-absorbing metal such as gold or titanium, and a suspended layer of bioink composed of hydrogel, cells, and bioactive agents [23]. When the laser beam pulses are focused onto the ribbon, the ink materials will be ejected from the ribbon and propelled towards the receiving substrate [4, 23]. This laser-assisted transfer of bioink has a high degree of precision and resolution (10–100 μm) for printing 2D and 3D patterns, using bioinks of high cell densities and viscosity without subjecting mechanical stress to the cells, [24]. The high resolution makes laser-assisted bioprinting a promising technique for hydrogels; however, the high cost of printing systems and lack of availability of commercial 3D laser bioprinters limit its explorative use in printing for clinical applications [3].

7.4.2 Extrusion-Based Printing

Extrusion-based printing relies on the application of pressure to create 3D patterns and constructs via computer-controlled layer-by-layer deposition. The material used as the bioink is in the form of molten or semi-molten polymers, polymer solutions, pastes, or dispersions [4]. The extrusion print head, which consists of a nozzle and a cartridge, moves via a computer-controlled process along three dimensions (*x*-, *y*-, and *z*-directions) [4]. The hydrogel ink, loaded into the cartridge, is extruded through a micro-sized nozzle or a microneedle in the form of a continuous filament onto a stationary platform via pneumatic, plunger, or screw-based pressure [25]. The accommodation of an extensive range of biomaterial types, room temperature processing, direct incorporation of cells into the bioink, and uniform cell distribution throughout the 3D-printed constructs are advantages of extrusion-based printing [25]. A crosslinking procedure is often required for extrusion-based techniques to maintain the shape of the 3D-printed hydrogel constructs, either during or after the printing process. The simplicity, predictability, and well-defined processing methods of extrusion-based techniques make it popular for hydrogel-based 3D printing [5, 6]. Extrusion-based bioprinting accommodates for the deposition of hydrogel bioinks with high cell densities, although the pressure during the extrusion process may affect cell viability [3].

7.4.3 Inkjet-Based Printing

The inkjet method of 3D printing is a non-contact technique akin to the conventional desktop inkjet printer. Inkjet printers can be divided into drop-on-demand or continuous ejection systems, where the droplet actuation mechanism is achieved through either a thermal or piezoelectric actuator head [4, 19, 25]. In the drop-on-demand ejection system, the actuator creates pulses to eject individual bioink droplets with predefined volumes [4]. In the continuous ejection system, continuous hydrogel ink is generated under pressure as a jet that can be further broken up into droplets where electrical signals control the direction of its movement [4]. Inkjet-based printing technology with thermal actuators use electrical pulses that generate bubbles to create pressure to eject hydrogel ink droplets onto a substrate, whereas a piezoelectric actuator generates pulses that create transient pressure for ejecting bioink droplets [25]. Inkjet printing has the advantage of high printing speed, low cost, and wide availability of printers; however, this technique is associated with varying cell encapsulation, and subjecting cells to thermal and mechanical stresses [3].

7.5 Printability and Printing Parameters

Printability is defined as the ability to form and maintain 3D constructs from bioink using any form of printing technique [26]. Printability affects the structure of the resulting 3D construct, its morphology, and mechanical and biological properties (Figure 7.3) [26, 27]. Extrusion based-printing favours the high encapsulation density of live cells

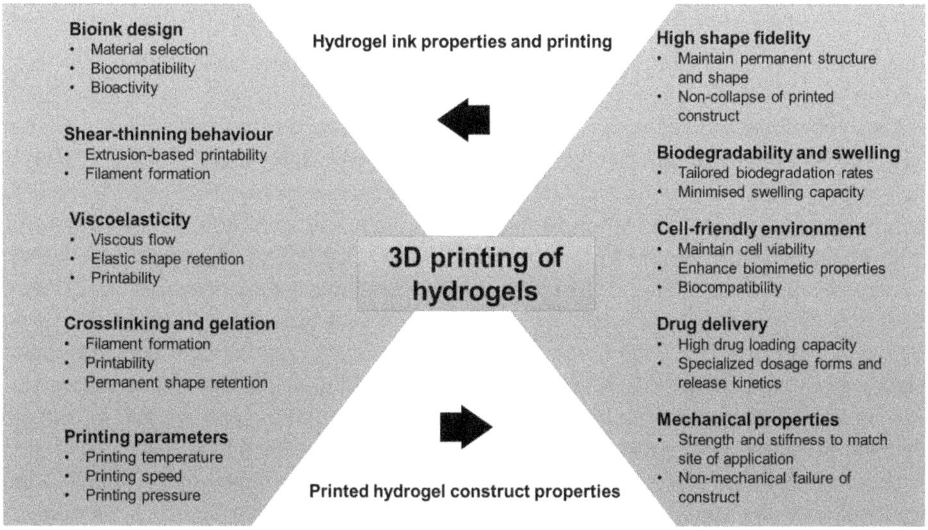

Figure 7.3 *Hydrogel ink properties that influence the final properties of the printed construct.*

and for this reason it is commonly used for 3D printing of hydrogel structures and therefore is discussed in the subsequent sections. In extrusion-based printing, hydrogels can be chemically or physically crosslinked to produce defined 3D structures. However, the crosslinking process is time-dependent and may result in the overflow and spreading of hydrogel causing it to deviate from its intended computer-aided design. Printability is an index used to measure the difference between the designed and the fabricated scaffold in the printing process [27]. Hydrogels of poor printability may damage encapsulated cells, collapse under the weight of previously deposited layers (which compromises the ability to stack subsequent layers of ink), or even fuse with those layers (which compromises the resolution and maintenance of the desired pore structure), thus failing to form an appropriate 3D structure [26, 28]. Therefore, 3D printability of a hydrogel material is defined as the ability of a hydrogel to form and maintain a reproducible 3D structure with dimensional integrity (also referred to as shape fidelity) [27]. The key categories and its factors affecting printability are: 1) bioink design; 2) crosslinking technique; and 3) printing parameters.

7.5.1 Bioink Design

The definition of a bioink has been proposed as 'a formulation of cells suitable for processing by an automated biofabrication technology that may also contain biologically active components and biomaterials' [29]. The term bioink refers to the cellular component present within the 3D-printed hydrogels, hence the prefix 'bio' that reflects the biological nature of the formulation. Polymer and hydrogel solutions that lack the cellular component are simply referred to as biomaterial inks [29]. This distinction is important as this biological cellular component significantly changes the hydrogel ink properties, its processing parameters, and 3D-printing conditions. The composition of the hydrogel bioink, its rheological properties, and its physical properties such as the contact angle, surface tension, and roughness should be considered for printability and cell encapsulation [26].

7.5.1.1 Materials Selection, Concentration and Viscosity

Polysaccharide-based and proteinaceous hydrogels are frequently used as bioinks because of their favourable effects on cell viability and interactions, particularly for applications in tissue engineering and regenerative medicine. Natural hydrogels (such as alginate, agarose, collagen, fibrin, gelatin, hyaluronic acid, and silk) are frequently used as bioink materials due to their structural similarity to native ECM. Hydrogel materials of collagen, fibrin, and gelatin contain inherent signalling molecules to promote cell adhesion. Natural hydrogels may be combined with synthetic materials to improve mechanical properties. Synthetic materials, which lack bioactive molecules, can be combined or functionalised with natural materials to impart hydrophilicity, bioactivity for cell interaction, and enzyme-mediated biodegradation [30]. Hydrophilicity of the ink, indicated by contact angle and surface tension, improves the biological properties of the hydrogel making it suitable for encapsulating cells and for promoting cellular interactions. The material of the hydrogel ink and the deposition surface of the printing substrate should also be considered, as a substrate of high surface energy compared to surface tension of the hydrogel will result in spreading of the ink, thus affecting printability and resolution [26]. To overcome ink spreading, the printing substrate or platform can be coated with a hydrophobic material to alter its wettability, provide a low-energy surface, and a higher contact angle of the ejected ink [31]. Furthermore, composition of the hydrogel ink determines its crosslinking technique, and this must be considered during material selection.

Hydrogel-based inks must have the ability to be extruded through narrow-sized nozzles or hollow needles in extrusion-based printing and maintain their shape stability after printing. The most important parameter defining the reproducibility and printability of hydrogels is their concentration and the accurate quantification of individual materials contained in a hydrogel mix [32]. The commonly used hydrogels for 3D printing include alginate, gelatin, and GelMA in concentrations of 2–4% w/v, 5% w/v, and 10% w/v, respectively, as these concentrations provide the optimal viscosity range for printing. Viscosity is described as the resistance of a fluid to flow under application of stress. Viscosity, determined by the molecular weight and concentration of the hydrogel solution, is another important parameter for printability and cell encapsulation efficiency [30, 32]. The greater the viscosity of a hydrogel ink, the greater the internal pressure (shear stress) of the hydrogel during extrusion, and the greater the potential for damaging suspended cells in the bioink. Therefore, hydrogels with shear-thinning properties are used to prevent cell damage by high shear stresses, to facilitate filament deposition during printing, and to maintain shape fidelity after printing [30, 32]. To obtain an optimal balance between good shape fidelity (harder hydrogels) and good printability (softer hydrogels), hydrogel composition, concentration, viscosity, and stiffness need to be modified [32].

7.5.1.2 Rheological Properties

The rheological properties of a hydrogel ink have the greatest influence on its printability. Rheology describes the deformation and flow properties of a material upon applied stress. A hydrogel ink undergoes several transitions in its rheological flow during the entire extrusion printing process, from its initial resting state pre-printing to its final resting state post-printing. During the printing process, the hydrogel ink is forced through the nozzle (increased shear stress), it takes a new shape, and reaches a new resting state from its original state (decreased shear stress) [30]. The main rheological properties that describe these transitions are: 1) viscosity; 2) viscoelastic shear moduli; 3) elastic recovery; and 4) yield stress [28]. Here, viscosity is defined as the ratio of shear stress to shear rate. Hydrogel fluids with a linear relationship (direct proportion) to shear stress and shear rate are termed Newtonian fluids, where viscosity remains constant despite application of shear force. Hydrogel fluids that diverge from linearity of shear stress to shear rate (inverse proportion) are termed non-Newtonian fluids, where viscosity can either increase or decrease upon the application of shear force.

7.5.1.3 Shear-Thinning

For extrusion printing of hydrogels, the ink should ideally be shear-thinning, exhibiting non-Newtonian behaviour, where the viscosity of the hydrogel fluid decreases upon increased application of shear stress. During extrusion, shear rate increases and so a hydrogel ink must decrease in viscosity to allow passage of the ink through the nozzle in the form of a filament. After extrusion, shear rate decreases resulting in increased viscosity of the deposited hydrogel ink and shape stability. A high zero-shear viscosity of the printed hydrogel structure assists with shape preservation to prevent its collapse post-printing [30]. In hydrogels, shear-thinning behaviour is attributed to shear-induced polymer disentanglement, whereas permanent shape stability is acquired by crosslinking post-printing [30]. Hybrid or composite hydrogel inks (such as nanocomposite hydrogels) containing colloidal dispersions of solid particles undergo shear-induced disruption of the interactions between solid particles. The high viscoelastic nature of colloidal hydrogel inks at rest is achieved through re-establishment of the interactions between the suspended solid particles to provide shape fidelity of the 3D-printed structure [30].

7.5.1.4 Viscoelasticity and Yield Stress

Hydrogel inks for extrusion printing must exhibit both flow and shape-retention properties. While passing through the nozzle, the hydrogel should flow easily with minimal internal resistance, especially for cell-suspended bioinks. After deposition of the hydrogel ink, its flow should cease immediately with instantaneous return of internal forces to resist deformation and retain shape elasticity [30]. Viscoelasticity is the property of viscous flow and elastic shape retention. It is measured in terms of storage (elastic) modulus (G') and loss (viscous) modulus (G''). Storage modulus (G') indicates the ability of a material to store energy elastically during deformation and is associated with elastic shape retention of a hydrogel. Loss modulus (G'') is a measure of the energy dissipated or lost and associated with viscous flow of a hydrogel. The shape retention property of 3D-printed hydrogel constructs is related to the yield stress, the stress at which the hydrogel material undergoes plastic deformation. Increased yield stress allows for improved filament formation and gel stiffness but may impede cell encapsulation efficiency. For shape fidelity and retention, the storage modulus (G') and yield stress correlate to the number of crosslinks and the degree of entanglements in the hydrogel ink network, as these interactions impart internal resistance to shape deformation [30].

7.5.1.5 Cell Encapsulation

Maintaining viability and proliferative ability of encapsulated cells within bioinks is well recognised. However, the impacts of cell encapsulation on the physico-chemical and physico-mechanical properties of the bioink is often overlooked, thus leading to challenges in the design of highly printable hydrogel bioinks. Cells within a bioink occupy a specific volume depending on the cell size and seeding density. The volume occupied by cells in a bioink impacts the viscoelastic properties, as cells have their intrinsic biomechanical properties that can affect printability and resolution of cell-laden hydrogels [33]. High cell densities can reduce the crosslinking efficiency of hydrogel bioinks by slowing the rate of gelation kinetics and physically obstructing the reactions between functional groups [30].

7.5.2 Crosslinking Techniques

Crosslinking is a critical parameter that determines the physico-chemical and mechanical properties of the resulting 3D-printed hydrogel construct and the viability of encapsulated cells. The crosslinking process solidifies the deposited hydrogel ink and ensures good shape fidelity and tuneable mechanical properties. Printability is affected by the type of crosslinking (physical, chemical, or photocrosslinking) and the degree of crosslinking. For improving printability, particularly in nozzle-based printing systems, the crosslinking and gelation kinetics should be precisely controlled to prevent nozzle blockages and irregularities in the 3D-printed constructs (Figure 7.4) [34]. Physically crosslinked hydrogels often produce mechanically weaker 3D-printed constructs than chemically crosslinked hydrogels but provide a cell-friendly environment for cell encapsulation. Thus, for hydrogel bioinks, balance needs to be achieved between printability, crosslinking degree, and cell viability. A lower crosslinking degree allows faster ink flowrates, whereas a higher crosslinking degree produces stiffer hydrogels that may quickly impede printability [34].

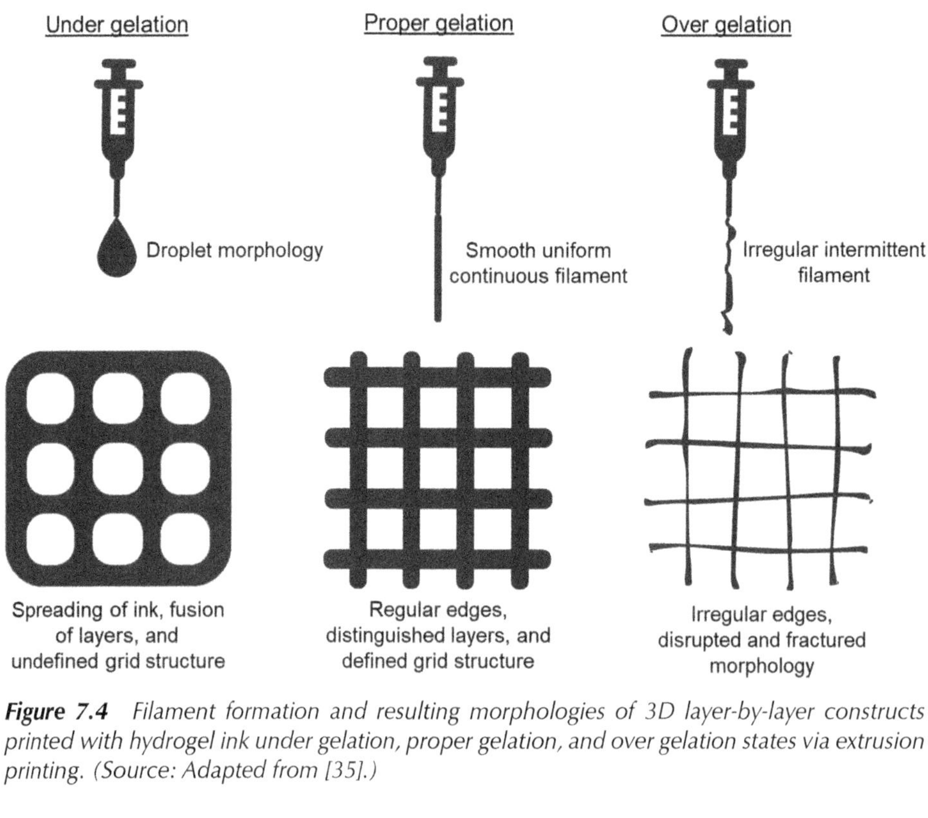

Figure 7.4 *Filament formation and resulting morphologies of 3D layer-by-layer constructs printed with hydrogel ink under gelation, proper gelation, and over gelation states via extrusion printing. (Source: Adapted from [35].)*

7.5.2.1 Thermal Crosslinking

Thermal crosslinking relies on the thermoresponsive behaviour of hydrogel inks to gel via heating or cooling of the material. It is a simple technique for inducing crosslinking; however, this technique may take longer than other crosslinking methods, the degree of crosslinking cannot be precisely controlled, and the application of heat or sudden transitions in temperature may negatively affect cell viability [34]. The effect of temperature-induced gelation, in terms of gelation time and density, is dependent on polymer concentration. The uncharged polymer chains of agarose forms random coils at 40 °C to transition into a fluid sol. Cooling the solution to 37 °C or lower initiates the formation of helical structures where the polymer chains form a 3D network to gel. The temperature setting on the 3D printers accommodates for the use of thermo responsive hydrogel inks.

7.5.2.2 Physical Ionic Crosslinking

Physical ionic crosslinking utilises a solution of divalent or multivalent ions, opposite to the ionic charge of the hydrogel ink, to induce crosslinks between the polymer chains resulting in gelation. Anionic polymers, such as alginate, are crosslinked with calcium ions from a solution of calcium chloride to induce gelation. Electrostatic interactions between two oppositely charged polymers may also result in crosslinking and gelation. There are three common ways of utilising ionic-crosslinkers to induce gelation of hydrogel inks: 1) bath-assisted printing; 2) aerosol spraying; and 3) pre-crosslinking [34]. Bath-assisted printing directly prints hydrogel ink into a solution of ionic-crosslinker. It provides rapid gelation, where the degree of crosslinking depends on the concentration of the ionic solution and the extend of exposure of the 3D-printed constructs to crosslinking. Aerosol spraying uses a fine mist of ionic-crosslinker solution directed on the printing nozzle for immediate gelation of the hydrogel ink as it exits the nozzle. Pre-crosslinking mixes in low concentrations of ionic-crosslinker with the hydrogel ink so that it becomes fully crosslinked after extrusion [34]. In nozzle-based printers, the degree of crosslinking and the resulting gelation of the hydrogel ink impacts printability in terms of filament formation and final layer-by-layer deposition of the 3D-printed hydrogels [35].

7.5.2.3 Chemical Crosslinking

A 3D network formation in chemically crosslinked hydrogels occur by irreversible covalent bonding between the chemical crosslinker molecule and functional groups on the polymer chains, where the crosslinker molecule is integrated into the material of the final 3D-printed construct. Chemical crosslinking can occur through various chemical reactions, such as Schiff base coupling, hydrazide-aldehyde coupling, Diels–Alder linkage, and azide-alkyne cycloaddition, initiated by heat or light [34]. The use of synthetic chemical crosslinkers, such as formaldehyde and glutaraldehyde, are less frequently used in 3D bioprinting of hydrogels bioinks due to their high cytotoxicity, despite their favourable mechanical properties on 3D-printed hydrogel constructs. The use of the natural non-toxic chemical crosslinker genipin is replacing the cytotoxic synthetic options in 3D printing of chitosan, gelatin, and collagen hydrogel inks [36, 37].

7.5.2.4 Photocrosslinking

Photocrosslinking requires a source of radiation (UV-light, visible light, or laser), a photo-curable resin, and a photo-initiator to induce chemical crosslinks [34]. When UV or visible light interacts with the photo-initiator, free radicals are generated to induce crosslinking and gelation in the hydrogel network via the following polymerisation mechanisms: 1) chain-growth mechanisms; 2) step-growth mechanisms; and 3) redox-based reactions [34, 38]. The printing techniques of extrusion, stereolithography, and laser-assisted printing are compatible with photocrosslinking. It is considered simple to operate, offering spatiotemporal and remote control, and rapid crosslinking processability at room temperature and physiological conditions [38, 39]. UV-light poses the risk of cell damage in bioinks and potential harm to the operator of the instrument; hence, visible light is preferred [39]. In addition, a high concentration of photo-initiator in the hydrogel ink may induce cellular toxicity, cell membrane disruption, and DNA damage to both encapsulated cells and host cells at the application site, particularly in the presence of free radicals generated by the photo-initiator [40, 41]. In the photocrosslinking technique for 3D printing of hydrogels, the selection of a minimally cytotoxic or non-cytotoxic photo-initiator is crucial for successful tissue engineering applications [41].

7.5.3 3D Printing Parameters

The printing parameters refer to the printer firmware settings that are controlled by the user. These parameters include temperature, printing speed, printing pressure, and nozzle size, which can all influence the printability and printing resolution of hydrogel inks [26, 32]. The printing parameters depend on the composition and rheology of the hydrogel ink. Therefore, the parameters must be precisely selected for each circumstance of 3D printing of hydrogel structures, where the ideal parameters may be identified via a process of trial and error.

7.5.3.1 Temperature

The two main components of extrusion-based printing that require temperature control are the cartridge temperature and the bed temperature [32]. The temperature setting of the cartridge (component where ink is loaded) and the bed (platform onto which ink is deposited) is used to facilitate the printing of hydrogel inks displaying temperature-dependent rheological properties, particularly thermoresponsive hydrogels. The cartridge and bed can be set to different temperatures to accommodate sol–gel phase transitions of thermoresponsive materials. For example, the cartridge can be set to 37–40 °C to induce a lower viscosity of alginate or gelatin inks, which then solidifies upon deposition onto a bed of lower temperature. Likewise, thermoresponsive inks that gel at physiological temperature can be loaded into a cartridge of 12–15 °C and deposited onto a heated bed of 37–40 °C to induce gelation [42]. The extrusion-based bioprinting process can also maintain normal cell culture incubation temperatures when the cartridge or bed is set to 37–40 °C, thus limiting cell shock from abrupt temperature changes [32].

7.5.3.2 Pressure

The printing pressure is a critical parameter to modulate the effect of high shear rate on the cell viability of bioinks. Excessive extrusion pressure may lead to poor printability and cell damage [26]. Sufficient pressure is required to initiate an easy, yet constant flow of the hydrogel ink through the nozzle and to ensure extrusion of a continuous filament in the layer-by-layer deposition of ink. The nozzle size is also related to printing pressure where increasing pressure is required for decreasing nozzle sizes, and thus the selection of nozzle size for enhanced printing resolution must be considered in accordance with the pressure setting [26].

7.5.3.3 Speed

The printing speed is the movement of the nozzle along the *x*–*y*-direction and is related to the total printing time [32]. The printing speed must be carefully adjusted to the hydrogel ink flowrate. At a constant flowrate of ink, lower printing speeds result in poor printability due to ejection of more ink per unit time, whereas higher printing speeds can lead to discontinuous filament formation and a disrupted layering of ink [26].

7.6 Clinical Translation

The full implementation of 3D printing in clinical practice requires certain safety, technological, regulatory, and manufacturing issues to be addressed, especially regarding on-demand personalised 3D printing of pharmaceutical products. These issues pertain to 3D printing of medicines in general, but also apply to 3D printing of hydrogels for applications in tissue engineering, wound healing dressings, and pharmaceutical drug delivery dosage forms.

7.6.1 Regulatory Considerations

For 3D hydrogels to be accepted for pharmaceutical and clinical use, detailed regulations are required to manage concerns of safety, efficacy, intellectual property, ethics, and patient rights and privacy [43]. Currently there is a lack of such specific guidance and regulations. Unlike mass production of medicines, 3D printing intended for personalised medicine will require new regulations regarding its manufacture and production in pharmacies, hospitals, doctors' offices, or perhaps the patients' homes [43]. The regulatory process of 3D-printed hydrogels should encompass three parts: 1) the type of 3D printer; 2) the composition of hydrogel ink; and 3) the finished 3D-printed construct [44]. The major concern is quality control relating to safety and efficacy. Regulations and guidance, incorporating good manufacturing practice, are required for standardising of the following: 1) printer specifications and printing techniques; 2) critical printing parameters; 3) the availability and use of biodegradable and biocompatible hydrogel inks (non-living and living materials); 4) sterilisation processes for implantable tissue engineering hydrogel constructs, sterile wound dressings, and viability of encapsulated cells; 5) assessment protocols for validating hydrogel properties and confirming *in vitro*/*in vivo* performance; and 6) approval for commercialisation of 3D-printed products [11, 44–46].

7.6.2 Manufacturing Considerations

Hydrogel-based 3D printing should comply with good manufacturing practices, whether large-scale or small-scale. A large-scale hydrogel printing setup may be more complex compared to small-scale on-demand printing in a pharmacy or hospital. Scale-up manufacturing will need to facilitate rapid change in printing parameters, hydrogel ink, and machine parts within a single run [44]. Currently, the large-scale production of 3D-printed hydrogels is far-off from reality because of the lack of ready-made printing equipment that is specifically designed for printing hydrogel inks of varying physico-mechanical properties, rheological properties, crosslinking mechanisms, and high drug loading, including components for printing living cells and thermolabile or water-sensitive drugs within hydrogels [43, 44]. The availability of several printing technologies may produce significant differences in the design, mechanical attributes, quality, reproducibility, and overall finish of 3D-printed hydrogels. To reduce end-product variability in quality, a quality-by-design (QbD) approach is needed to optimise quality control through understanding the critical material attributes (CMAs), critical process parameters (CPPs), as well as in-process and finished product testing parameters for assuring quality consistency [46].

7.6.3 Limitations and Future Direction

The main limitations in 3D printing of hydrogels are the low printing speeds and low printing resolution to facilitate the on-demand printing of complex biomimetic tissues and devices for clinical applications. The current hydrogel-based 3D-bioprinting systems cannot fully replicate the finer architectures, complexities, and functionalities of natural tissues and organs [47]. The low printing speed will lead to time constraints in fabricating products of complex designs, refined finishes, and high reproducibility [44, 45, 48]. Printing technology capable of printing with multiple compositions of hydrogel inks and bioinks will advance 3D-printed tissue engineering applications. The growing emergence of bioprinting companies indicates positive prospects for newer and improved printing technologies for biomaterials, hydrogel inks, and pharmaceutical active ingredients. The use of multiple materials within a single construct will improve the biomimetic properties and range of functionalities of 3D-printed structures [47]. To achieve the clinical translation of hydrogel-based 3D printing, proper technological setups, training of personnel to identify product defects and quality, and regulations overseeing the production process is required. To assist with in-house quality control measures, non-destructive spectroscopic analytical methods (such as near infrared and Raman spectroscopy) could be used to confirm solid-state characteristics, composition, and drug distribution of 3D-printed hydrogels, whereas textural analysis could ensure appropriate mechanical properties, before dispensing the product to the patient [11, 45].

7.7 Conclusions

The combination of selecting a hydrogel printing technique, the design of hydrogel inks and bioinks, and the process of crosslinking determines the success of printing viable 3D hydrogel constructs for tissue engineering applications, wound healing dressings, and pharmaceutical dosage forms. An important aspect for achieving high printability is proper filament formation for the precise deposition of hydrogel ink and the production of defined structures of excellent shape fidelity. Printability is largely dependent on the composition and rheological properties of the hydrogel ink to improve its compatibility and cell processing abilities with the currently available array of 3D-printing techniques and equipment. This chapter overviewed the various printing techniques and the factors affecting printability of hydrogel inks and outlined the main limitations to scale-up manufacturing and clinical translation. Large-scale manufacture of 3D-printed hydrogels may not be possible yet, but the prospect of small-scale on-demand 3D printing of hydrogels for customised wound dressings and paediatric-friendly or geriatric-friendly dosage forms is likely. To fully exploit the potential of this unique fabrication technique, regulations and guidelines need be devised to oversee product quality, safety, and efficacy of 3D printing in clinical practice. In addition, regulatory agencies need to be prepared in terms of understanding the technological details and pathways for product review and approval for commercialisation [46].

References

[1] Unagolla, J.M. and Jayasuriya, A.C. (2020). Hydrogel-based 3D bioprinting: a comprehensive review on cell-laden hydrogels, bioink formulations, and future perspectives. *Applied Materials Today* **18**: 100479.

[2] Kessel, B., Lee, M., Bonato, A. et al. (2020). 3D bioprinting of macroporous materials based on entangled hydrogel microstrands. *Advanced Science* **7** (18): 2001419.

[3] He, Y., Yang, F., Zhao, H. et al. (2016). Research on the printability of hydrogels in 3D bioprinting. *Scientific Reports* **6** (1): 29977.

[4] Li, J., Wu, C., Chu, P.K., and Gelinsky, M. (2020). 3D printing of hydrogels: rational design strategies and emerging biomedical applications. *Materials Science and Engineering: R: Reports* **140**: 100543.

[5] Advincula, R.C., Dizon, J.R.C., Caldona, E.B. et al. (2021). On the progress of 3D-printed hydrogels for tissue engineering. *MRS Communications* **11** (5): 539–553.

[6] Soleymani Eil Bakhtiari, S., Bakhsheshi-Rad, H.R., Karbasi, S. et al. (2021). 3-dimensional printing of hydrogel-based nanocomposites: a comprehensive review on the technology description, properties, and applications. *Advanced Engineering Materials* **23** (10): 2100477.

[7] Long, J., Etxeberria, A.E., Nand, A.V. et al. (2019). A 3D printed chitosan-pectin hydrogel wound dressing for lidocaine hydrochloride delivery. *Materials Science and Engineering: C* **104**: 109873.

[8] Tsegay, F., Elsherif, M., and Butt, H. (2022). Smart 3D printed hydrogel skin wound bandages: a review. *Polymers* **14** (5): 1012.

[9] Cheng, Y., Qin, H., Acevedo, N.C. et al. (2020). 3D printing of extended-release tablets of theophylline using hydroxypropyl methylcellulose (HPMC) hydrogels. *International Journal of Pharmaceutics* **591**: 119983.

[10] Acosta-Vélez, G.F. and Wu, B.M. (2016). 3D pharming: direct printing of personalized pharmaceutical tablets. *Polym Science* **2** (1): 11.

[11] Seoane-Viaño, I., Januskaite, P., Alvarez-Lorenzo, C. et al. (2021). Semi-solid extrusion 3D printing in drug delivery and biomedicine: personalised solutions for healthcare challenges. *Journal of Controlled Release* **332**: 367–389.

[12] Madduma-Bandarage, U.S.K. and Madihally, S.V. (2021). Synthetic hydrogels: synthesis, novel trends, and applications. *Journal of Applied Polymer Science* **138** (19): 50376.

[13] Singhal, R. and Gupta, K. (2016). A review: tailor-made hydrogel structures (classifications and synthesis parameters). *Polymer-Plastics Technology and Engineering* **55** (1): 54–70.

[14] Lin, C.C. and Metters, A.T. (2006). Hydrogels in controlled release formulations: network design and mathematical modeling. *Advanced Drug Delivery Reviews* **58** (12–13): 1379–1408.

[15] Chen, F.M. and Liu, X. (2016). Advancing biomaterials of human origin for tissue engineering. *Progress in Polymer Science* **53**: 86–168.

[16] Saroia, J., Yanen, W., Wei, Q. et al. (2018). A review on biocompatibility nature of hydrogels with 3D printing techniques, tissue engineering application and its future prospective. *Bio-Design and Manufacturing* **1** (4): 265–279.

[17] Stratton, S., Shelke, N.B., Hoshino, K. et al. (2016). Bioactive polymeric scaffolds for tissue engineering. *Bioactive Materials* **1** (2): 93–108.

[18] Tamay, D.G., Dursun Usal, T., Alagoz, A.S. et al. (2019). 3D and 4D printing of polymers for tissue engineering applications. *Frontiers in Bioengineering and Biotechnology* **7**: 164.

[19] Billiet, T., Vandenhaute, M., Schelfhout, J. et al. (2012). A review of trends and limitations in hydrogel-rapid prototyping for tissue engineering. *Biomaterials* **33** (26): 6020–6041.

[20] Martinez, P.R., Goyanes, A., Basit, A.W., and Gaisford, S. (2017). Fabrication of drug-loaded hydrogels with stereolithographic 3D printing. *International Journal of Pharmaceutics* **532** (1): 313–317.
[21] Xing, J.F., Zheng, M.L., and Duan, X.M. (2015). Two-photon polymerization microfabrication of hydrogels: an advanced 3D printing technology for tissue engineering and drug delivery. *Chemical Society Reviews* **44** (15): 5031–5039.
[22] Hee, K.S. and Chu, C.C. (2009). Visible light induced dextran-methacrylate hydrogel formation using (−)-riboflavin vitamin B_2 as a photoinitiator and *L*-arginine as a co-initiator. *Fibers and Polymers* **10** (1): 14–20.
[23] Ventura, R.D. (2021). An overview of laser-assisted bioprinting (LAB) in tissue engineering applications. *Medical Laser: Engineering, Basic Research, and Clinical Application* **10** (2): 76–81.
[24] Hopp, B., Smausz, T., Kresz, N. et al. (2005). Survival and proliferative ability of various living cell types after laser-induced forward transfer. *Tissue Engineering* **11** (11–12): 1817–1823.
[25] Li, J., Chen, M., Fan, X., and Zhou, H. (2016). Recent advances in bioprinting techniques: approaches, applications and future prospects. *Journal of Translational Medicine* **14** (1): 271.
[26] Naghieh, S. and Chen, X. (2021). Printability: a key issue in extrusion-based bioprinting. *Journal of Pharmaceutical Analysis* **11** (5): 564–579.
[27] Naghieh, S., Sarker, M., Sharma, N.K. et al. (2019). Printability of 3D printed hydrogel scaffolds: influence of hydrogel composition and printing parameters. *Applied Sciences* **10** (1): 292.
[28] Ribeiro, A., Blokzijl, M.M., Levato, R. et al. (2017). Assessing bioink shape fidelity to aid material development in 3D bioprinting. *Biofabrication* **10** (1): 014102.
[29] Groll, J., Burdick, J.A., Cho, D.W. et al. (2018). A definition of bioinks and their distinction from biomaterial inks. *Biofabrication* **11** (1): 013001.
[30] Schwab, A., Levato, R., D'Este, M. et al. (2020). Printability and shape fidelity of bioinks in 3D bioprinting. *Chemical Reviews* **120** (19): 11028–11055.
[31] Vafaei, S., Tuck, C., Ashcroft, I., and Wildman, R. (2016). Surface microstructuring to modify wettability for 3D printing of nano-filled inks. *Chemical Engineering Research and Design* **109**: 414–420.
[32] Mancha Sánchez, E., Gómez-Blanco, J.C., López Nieto, E. et al. (2020). Hydrogels for bioprinting: a systematic review of hydrogels synthesis, bioprinting parameters, and bioprinted structures behavior. *Frontiers in Bioengineering and Biotechnology* **8**: 776.
[33] Schwartz, R., Malpica, M., Thompson, G.L., and Miri, A.K. (2020). Cell encapsulation in gelatin bioink impairs 3D bioprinting resolution. *Journal of the Mechanical Behaviour of Biomedical Materials* **103**: 103524.
[34] GhavamiNejad, A., Ashammakhi, N., Wu, X.Y., and Khademhosseini, A. (2020). Crosslinking strategies for 3D bioprinting of polymeric hydrogels. *Small* **16** (35): 2002931.
[35] Ouyang, L., Yao, R., Zhao, Y., and Sun, W. (2016). Effect of bioink properties on printability and cell viability for 3D bioplotting of embryonic stem cells. *Biofabrication* **8** (3): 035020.
[36] Rajabi, M., McConnell, M., Cabral, J., and Ali, M.A. (2021). Chitosan hydrogels in 3D printing for biomedical applications. *Carbohydrate Polymers* **260**: 117768.
[37] Yu, Y., Xu, S., Li, S., and Pan, H. (2021). Genipin-cross-linked hydrogels based on biomaterials for drug delivery: a review. *Biomaterial Science* **9** (5): 1583–1597.
[38] Qin, X.H., Ovsianikov, A., Stampfl, J., and Liska, R. (2014). Additive manufacturing of photosensitive hydrogels for tissue engineering applications. *BioNanoMaterials* **15** (3–4): 49–70.
[39] Zheng, Z., Eglin, D., Alini, M. et al. (2021). Visible light-induced 3D bioprinting technologies and corresponding bioink materials for tissue engineering: a review. *Engineering* **7** (7): 966–978.
[40] Williams, C.G., Malik, A.N., Kim, T.K. et al. (2005). Variable cytocompatibility of six cell lines with photoinitiators used for polymerizing hydrogels and cell encapsulation. *Biomaterials* **26** (11): 1211–1218.
[41] Choi, J.R., Yong, K.W., Choi, J.Y., and Cowie, A.C. (2019). Recent advances in photo-crosslinkable hydrogels for biomedical applications. *BioTechniques* **66** (1): 40–53.
[42] Celikkin, N., Simó Padial, J., Costantini, M. et al. (2018). 3D printing of thermoresponsive polyisocyanide (PIC) hydrogels as bioink and fugitive material for tissue engineering. *Polymers* **10** (5): 555.

[43] Chen, G., Xu, Y., Chi Lip Kwok, P., and Kang, L. (2020). Pharmaceutical applications of 3D printing. *Additive Manufacturing* **34**: 101209.
[44] Parhi, R. (2021). A review of three-dimensional printing for pharmaceutical applications: quality control, risk assessment and future perspectives. *Journal of Drug Delivery and Science Technology* **64**: 102571.
[45] Lamichhane, S., Bashyal, S., Keum, T. et al. (2019). Complex formulations, simple techniques: can 3D printing technology be the Midas touch in pharmaceutical industry? *Asian Journal of Pharmaceutical Sciences* **14** (5): 465–479.
[46] Beg, S., Almalki, W.H., Malik, A. et al. (2020). 3D printing for drug delivery and biomedical applications. *Drug Discovery Today* **25** (9): 1668–1681.
[47] Heinrich, M.A., Liu, W., Jimenez, A. et al. (2019). 3D bioprinting: from benches to translational applications. *Small* **15** (23): 1805510.
[48] Parulski, C., Jennotte, O., Lechanteur, A., and Evrard, B. (2021). Challenges of fused deposition modeling 3D printing in pharmaceutical applications: where are we now? *Advanced Drug Delivery Reviews* **175**: 113810.

8

Analytical Characterisation of 3D-Printed Medicines

Ana Luiza Lima[1], Lívia L. Sá-Barreto[2] and Marcilio Cunha-Filho[1]

[1] *Laboratory of Food, Drugs, and Cosmetics (LTMAC), School of Health Sciences, University of Brasilia, Brasília, Brazil*

[2] *Faculty of Ceilandia, University of Brasilia (UnB), Brasília, Brazil*

8.1 Introduction

Since the industrial revolution of the eighteenth century, no technological advance has caused as much anticipation in the pharmaceutical sector as the use of additive manufacturing to produce medicines. At the time, the jump in the production scale resulted in unexpected stability problems and quality deviations that could only be overcome with sophisticated analytical instrumentation that notably emerged in the 1950s, based on spectroscopy and chromatography [1].

In less than a decade, additive manufacturing, also called 3D printing, has demonstrated that it can revolutionise the pharmaceutical segment by producing personalised medicines for various delivery routes with different drug dosages and release characteristics, in addition to benefits in cost and easy operation [2, 3]. In particular, such technology makes it possible to unlimitedly expand the existing types of dosage forms, which were constrained by the limitations of conventional industrial production [4]. However, for this prediction to be fulfilled and the full potential of 3D technology to be properly exploited, it is necessary, once again, to resort to cutting-edge analytical techniques that can guarantee the quality parameters expected for a pharmaceutical product.

3D Printing of Pharmaceutical and Drug Delivery Devices: Progress from Bench to Bedside, First Edition.
Edited by Dimitrios A. Lamprou, Dennis Douroumis and Sheng Qi.

In this innovative scenario, it is crucial to consider that additive manufacturing encompasses a wide variety of technologies in which new materials and processing conditions, including stress sources, such as high temperature, light, and shear [5], challenge the pharmaceutical quality standards in place. According to the American Society for Testing and Materials (ASTM), the most promising techniques for pharmaceuticals are listed in Figure 8.1, along with the most common materials and the processing stress source [6, 7].

Thus, to guide the development of 3D pharmaceuticals, it is important to properly operate, interpret, and apply appropriate analytical techniques considering each personalised drug product. More than ever, the quality of pharmaceutical products made by 3D printing needs to be built through clear analytical protocols, from the choice of materials that will constitute the printed dosage form by preformulation studies, through in-process controls that adequately measure the intermediate products generated from the still partial production process, and finally, the use of analytical characterisations adapted to the final products that can be of an unimaginable number [8–10]. Figure 8.2 illustrates the characterisation strategy required for the design and production of 3D pharmaceuticals.

Hence, in the current stage of still preliminary development of 3D printing of pharmaceuticals, one of the main challenges is to ingeniously combine the conventional analytical tools used in pharmaceutical production with the instrumentation typically used in

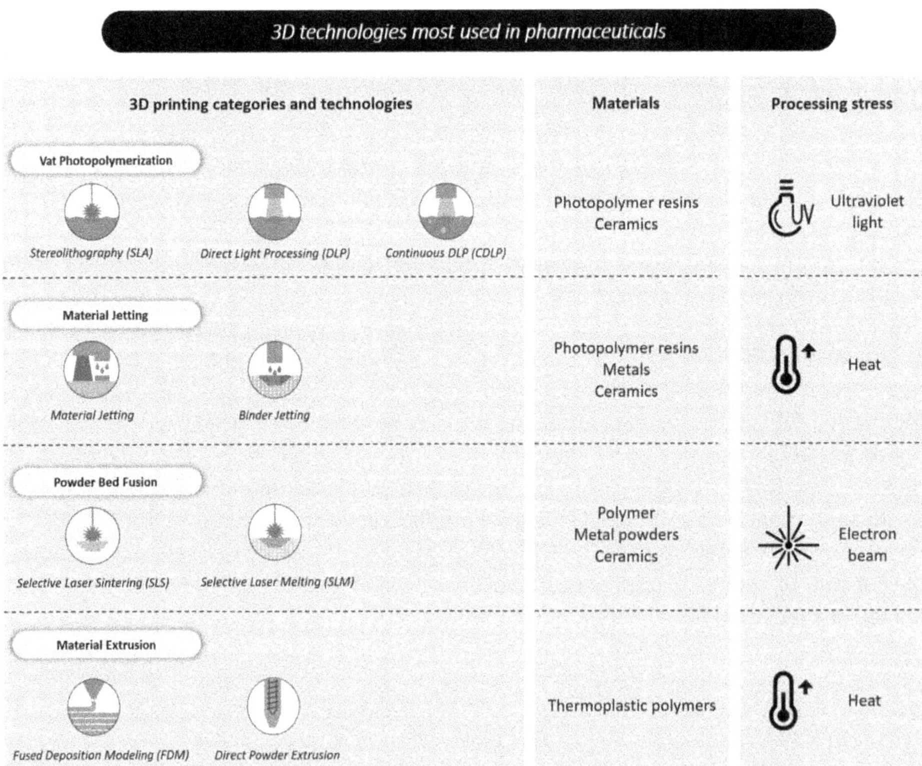

Figure 8.1 *3D-printing categories and the respective technologies most used to develop pharmaceuticals, along with the most explored materials and the processing stress source.*

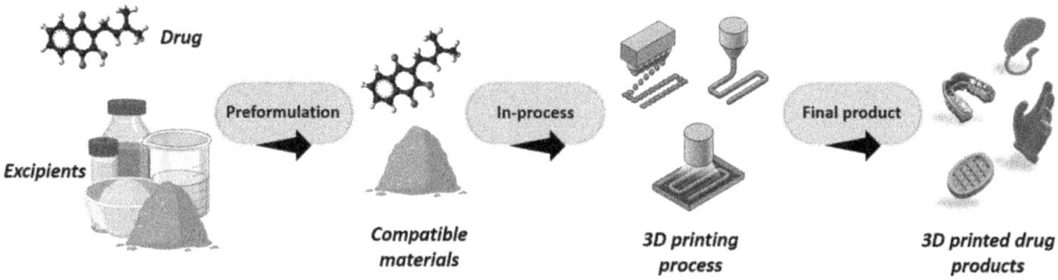

Figure 8.2 Schematic illustration for designing and developing 3D pharmaceutical devices, including the steps of preformulation, in-process control, and final product characterisation.

evaluating 3D objects in different areas. Notably, the developers should keep in mind the necessary adaptation to the printing technology and the drug product to be developed.

8.2 Preformulation

Preformulation studies are the first stage for pharmaceuticals development. At this phase, an in-depth evaluation of the drug–excipient interactions is performed to assess their compatibility and feasibility [11]. An adequate choice of formulation components at this early stage can be decisive in obtaining a drug product that meets the expected performance and health agencies' requirements. For this, instrumentation capable of measuring the physical-chemical aspects of the samples must be used. Especially in the case of 3D products, analytical techniques sensitive to changes in certain physical properties can provide valuable insights into the miscibility of materials and even their printability.

8.2.1 Thermal Analysis

Thermal analysis includes several techniques in which an analytical signal is measured as a function of a controlled temperature programme. According to the thermal events observed, it is possible to monitor physico-chemical interactions and potential instabilities between drug and excipients, which will guide formulations' components screening [12].

The most common analytical instrumentation responsible for thermal analysis in a preformulation protocol includes differential scanning calorimetry (DSC), differential thermal analysis (DTA), and thermogravimetric analysis (TGA) [13, 14]. DSC and DTA are based on the measurement of the samples' heat flow and can identify physical transitions such as melting, crystallisation, and glass transition (Figure 8.3A). Such analyses are of particular interest for materials used in 3D printing, in most cases, polymers and resins used to form the 3D matrix [15].

The DSC allows more accurate measurements than DTA, as it works with independent heating systems between the test sample and the reference [16, 17]. The greater precision of this instrument can be particularly decisive in the evaluation of drug–polymer interactions in 3D-printed systems. In fact, depending on the printing conditions and the proportion of the components, subliminal variations in the glass transition perceived by the DSC could lead to important repercussions on printability.

Moreover, the crystallinity of the drug evaluated in the presence of each excipient obtained by the DSC can also play an important role in preformulation evaluations

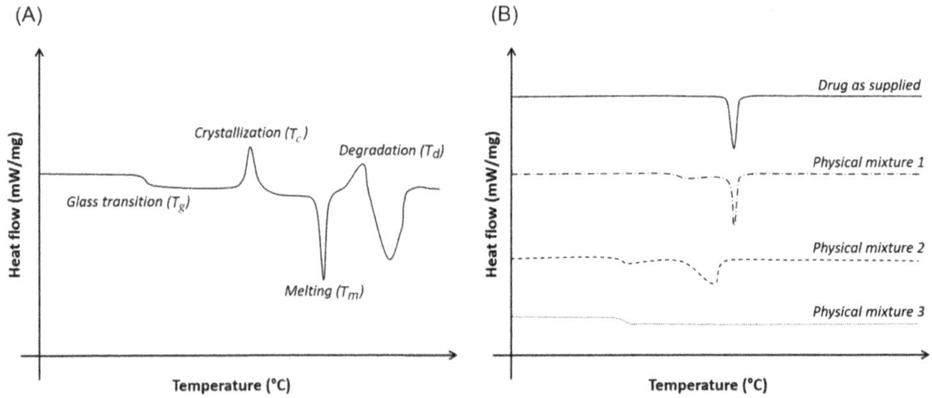

Figure 8.3 *Hypothetical DSC curves illustrating: (A) the possible thermal events that can be measured in this analytical technique, such as glass transition (T_g), a kinetic phenomenon in which the amorphous regions change from a hard glassy state with limited molecular motion to a flexible soft state with intense molecular motion at a given temperature region; crystallisation temperature (T_c), thermal zone in which heat is released from the sample and the molecular chains have sufficient energy to form ordered arrangements and create a crystal structure; melting temperature (T_m), thermal zone in which heat is absorved from the sample to liquefy the solid structure; degradation temperature (T_d), thermal zone in which heat is used to permanently break chemical bonds, destructuring the original molecule. (B) Some of the responses that can be found in a preformulation test for drug–polymer mixtures, wherein physical mixture 1 represents no change in melting peak or polymer T_g, indicating no interaction between components (18); physical mixture 2 exemplifies anticipation and decrease in melting enthalpy along with the reduction in polymer T_g, indicating a partial dispersion of the drug (19); physical mixture 3 represents the disappearance of the melting event and the reduction in polymer T_g, indicating total drug solubilisation (18).*

involving materials for 3D printing. Thereby, the changes observed in the drug's melting peak are related to some interaction degree between the components. Particularly, anticipation or decrease in the melting peak is related to its interaction with the formulation components, which may occur due to partial or total solubilisation of the drug in the pharmaceutical vehicle, as schematically represented in Figure 8.3B [18, 19].

Most 3D pharmaceutical products are designed with few excipients and a considerable amount of an organic matrix; thus, monitoring drug solubilisation is essential to ensure a final product with adequate pharmaceutical performance [20]. Additionally, in 3D techniques that explore curing or high temperatures (e.g., vat photopolymerisation, material jetting, a material extrusion), changes in polymer's glass transition must be monitored in polymer–drug mixtures in order to assist not only the choice of the material but also the processing conditions.

TGA, in turn, is a technique used to assess the decomposition profile of the drug and excipients based on accurate mass loss measurements during such deleterious events. The decomposition profile of a compound under certain analysis conditions occurs according to reproducible events. Consequently, changes in these weight loss patterns, which anticipate or intensify decomposition kinetics, can suggest loss of stability and incompatibilities. Indeed, as observed in DSC, changes in the TGA curve are associated with component interaction. Both thermal responses can indicate some sign of incompatibility or may produce unstable 3D products. The thermal simulations of such analyses are especially realistic for 3D-printing technologies that exploit high temperatures, such as Fused deposition modelling (FDM), Stereolithography (SLA), and Digital light processing (DLP) [21, 22]. Establishing analysis programmes that can accurately simulate the particular printing conditions can be the key to better taking advantage of these analytical tools.

Figure 8.4 shows hypothetical TGA curves and their first derivative, which enables the visualisation of all decomposition phases. Therefore, an adequate response for a preformulation study of 3D pharmaceuticals would be the alteration of the decomposition profile, indicative of interaction but without degradation in the heat zone chosen for processing. The chart highlights the usual temperature range for 3D printing by extrusion material (150–250 °C) [23].

In fact, small percentages of mass loss in slow decomposition kinetics are acceptable (physical mixture 1 and 2; Figure 8.4). As TGA promotes an accelerated degradation, a conventional exposure to the sample's initial decomposition temperature will not necessarily compromise the material's thermal stability, mainly in cases of slow kinetic degradation and reduced processing time, as in 3D printing. On the other hand, interactions that lead to intense mass loss at lower temperatures are not stable to fit the thermal energy power source required in additive manufacturing (physical mixture 3; Figure 8.4) [10, 24].

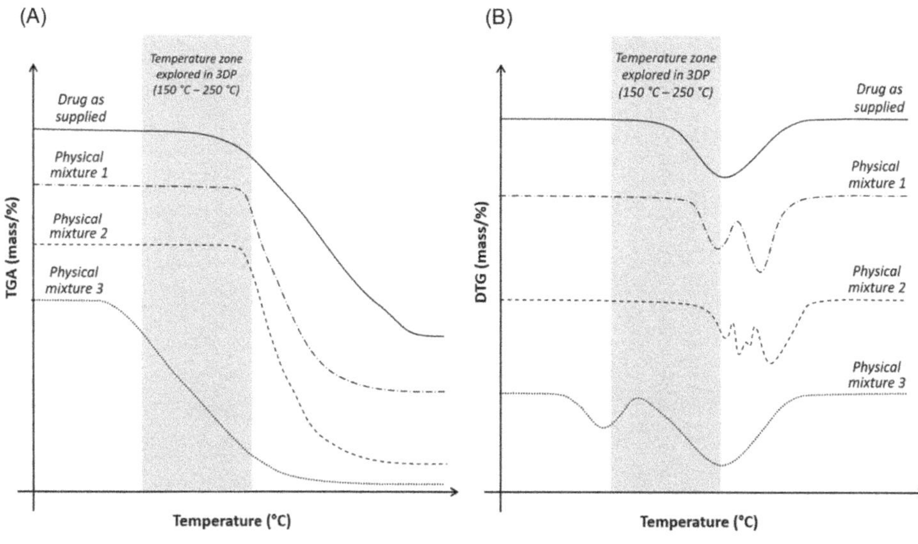

Figure 8.4 Hypothetical TGA curves illustrating (A) possible decomposition profiles and (B) its first derivative (DTG) showing the peaks corresponding to the different phases of drugs' decomposition and three different physical mixtures. The shaded area indicates the highest temperatures used in 3D printing (3DP). Physical mixture 1 represents two decomposition phases, in which the sum of two degradation events is observed, indicating little interaction between the components or interactions that do not affect the thermal degradation profile. Physical mixture 2 exemplifies several decomposition steps, showing that the drug–excipient interaction modified the materials degradation kinetics. Physical mixture 3 represents an interaction between the components capable of drastically modifying the decomposition profile of the materials, anticipating and accelerating their degradation.

8.2.2 X-Ray Powder Diffraction (XRPD)

The evaluation of the crystallinity of a drug in the 3D-printed object can be decisive for its correct functioning; in this sense, X-ray powder diffraction (XRPD) is a simple and highly recommended technique. The principle of this analytical instrumentation is based on the orderly and patterned arrangement of the atomic structure of crystals. Thereby, an X-ray beam impinges the solid sample at specific angles, and then they are scattered according to the organisation of electrons in the atoms, defining the diffraction patterns [25]. This signal is converted into peaks, where its position is used to determine the size and shape of the crystal, while the intensity of the diffraction peak determines the atomic position within the crystal [26].

One of the great advantages of producing pharmaceuticals by 3D printing is the ability of this technology to render and maintain the amorphous profile of the drug, conditioning its release to the embedded matrix [27]. In this way, XRPD should be applied in preformulation studies of 3D pharmaceutical dosages in order to anticipate the amorphisation of the drug and even to check the possible formation of new crystalline phases [24, 28].

The crystalline profile commonly observed in drugs is shown in Figure 8.5. The disappearance of characteristic peaks of the original crystal and the appearance of new signals

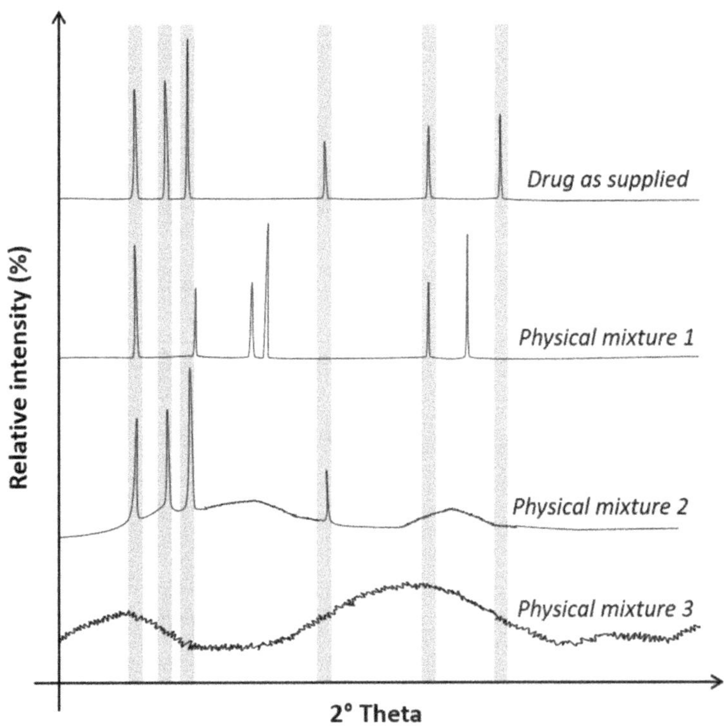

Figure 8.5 Hypothetical XRPD diffractograms of the drug and three different physical mixtures. The characteristic peaks of the drug original crystal are shaded. Physical mixture 1 describes the disappearance of peaks corresponding to drug crystal structure and the appearance of new peaks different from its original crystalline form, suggesting the formation of a new polymorph of the drug. Physical mixture 2 represents the presence of some background noise with the disappearance of some peaks, indicating partial amorphisation of the drug. Physical mixture 3 exemplifies the disappearance of all peaks corresponding to the drug's crystalline form and an intense baseline noise, indicating complete amorphisation of the drug.

on the diffractogram are unmistakable signs of changes in the crystal organisation, i.e., polymorphism (physical mixture 1; Figure 8.5) [29]. Moreover, changes in baseline, such as the intense presence of noise with non-peak identification, indicate a partially or entirely drug amorphisation (physical mixtures 2 and 3; Figure 8.5) [28, 30].

Therefore, this technique should be explored together with thermal analysis to identify pharmaceutical matrices and excipients that can amorphise the drug and do not trigger the appearance of polymorphs. When observed in physical mixtures, this response is beneficial to developing 3D pharmaceuticals, since additive manufacturing processing tends to intensify the drug–excipient interactions, stabilising and enhancing the previously observed amorphous character [31]. This was followed, for example, with indomethacin, which remained amorphous in the polymer Kollidon® VA 64 after processing by SLS 3D printing [27] and with griseofulvin that even with the intense stress of FDM 3D printing remained amorphous in tablets produced using hydroxypropyl cellulose [32].

8.2.3 Infrared Spectroscopy

The analytical techniques based on vibrational spectroscopy in the infrared region include near-infrared (NIR) and mid-infrared (MIR). Particularly, MIR spectrometry, which uses lower wavelengths (400–4,000 cm^{-1}), can recognise hydrophilic and hydrophobic functional groups. Notably, the region of 900–1,500 cm^{-1} is ideal for identifying the chemical structure of drugs [33], making this technique one of the most demanded analytical tests for a preformulation study [34, 35].

The identification of functional groups is based on detecting interatomic chemical bonds. First, infrared incident light passes through the sample, which triggers the vibration of specific molecular bonds due to infrared energy absorption. Then, this signal is transmitted to a detector and is converted into a spectrum (Figure 8.6) [36]. A mathematical model is required to decode the analytical signal. The most used procedure for this purpose is Fourier transformation [37, 38].

Therefore, FTIR is a straightforward methodology to be applied as a compatibility screening tool, since the vibrational alterations detected by this method allow the identification of possible intermolecular interactions between the formulation components. The reduction in peak intensity or the appearance or disappearance of some peaks indicates interactions between the drug and the excipient [33, 37].

Figure 8.7 shows hypothetical FTIR spectra in which alterations in the peaks correspond to the characteristic functional groups of the molecule, signaling the chemical interaction between the components. Moreover, based on this analysis, not only polymer–drug

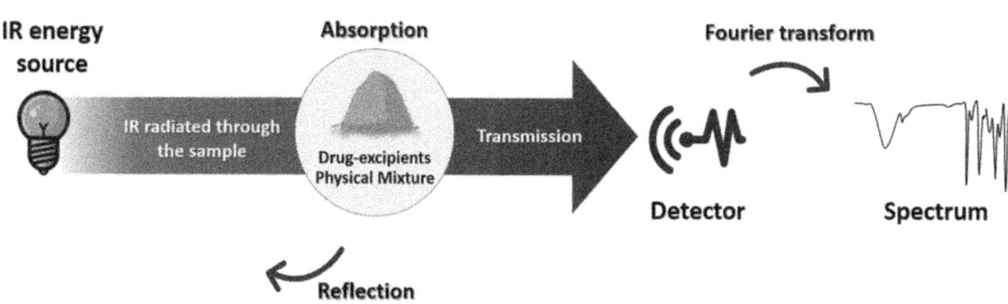

Figure 8.6 Schematic illustration of spectrum detection by infrared spectroscopy, from infrared light radiation to sample response and data detection.

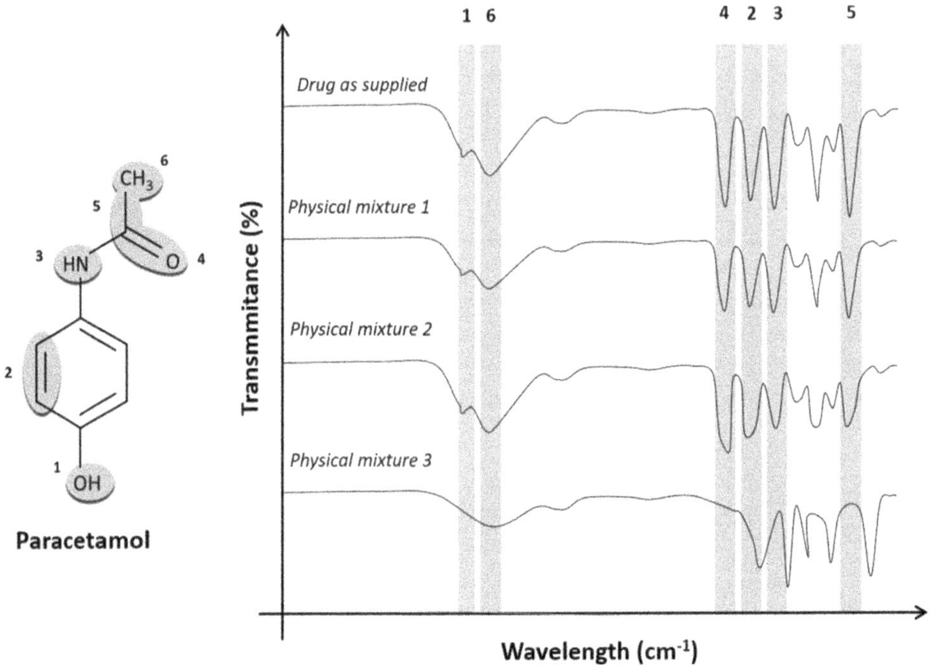

Figure 8.7 Hypothetical FTIR spectra of the drug and three different physical mixtures. The shaded peaks correspond to the characteristic functional groups of the drug paracetamol. Physical mixture 1 represents practically unchanged drug spectra, indicating little or no interaction between the components. Physical mixture 2 exemplifies an enlargement and reduction of the peak intensity, indicating changes in the molecular structure due to inter- or intramolecular interactions, maintaining the chemical stability of the drug. Physical mixture 3 represents the disappearance of peaks corresponding to the drug, indicating its instability and incompatibility with the excipient.

interaction is inferred by also its chemical stability in order to assist the screening of the best excipients [24]. Indeed, the disappearance of a peak following the non-appearance of a new one that could correspond to a new bond with that atom indicating degradation and loss of molecular structure [39].

8.2.4 Hot-Stage Microscopy (HSM)

Organoleptic changes such as colour and shape, even if unrelated to a chemical modification of the sample, constitute quality deviations and must be anticipated in preformulation studies. Hot-stage microscopy (HSM) is a widely used technique to assess sample morphology changes during heating. Therefore, it is a valuable tool to complement the other analyses herein presented, in which melting, solubilisation, crystallisation, and possible degradation or polymorphism events can be noticed [13]. In this characterisation, the sample is placed in a small hot-stage cell with a microscope that allows the visualisation of the morphology of thermal events (Figure 8.8) [40].

Some changes resulting from heat exposure are not observed in the thermal analysis, since its low concentration does not generate an analytical signal [41, 42]. Thereby, the evaluation of drug behaviour through HSM becomes even more interesting, especially for

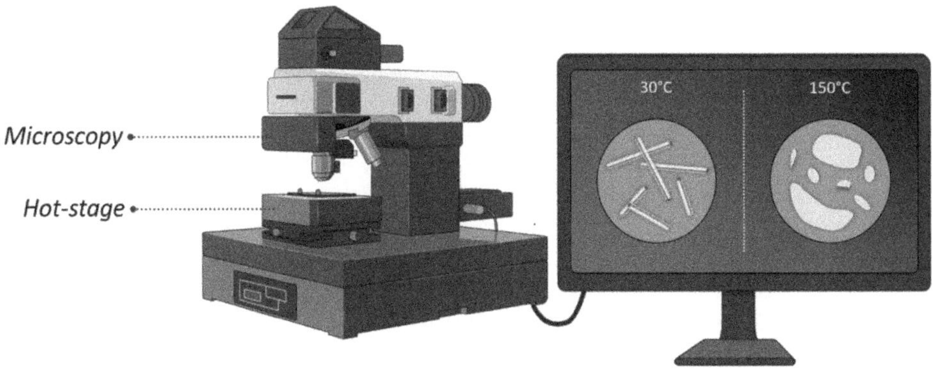

Figure 8.8 *Schematic illustration of hot-stage microscopy, showing the morphological analysis of the crystal melting.*

the preformulation of 3D products developed by categories that explore temperature variation, such as material extrusion, material jetting, and vat photopolymerisation.

Furthermore, some events observed by DSC, TGA, and even XRPD that are difficult to understand can have their interpretation assisted with the images obtained by HSM [43]. Thus, this analysis can be a complementary tool to other analytical instrumentations [44].

8.2.5 Customizsd Sample Preparation for the Preformulation Protocol

Additive manufacturing uses different materials and energy sources than conventional technology to produce pharmaceutical products. Therefore, innovative steps can be introduced in the preparation of samples trying to simulate the processing stress that the material will be exposed to when printed, e.g., using heating or light (Figure 8.1). This will provide a proper application of the above-mentioned analytical techniques in a preformulation study of 3D pharmaceutical dosage forms.

In this direction, a protocol for preformulation studies for FDM 3D printing proposed heating the samples for 5 minutes at a predetermined temperature in order to simulate the thermal stress of the hot-melt extrusion (HME) process, followed by cooling at room temperature and then heating again at a specific temperature for 2 minutes, aiming to simulate the heating of the print step [10]. Moreover, another work performed a similar double-heating protocol for HME-FDM systems whose extrusion process is more accelerated, in which the samples were heated for 2 minutes at each chosen temperature [24]. In both studies, the time–heat exposure was classified as greater than residence processing time to anticipate, in the worst possible scenario, the repercussion of physical-chemical changes in the sample due to thermal stress.

8.3 In-Process Characterisations

The 3D printing of drug products involves mostly continuous processes with few production steps. However, particularly in the case of FDM, the drug-loaded filaments are intermediate products whose viscoelastic and mechanical characteristics are decisive for correct printing and should therefore be closely monitored. Accordingly, establishing in-process control protocols that fulfill this purpose is mandatory as part of the quality assurance system for pharmaceutical production.

8.3.1 Mechanical Analysis

Mechanical evaluation is commonly used on filaments for FDM 3D printing, since it is crucial information to assist printability requirements and successful feedability. In fact, most pharmaceutical polymers are not designed to suit 3D technology; in this way, extruded polymeric filaments without other adjuvants are most likely too hard or too soft, which hamper the traction and printing processes [45]. Therefore, in order to choose the ideal filament for printing, mechanical properties are assessed before the printing step, as an in-process control [46].

This evaluation can be performed using a universal testing machine or a texture analyser. Both instruments equipped with the proper geometry cell can perform three-point bending, resistance, and elongation tests, which aim to measure the maximum stress and force required to fracture the filament (Figure 8.9) [47–49].

The three-point bending and the resistance tests use compressive forces. The first test is performed on a selected piece of the filament, while the resistance test is carried out throughout the entire sample (Figure 8.9). This difference makes bending tests more recommendable since the reduced sample volume leads to a lower probability of finding errors and generating large statistical deviations, especially in the case of sensitive materials such as brittle filaments [46].

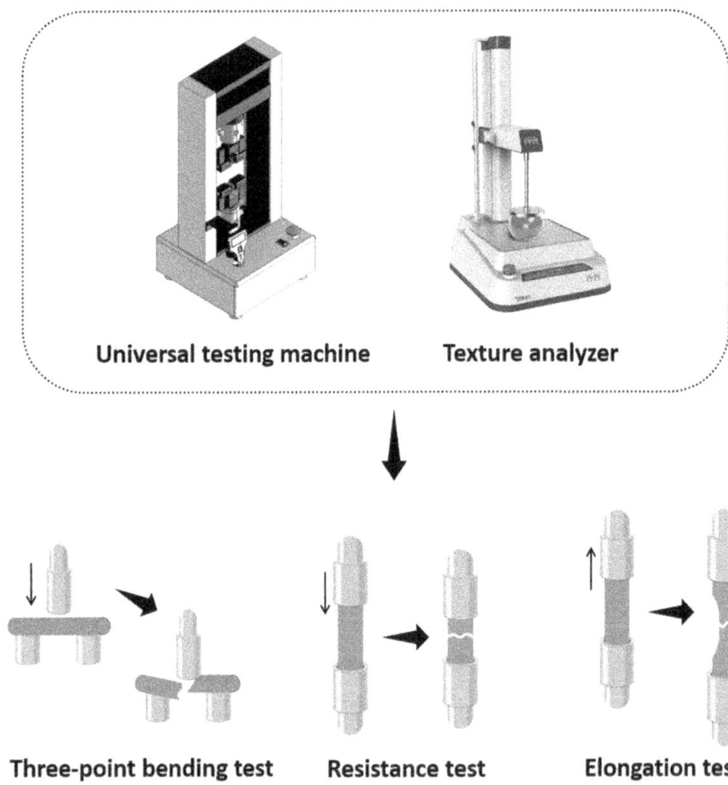

Figure 8.9 Schematic illustration of mechanical evaluation performed in a universal testing machine or a texture analyser. The different analyses commonly performed to measure the maximum stress and force required to fracture the filament are represented.

On the other hand, elongation tests, also known as tensile tests, use extensional forces to stretch the filament and measure the fracture point (Figure 8.9). Considering that the filaments are pulled by the gears of an FDM 3D printer, the majority of force applied under the samples are tension forces. Thus, this type of test would be more recommendable due to the high similarity with the mechanical feeding process of the FDM 3D printer [50]. However, the analytical deviation of this test is very high due to the force being performed throughout the entire sample, as explained before, which requires an increased number of replicates [46].

Regardless of the experimental chosen protocol, the result is represented in force-displacement or stress-strain graphs. The latter is the most used, as it provides extensive information about the material's mechanical properties (Figure 8.10A). In stress-strain diagrams, the first part of the curve is formed by a linear growth (elastic deformation), followed by a necking behaviour, in which small changes in stress cause pronounced variation in strain (plastic deformation) (Figure 8.10A) [51].

The difference between these two deformations lies in the material's ability to recover. When stress is employed in a solid material, its chemical bonds stretch [52]. Then, when the force is removed, the intra-atomic bonds can relax and recover, return to their original form (elastic deformation), or be irreversibly destroyed and maintain their new conformation (plastic deformation) [53].

Breaking stress higher than 8.5 MPa has been described for filaments capable of supporting the traction of a FDM 3D printer. Hence, this value is considered the lower stress limit to FDM materials [54]. Moreover, the area under the curve provides information about the material's toughness, i.e., larger areas correspond to ductile materials and smaller areas to brittle materials (Figure 8.10A) [55, 56]. In general, ductile samples have elastic and plastic deformation, while brittle materials only deform elastically. Additionally, this variable is related to fracture stress. Still, it is not totally dependent on it, as it is possible to obtain a ductile material that fractures at small stress values [57].

In this sense, to choose the ideal filament for printing, it is necessary to observe not only the maximum fracture point but all the deformation behaviour until filaments break. In

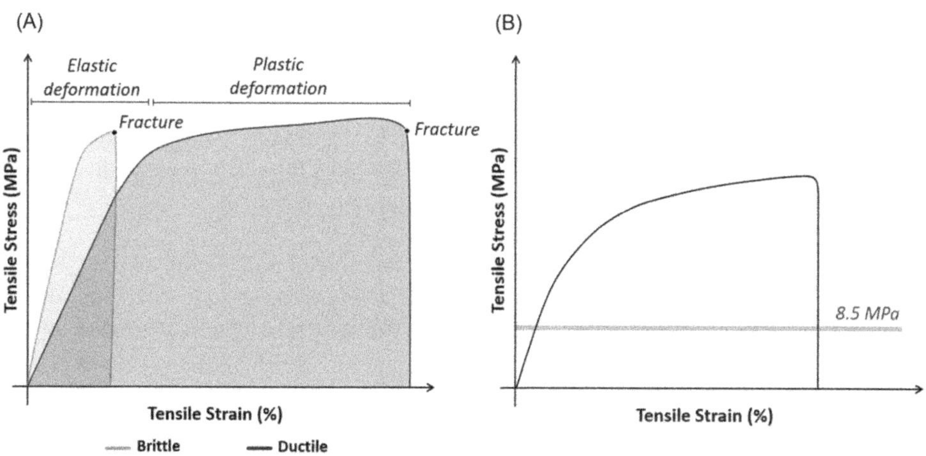

Figure 8.10 *Hypothetical stress-strain curves from the mechanical evaluation (A) indicating elastic and plastic deformation, toughness (area under the curve), and fracture point of brittle (fragile) and ductile (resistant) materials. (B) Example of the ideal curve for polymeric filaments used in FDM 3D printers.*

general, ideal filaments to support the traction and properly feed a FDM 3D printer have elastic deformation up to the limit stress values (8.5 MPa) and that are not highly ductile, since the polymer matrix needs to be able to deform to create the desired 3D structure, and then recover, aiming to solidify the final drug product (Figure 8.10B) [54].

8.3.2 Rheological Analysis

As mentioned before, pharmaceutical 3D objects are developed based on resins and polymers (Figure 8.1); hence, the formulations are extremely viscoelastic. In this way, rheological evaluation, especially oscillatory shear rheology, measured by a rheometer, is an indispensable tool to guide the production of 3D products, since it provides information on the deformation profile of viscoelastic compounds in different thermal-shear conditions [58].

In FDM 3D printing, the viscoelastic behaviour of the filament needs to provide adequate flow to the printer nozzle; however, its complex viscosity cannot be too low, otherwise the material will overly deform and could clog the printer [59]. On the other hand, for light-based vat photopolymerisation, the complex viscosity of the formulation needs to be as low as possible to allow adequate recoating of the liquid monomer on the polymerised underlayer [60]. Still, lower viscosity values have been described as a desirable condition for material jetting to achieve the appropriate spray of droplets [61].

To correctly assess the formulation's processability and characterise its components, it is necessary to ensure that the molecular structure will not be damaged or destroyed during the analysis, allowing reliable results. This chemical stability is difficult to be established with solid materials since the extreme rigidity can make materials brittle, leading to fragile intramolecular bonding forces, which can break during oscillation tests and disable the evaluation, especially at temperatures in which the material remains solid [62].

Thus, the first step of the evaluation is to establish linear viscoelastic response (LVR) and non-linear measurement intervals [63]. The strain is the rheological variable that will control this phenomenon, representing the amplitude in which the deformation can occur. Thus, changes in the strain inside the LVR of the material do not modify the rheological response. In contrast, changes in the strain outside the LVR imply a non-constant deformation with a risk of molecular destruction, making the analysis unfeasible [64].

The determination of the LVR is performed by an amplitude (strain) sweep test, in which the applied amplitude is increased at a fixed frequency, allowing to observe a transition between the linear and non-linear regimes (Figure 8.11) [65]. The LVR is different for each material, being extensive for liquid compounds and short for solid compounds, making high strain values easily accepted by fluids, while the molecular control of elastic materials is more demanding [66, 67].

Methods that work with low strain values are known as the small amplitude oscillatory shear (SAOS), while those that work with high strain values are known as the large amplitude oscillatory shear (LAOS) [68]. As the LVR of solid materials is always formed by low strain values, the SAOS methodology is often chosen to characterizs pharmaceutical systems developed by extrusion or 3D printing.

For materials characterisation, rheological measurements of viscosity, modulus, and tangent of angle δ are evaluated as a function of frequency and temperature variations [58, 66]. Amongst all the numerous analyses that can be performed on an oscillatory rheometer, the frequency sweep test is the most popular [69–71].

In this experimental technique, temperature and strain are constant, and an increasing frequency is applied to the material in order to evaluate their moduli (G' and G'') and

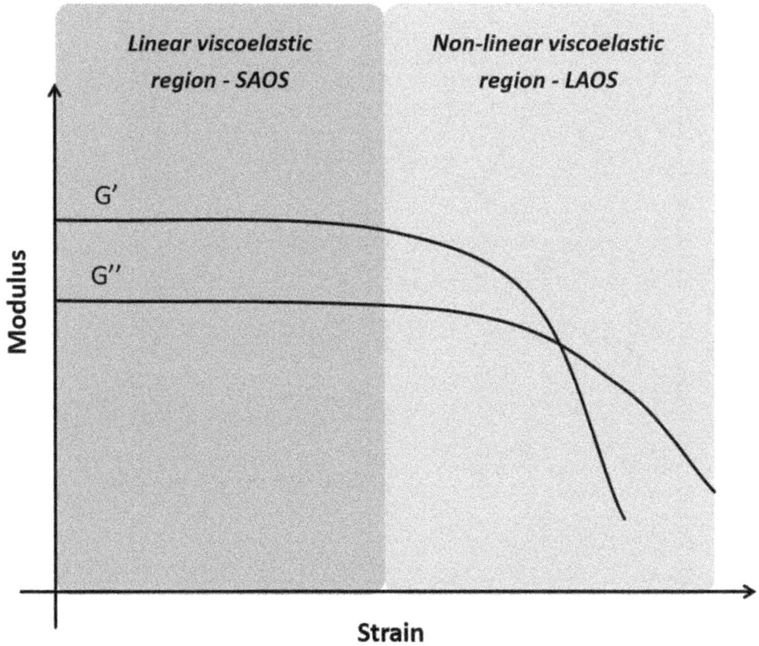

Figure 8.11 *Hypothetical graphical representation of an amplitude sweep analysis in which the loss modulus (G") and storage modulus (G') are plotted. The dark grey region indicates the linear viscoelastic response and the strain values that form the SAOS method. In contrast, the light gray region indicates the non-linear viscoelastic response and the strain values that form the LAOS method.*

complex viscosity (Figure 8.12) [72, 73]. This provides the time-dependent behaviour of a sample, since low frequencies simulate slow motion or slow processing conditions (dark grey region Figure 8.12), and high frequencies simulate fast motion or fast processing conditions (light grey region. Figure 8.12) [74, 75].

When applied to FDM, frequency sweeps are proven methods to choose the printing speed, as they can collect information on the behaviour and inner structure of the formulation under predefined procedure parameters (shear and temperature are constant) and assess the time required for the material to deform optimally [76]. Additionally, this experimental tool is useful to choose the processing temperature in material jetting 3D printing [61]. Several possible processing temperatures are selected in most research, and the frequency sweep analysis is performed in these different isotherms. Therefore, the condition that provides adequate deformation in a short time and low temperature is considered ideal for printing [77–79].

Rheological temperature ramp analysis is a widely explored test in materials science for evaluating polymers suitable for 3D printing [80, 81]. Indeed, such oscillatory assay has been increasingly used to characterise polymer-based materials for SLA, material jetting, and FDM [66, 82, 83]. In this experiment, rheological patterns are monitored according to a temperature ramp (Figure 8.13).

In practice, the viscoelastic behaviour of the material is evaluated as a function of temperature in order to assess its deformation capacity at different heat profiles to enable the

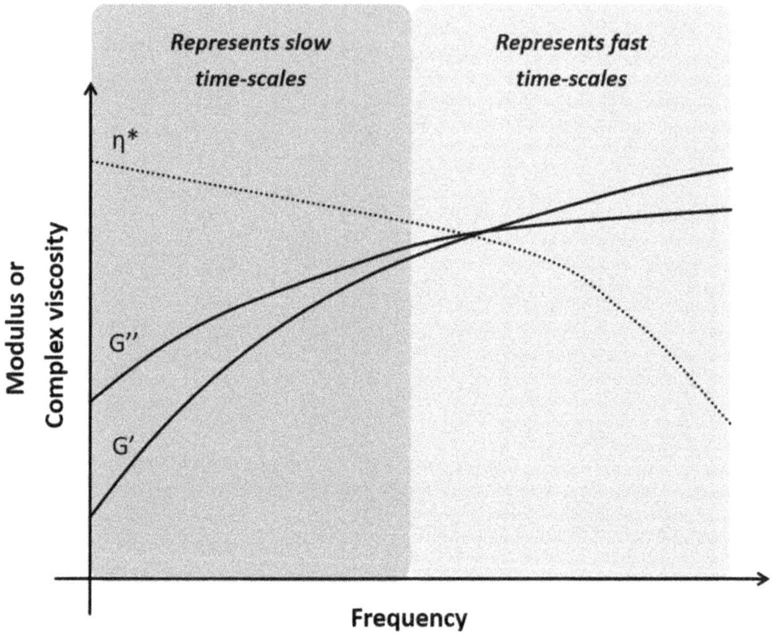

Figure 8.12 *Hypothetical graphical representation of a frequency sweep analysis in which the loss modulus (G''), storage modulus (G'), and complex viscosity (η*) are plotted. The dark grey region indicates the low frequency values used to represent slow processing conditions, while the light grey region indicates the high frequency values used to describe fast processing conditions.*

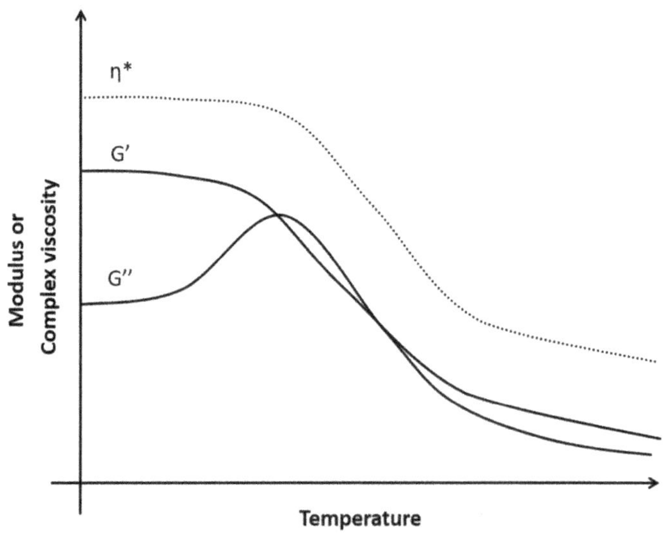

Figure 8.13 *Hypothetical graphical representation of a temperature ramp analysis loss modulus (G"), storage modulus (G'), and complex viscosity (η*) are plotted.*

ideal definition of thermal and stress for the printing process [84]. For pharmaceutical systems developed by FDM, the elastic deformation (marked by G′) cannot be too high, otherwise, the polymer will not deform; consequently, the polymer-drug interaction is impaired. However, if the viscous behaviour (marked by G″) is greater, the deformation will be so intense that the polymer will not recover and solidify after extrusion or printing [58].

Moreover, for assessing the complete deformation of the formulation, complex viscosity and complex modulus are monitored as they gather information about the resistance of flow and deformation, respectively, considering the viscoelasticity of the material [10]. Additionally, for characterisation purposes, this analysis is commonly performed to assess the glass transition region (T_g) and melting point (T_m) of the polymer [85, 86].

For a solid polymer, the glass transition range starts at the first signs of materials softening, and it ends when the molten state has been completely reached (grey dotted line Figure 8.14) [87]. In this way, the T_g temperature is established approximately in the middle of the T_g range, and it can change according to the proposed mathematical method of interpretation: temperature onset of G′ decay [88], peak preceding G″ decay [89], or peak of tan δ [90]. Since tan δ considers both viscous and elastic deformation, it is the most described method for calculating T_g temperature.

The melting point in a rheological temperature ramp analysis is classified as the crossover temperature (G′ = G″) [85]. Figure 8.14A shows the viscoelastic profile of an amorphous polymer, indicating its T_g temperature, while Figure 8.14B shows the viscoelastic profile of a crystalline polymer, showing T_g and T_m.

Conclusively, besides supporting the identification of the best process conditions, a temperature ramp analysis can also evaluate a crystalline profile, decomposition, compatibility, and capacity for molecular interaction [28, 46]. Moreover, the intrinsic characterisation proposed by this rheological test can provide information during and after the material is stressed, since the rheometer is able to return to ambient conditions while continuing to monitor the rheological variables.

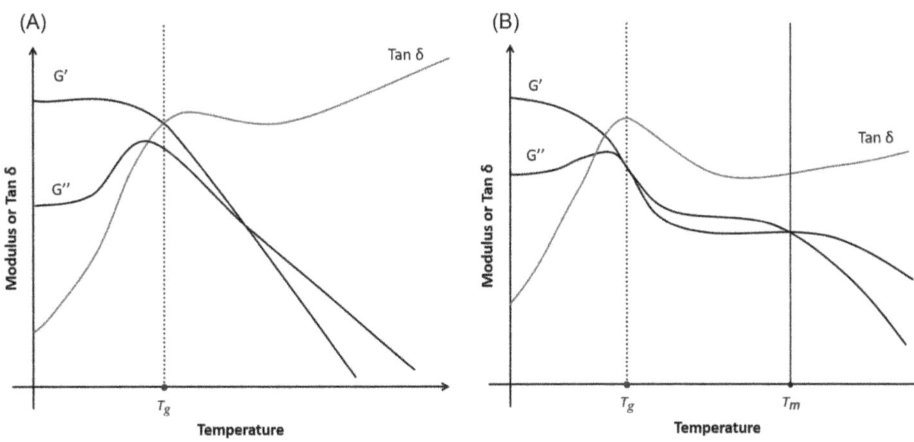

Figure 8.14 *Hypothetical graphical representation of a temperature ramp analysis in which the loss modulus (G''), storage modulus (G'), and tangent of the angle (Tan δ) are plotted. Tan δ peak indicates the glass transition temperature (T_g) of (A), an amorphous polymer, and (B) a crystalline polymer, in which the melting point (T_m) is also represented and indicated by the modulus crossover.*

In this way, this analytical tool can identify the influence of processing conditions on product stability, predicting possible degradation, incompatibility, and vehicle crystallisation caused by extrusion or printing. Hence, oscillatory rheology is a sensitive and comprehensive instrument that guarantees 3D drug products with the expected pharmaceutical requirements.

8.3.3 Drug Characterisation

Considering that intermediate products for 3D printing may require thermal and mechanical stress of the formulation for its correct miscibility with the drug, it is important to evaluate these samples by stability-indicating methods to be properly validated. This is highly recommended in the case of FDM 3D printing that employs drug filaments produced by hot-melt extrusion [91].

For this, chemical assays commonly used for drug determination can be perfectly applied as long as their selectivity is evaluated. Liquid chromatography equipment with UV, fluorescent, or even mass spectrometry detectors are very required options. However, all these destructive methods are time-consuming and demand complex sample preparation. Thus, analytical alternatives such as NIR and Raman spectroscopy can be used to evaluate drug content and uniformity, as they are simple, fast, and non-destructive techniques [92–94]. In practice, these spectroscopic techniques analyse samples at different concentrations, and thus calibration models such as partial least squares are established to determine drug content [95–97].

8.4 Final Product

3D-printed medicines produced on a large scale or personalised in compound pharmacies and hospitals must undergo the quality control tests established in the pharmacopeias. Additionally, some additional tests with greater adherence to the characterisation of these new devices should be run in the routine evaluation of the 3D drug product.

8.4.1 Morphological Analysis

One of the great advantages of applying 3D technology in the development of pharmaceuticals is the ability to personalise the object with specific sizes and dimensions to produce devices with modified release and better performance. Therefore, the evaluation of the morphological structure of the final product is crucial to guarantee its stability and effectiveness. Accordingly, image analyses tools are highly demanded, including stereomicroscopy, for an overview of the macrostructure, and scanning electron microscopy (SEM), to analyse in more detail the morphological specifications [41, 98].

Stereomicroscopy associates the reflection of light with different lenses leading to the visualisation of the 3D object with magnifications of up to 50x (Figure 8.15A) [99]. Additionally, polarised light can be used to facilitate the identification of crystalline domains [100, 101]. On the other hand, instead of light, SEM methodology uses an electron beam that interacts with the sample's surface, scanning it and providing a detailed image of the object [102]. This type of interaction occurs due to the electrical conductivity of the object's surface. However, as pharmaceutical materials are non-conductive, it is necessary to pre-coat the sample, which is typically performed using gold [103, 104]. In this way, SEM provides images with magnifications up to a nanometric scale [105], identifying any morphological irregularity of the pharmaceutical product (Figure 8.15B).

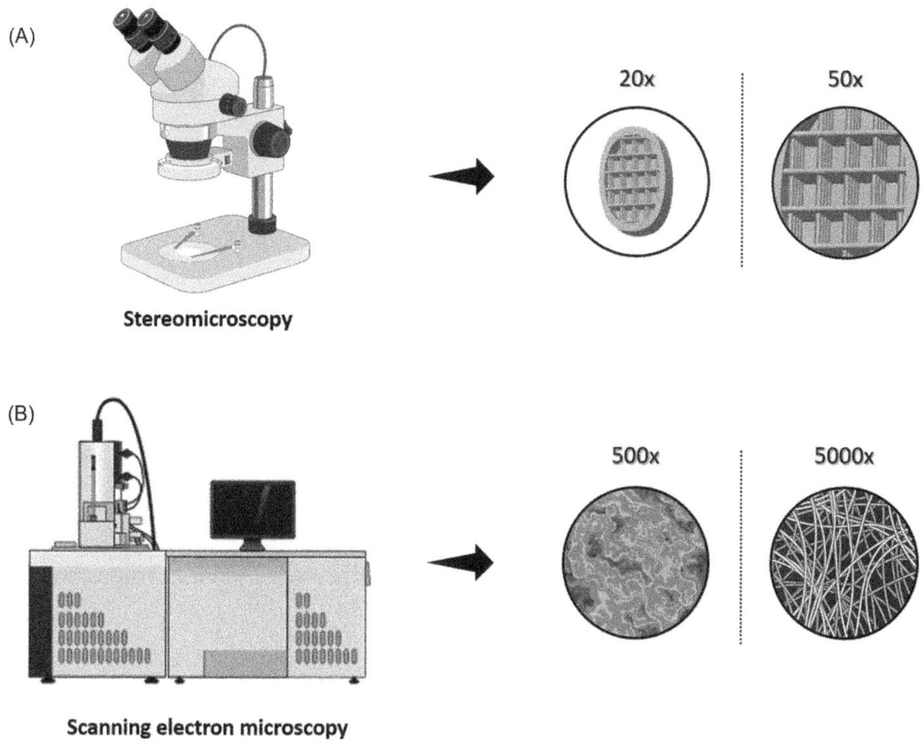

Figure 8.15 *Schematic illustration of (A) stereomicroscope and (B) scanning electron microscope, along with hypothetical images that these two techniques can provide.*

Both techniques can be used to accurately determine the dimensions of 3D-printed drug objects, ensuring that the drug device is printed as intended. The morphological precision of 3D-printed drugs is essential to ensure their customised use in the patient, as well as their perfect performance as modified drug delivery systems.

8.4.2 X-Ray Computed Microtomography (XμCT)

X-ray computed microtomography (XμCT), also known as micro-computed tomography (micro-CT) is a radiographic imaging technique that uses X-rays to create cross-sections of a solid object to produce 3D virtual images of the internal structure [106]. For that, the sample is placed on a rotation stage between a detector and an X-ray source that is transmitted through aiming to produce 2D image of one section of the object. Then, this procedure is repeated while the sample completes a 360° rotation on its axis so that multiple images are detected from all the different sections of the sample (Figure 8.16) [107].

After this procedure, a specialised computer software performs a tomographic reconstruction based on the 2D images, and the correlation is made between pixel intensity and object density [108]. Later, a 3D image of the object is built, making it possible to analyse the material's internal structure without destroying it.

This technique is valuable within the microstructural characterisation of 3D pharmaceutical devices to guide the comprehension of its dissolution profile and drugs release [109, 110]. Furthermore, this assay allows that possible errors and morphological deviations

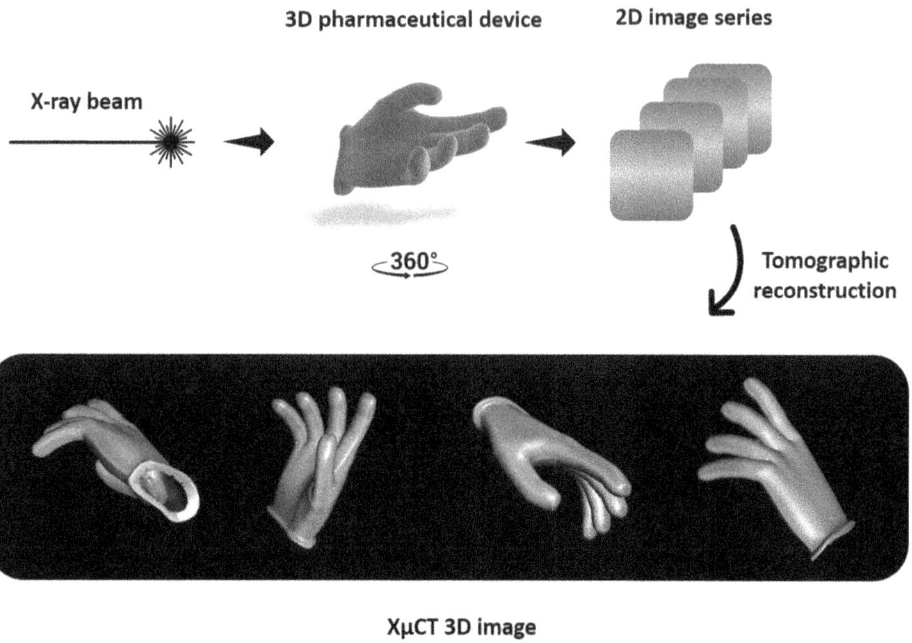

Figure 8.16 Schematic illustration of 3D images of a pharmaceutical device (drug-loaded gloves) obtained by X-ray computed microtomography (XµCT).

caused by 3D printing can be accurately identified, as the obtained image can be compared with the original CAD file [111].

Therefore, compared to the instrumentation commonly used to assess samples morphology, this technique has several advantages. XµCT is more accurate than stereomicroscopy, and analyses the entire object non-destructively, in addition to not requiring a complex sample preparation protocol, unlike SEM.

8.4.3 Terahertz Pulsed Imaging (TPI)

Terahertz pulsed imaging (TPI) is another valuable technique for non-destructive characterisation of the inner structures of a solid sample. Terahertz radiation, also called far-infrared, is an electromagnetic spectrum between the mid-infrared and microwave regions (2–133 cm^{-1}, 0.1–4 THz) that generates large-mass intramolecular oscillations, phonon lattice dynamics, and intermolecular bond vibrations [112]. In fact, these interactions are different in amorphous and crystalline materials, and as terahertz radiation easily passes through amorphous structures, it is ideal for penetrating most pharmaceutical excipients, especially polymers, which makes TPI technology relevant for the characterisation of 3D pharmaceutical dosage forms [113, 114].

In this technique, an automated scanner emits terahertz beams in a pulsed way through the sample, which is reflected in different time bands, depending on the composition and density of each material as well as the depth of the external layer. Then, the radiation is detected, and the variation in the spectrum's detection time provides information on the sample's internal structure [112, 115].

In view of this technology, it is possible to calculate the thickness of the shell and the infill of 3D pharmaceutical products [109]. TPI can also be a valuable tool for evaluating

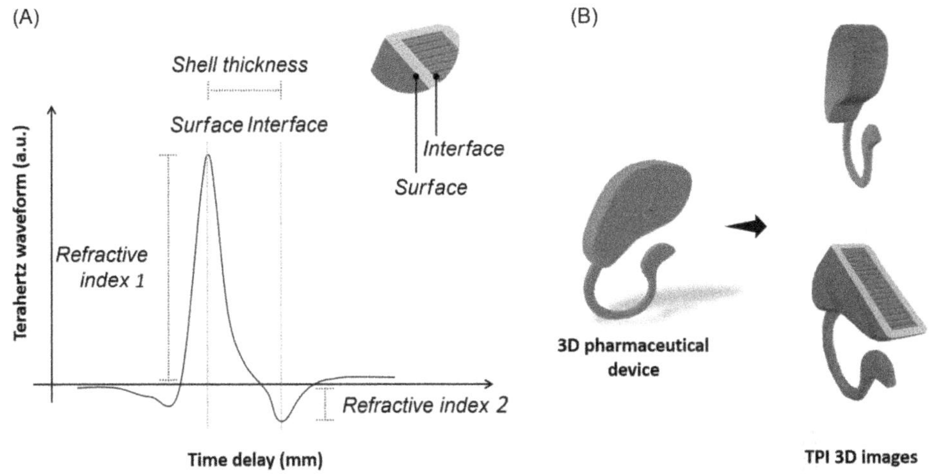

Figure 8.17 (A) Hypothetical graphical representation of the terahertz waveform of a pharmaceutical device (drug-loaded hearing aid), in which the peak intensity corresponding to refractive index 1 represents the device surface (external shell density), and refractive index 2 illustrates the device interface (the region between shell and infill). The variation between the two peaks is the shell thickness. (B) Schematic illustration of 3D images obtained by terahertz pulsed imaging.

drug product density and dissolution kinetics [116]. Moreover, as the radiation has low energy, there is no risk of heating or induction of photochemical reactions, which preserves the sample during analysis [117].

Figure 8.17A corresponds to the graph obtained from this analysis, in which the detected terahertz waveform is plotted as a function of time, generating peaks that correspond to surface and interface dimensions [112]. Additionally, as the delay time is used to calculate the depth of the layers, it is possible to build a 3D image from all the data obtained (Figure 8.17B) [109, 117].

Both XμCT and TPI are powerful techniques to provide information on the microstructure and drug performance of 3D pharmaceutical dosage forms. However, as XμCT analysis is very slow (~1 hour [109]), it would be difficult to use this instrument as quality control of final products. On the other hand, TPI analysis is performed in a few seconds, allowing multiple samples to be analysed rapidly. In addition to these data-acquisition time differences, the XμCT's rotating technology scans the 3D structure, providing images with precise dimensions. At the same time, in TPI, the terahertz pulse focuses on parts of the sample, wherein the 3D image is obtained mathematically through the refractive index, which can produce errors and limitations [118, 119]. Therefore, the technique used for characterisation must be chosen on a cost–benefit basis.

8.4.4 Mercury Porosimetry

The technical printing parameters such as infill density, infill type, printing speed, or laser scanning speed, can potentially impact the 3D object porosity [120]. This property, in its turn, can modify the drug release rates, especially of oral dosage forms [121]. Thereby, it is a characteristic that must be monitored in the 3D final product.

Porosimetry is the technique that assesses the porosity of a sample. First, the material is immersed in a liquid forced into the sample's pores by increasing pressure. Therefore, the liquid used must have a high contact angle (greater than 90°), in conjunction with the ability to avoid wetting the sample. Hence, the liquid used for this test is mercury [122].

During the analysis, the pressure needed for the liquid to penetrate the pores is inversely proportional to its size; the smaller the pore, the greater the pressure used. Finally, the mercury porosimeter provides data of cumulative pore volume (Figure 8.18A) and pore size distribution (Figure 8.18B) as a function of pore diameter [123, 124]. In addition, simulations of the pore's arrangement in the 3D drug device can be performed by processing the porosimetry data in a specific software [125], as exemplified in Figure 8.18C.

8.4.5 Helium Pycnometry

Another critical property capable of modifying the drug release is the product's density, which is assessed using helium pycnometry [121]. This test is performed with helium, as it is considered an inert gas. Thus, it is not adsorbed or absorbed by the solid material. Additionally, it is a molecule sufficiently small to penetrate any discontinuous space of the 3D object [122].

Therefore, in this technique, the sample is placed in a chamber filled with helium, which is able to penetrate samples pores. Additionally, a second chamber is also filled with helium

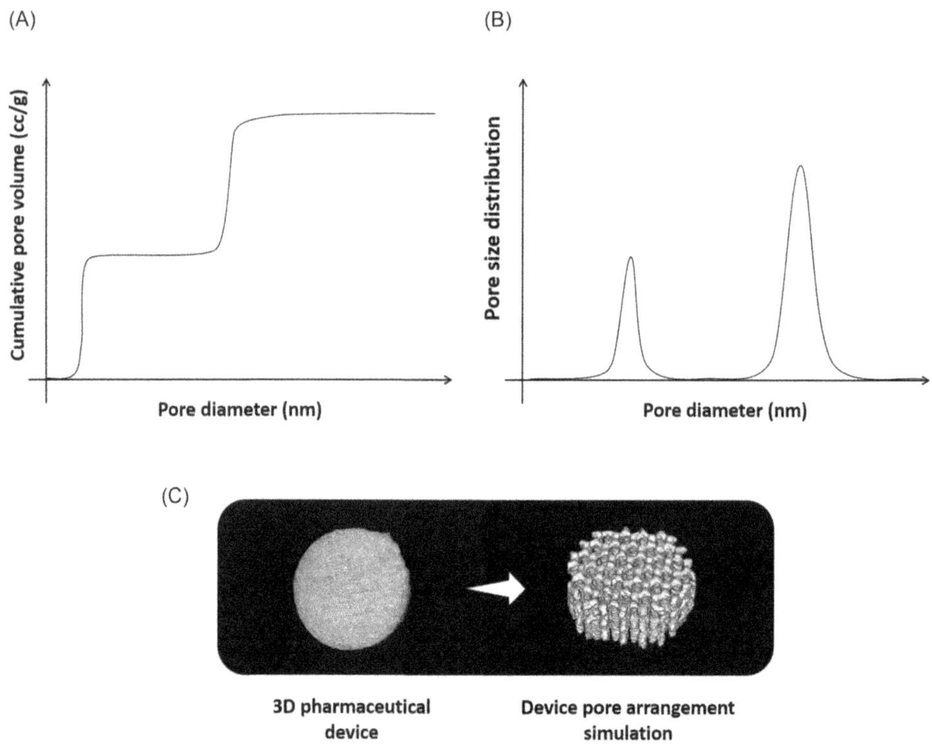

Figure 8.18 Hypothetical (A) cumulative pore volume; (B) pore size distribution curves obtained through mercury porosimeter; and (C) computer simulation of the pores arrangement of a pharmaceutical device (tablet), obtained by specialised software.

but with no solid content (reference chamber) [122]. Then, from the observed variation in the volume and pressure of the two chambers, it is possible to obtain a graph of the samples' skeletal density as a function of data acquisition cycles performed by the pycnometer [126]. Each pharmaceutical formulation has a unique compound composition, which leads to a printed object with unique and unpredictable densities. Therefore, these measurements are important for the development of 3D products as well as for monitoring their batch-to-batch properties.

8.5 Conclusions

As with the first major transformation of pharmaceutical production in the eighteenth century, the proper use of analytical testing from drug design to a final product could be critical to the success of additive manufacturing of pharmaceuticals. At the moment, although there is already a wide range of analytical tools available for this purpose, there are still few studies that explore such tools within the production of 3D medicines. It is expected that the intensive R&D efforts carried out in recent years can fill the existing gaps, and the analytical development applied to 3D printed medicines can subsidise the use of this technology for society in a safe way.

References

[1] Siddiqui, M.R., Al Othman, Z.A., and Rahman, N. (2017). Analytical techniques in pharmaceutical analysis: a review. *Arabian Journal of Chemistry* **10**: S1409–S1421.
[2] Seoane-Viaño, I., Trenfield, S.J., Basit, A.W., and Goyanes, A. (2021). Translating 3D printed pharmaceuticals: from hype to real-world clinical applications. *Advanced Drug Delivery Reviews* **174**: 553–575.
[3] Araújo, M.R.P., Sa-Barreto, L.L., Gratieri, T. et al. (2019). The digital pharmacies era: how 3D printing technology using fused deposition modeling can become a reality. *Pharmaceutics* **11**: 128.
[4] Pinho, L.A.G., Gratieri, T., Gelfuso, G.M. et al. (2021 Oct 19). Three-dimensional printed personalized drug devices with anatomical fit: a review. *Journal of Pharmacy and Pharmacology* **74** (10): 1391–1405.
[5] Mathew, E., Pitzanti, G., Larrañeta, E., and Lamprou, D.A. (2020). Three-dimensional printing of pharmaceuticals and drug delivery devices. *Pharmaceutics* **12**: 266.
[6] Goyanes, A., Allahham, N., Trenfield, S.J. et al. (2019). Direct powder extrusion 3D printing: fabrication of drug products using a novel single-step process. *International Journal of Pharmaceutics* **567**: 118471.
[7] ASTM F2792 – 12a (2013 Sept 09). *Standard terminology for additive manufacturing technologies*. Available from: http://www.additive3d.com.
[8] van den Heuvel, K.A., de Wit, M.T.W., and Dickhoff, B.H.J. (2021). Evaluation of lactose based 3D powder bed printed pharmaceutical drug product tablets. *Powder Technology* **390**: 97–102.
[9] Russi, L. and Del Gaudio, C. (2021). 3D printed multicompartmental capsules for a progressive drug release. *Annals of 3D Printed Medicine* **3**: 100026.
[10] Pinho, L.A.G., Lima, A.L., Sa-Barreto, L.L. et al. (2021). Preformulation studies to guide the production of medicines by fused deposition modeling 3D printing. *AAPS PharmSciTech* **22**: 263.
[11] Alves, G.L., Teixeira, F.V., da Rocha, P.B.R. et al. (2021). Preformulation and characterization of raloxifene-loaded lipid nanoparticles for transdermal administration. *Drug Delivery and Translational Research* **12** (3): 526–537.

[12] Teixeira, F.V., Alves, G.L., Ferreira, M.H. et al. (2018 Apr 1). Preformulation studies to guide the development of raloxifene lipid-based delivery systems. *Journal of Thermal Analysis and Calorimetry* **132** (1): 365–371.
[13] Noonan, T.J., Chibale, K., Bourne, S.A., and Caira, M.R. (2020). A preformulation co-crystal screening case study: polymorphic co-crystals of an imidazopyridazine antimalarial drug lead with the coformer succinic acid. *Journal of Molecular Structure* **1204**: 127561.
[14] Alhijjaj, M., Belton, P., Fábián, L. et al. (2018). Novel thermal imaging method for rapid screening of drug-polymer miscibility for solid dispersion based formulation development. *Molecular Pharmaceutics* **15** (12): 5625–5636.
[15] Al Rashid, A., Khan, S.A., Al-Ghamdi, S. G., and Koç, M. (2021). Additive manufacturing of polymer nanocomposites: needs and challenges in materials, processes, and applications. *Journal of Materials Research and Technology* **14**: 910–941.
[16] Leyva-Porras, C., Cruz-Alcantar, P., Espinosa-Solís, V. et al. (2020). Application of differential scanning calorimetry (DSC) and modulated differential scanning calorimetry (MDSC) in food and drug industries. *Polymers* **12**: 5.
[17] Seifi, H., Gholami, T., Seifi, S. et al. (2020). A review on current trends in thermal analysis and hyphenated techniques in the investigation of physical, mechanical and chemical properties of nanomaterials. *Journal of Analytical and Applied Pyrolysis* **149**: 104840.
[18] Rocha, J.L., Pires, F.Q., Gratieri, T. et al. (2022 Feb). Thermal analysis applied to the development of nanostructured lipid carriers loading propranolol using quality-by-design strategies. *Thermochimica Acta* **708**: 179143. Available from: https://linkinghub.elsevier.com/retrieve/pii/S0040603121002835.
[19] Delgado-Sánchez, C., Partal, P., Martín-Alfonso, M.J., and Navarro, F.J. (2021). Role of crystallinity on the thermal and viscous behaviour of polyethylene glycol-in-silicone oil (o/o) phase change emulsions. *Journal of Industrial and Engineering Chemistry* **103**: 348–357.
[20] Hamed, R., Mohamed, E.M., Rahman, Z., and Khan, M.A. (2021). 3D-printing of lopinavir printlets by selective laser sintering and quantification of crystalline fraction by XRPD-chemometric models. *International Journal of Pharmaceutics* **592**: 120059.
[21] Xinchang, Z. and Frank, L. (2021). Introduction to additive manufacturing. In: *Handbooks in Advanced Manufacturing, Additive Manufacturing*, 1–31. Elsevier.
[22] Palucci Rosa, R. and Rosace, G. (2021). Nanomaterials for 3D printing of polymers via stereolithography: concept, technologies, and applications. *Macromolecular Materials and Engineering* **306**: 2100345.
[23] Daminabo, S.C., Goel, S., Grammatikos, S.A. et al. (2020). Fused deposition modeling-based additive manufacturing (3D printing): techniques for polymer material systems. *Materials Today Chemistry* **16**: 100248.
[24] Silva, I.A., Lima, A.L., Gratieri, T. et al. (2021). Compatibility and stability studies involving polymers used in fused deposition modeling 3D printing of medicines. *Journal of Pharmaceutical Analysis* **12** (3): 424–435.
[25] v, L.A., Matruglio, A., Brett, D.J.L. et al. (2020). Using *in-situ* laboratory and synchrotron-based X-ray diffraction for lithium-ion batteries characterization: a review on recent developments. *Condensed Matter* **5**: 1–28.
[26] Ameh, E.S. (2019). A review of basic crystallography and X-ray diffraction applications. *International Journal of Advanced Manufacturing Technology* **105** (7–8): 3289–3302.
[27] Thakkar, R., Jara, M.O., Swinnea, S. et al. (2021). Pharmaceutics impact of laser speed and drug particle size on selective laser sintering 3D printing of amorphous solid dispersions. *Pharmaceutics* **13**: 1149.
[28] Lima, S.G.B., Pinho, L.A.G., Pereira, M.N. et al. (2018). Preformulation studies of finasteride to design matrix systems for topical delivery. *Journal of Pharmaceutical and Biomedical Analysis* **161**: 273–279.
[29] Aragon, F.F.H., Haeck, C.M., Morais, P.C., and Variano, B. (2021). Polymorphism characterization of segesterone acetate: a comprehensive study using XRPD, FT-IR and Raman spectroscopy. *International Journal of Pharmaceutics* **596**: 120234.
[30] Pinho, L.A.G., Lima, S.G.B., Malaquias, L.F.B. et al. (2018). Improvements of theobromine pharmaceutical properties using solid dispersions prepared with newfound technologies. *Chemical Engineering Research and Design* **132**: 1193–1201.

[31] Jamróz, W., Kurek, M., Czech, A. et al. (2018). 3D printing of tablets containing amorphous aripiprazole by filaments co-extrusion. *European Journal of Pharmaceutics and Biopharmaceutics* **131**: 44–47.

[32] Buyukgoz, G.G., Kossor, C.G., and Davé, R.N. (2021). Enhanced supersaturation via fusion-assisted amorphization during FDM 3D printing of crystalline poorly soluble drug loaded filaments. *Pharmaceutics* **13** (11): 1857.

[33] Petit, T. and Puskar, L. (2018). FTIR spectroscopy of nanodiamonds: methods and interpretation. *Diamond and Related Materials* **89**: 52–66.

[34] Hell, J., Prückler, M., Danner, L. et al. (2016). A comparison between near-infrared (NIR) and mid-infrared (ATR-FTIR) spectroscopy for the multivariate determination of compositional properties in wheat bran samples. *Food Control* **60**: 365–369.

[35] Ferreira, D.S., Galão, O.F., Pallone, J.A.L., and Poppi, R.J. (2014). Comparison and application of near-infrared (NIR) and mid-infrared (MIR) spectroscopy for determination of quality parameters in soybean samples. *Food Control* **35** (1): 227–232.

[36] Valand, R., Tanna, S., Lawson, G., and Bengtström, L. (2020). A review of Fourier Transform Infrared (FTIR) spectroscopy used in food adulteration and authenticity investigations. *Food Additives and Contaminants. Part A Chemistry, Analysis, Control, Exposure and Risk Assessment* **37**: 19–38.

[37] Thomas, S., Thomas, R., Zachariah, A.K., and Mishra, R.K. (2017). *Spectroscopic Methods for Nanomaterials Characterization* **2**. 73–94. Elsevier.

[38] Tiernan, H., Byrne, B., and Kazarian, S.G. (2020). ATR-FTIR spectroscopy and spectroscopic imaging for the analysis of biopharmaceuticals. *Spectrochimica Acta – Part A: Molecular and Biomolecular Spectroscopy* **241**: 118636.

[39] Ioakeimidis, C., Fotopoulou, K.N., Karapanagioti, H.K. et al. (2016). The degradation potential of PET bottles in the marine environment: an ATR-FTIR based approach. *Scientific Reports* **6**: 23501.

[40] Ashton, G.P., Harding, L.P., Midgley, G., and Parkes, G.M.B. (2020). Hot-stage microscopy: direct analysis in real-time mass spectrometry (HDM) as a novel tool for monitoring thermally-driven reactions on a small scale. *Analytica Chimica Acta* **1128**: 129–139.

[41] Lima, S.G.B., Pinho, L.A.G., Sa-Barreto, L.L. et al. (2021). Granules of finasteride and cyclodextrin obtained by hot-melt extrusion to target the hair follicles. *Powder Technology* **391**: 311–320.

[42] Lima, A.L., Pinho, L.A.G., Chaker, J.A. et al. (2020). Hot-melt extrusion as an advantageous technology to obtain effervescent drug products. *Pharmaceutics* **12** (8): 779.

[43] Chu, L., Grouve, W.J.B., van Drongelen, M. et al. (2021). Influence of the polymer interphase structure on the interaction between metal and semicrystalline thermoplastics. *Advanced Engineering Materials* **23** (2): 2000518.

[44] Kumar, A., Singh, P., and Nanda, A. (2020). Hot stage microscopy and its applications in pharmaceutical characterization. *Applied Microscopy* **50** (1): 12.

[45] Dumpa, N., Butreddy, A., Wang, H. et al. (2021). 3D printing in personalized drug delivery: an overview of hot-melt extrusion-based fused deposition modeling. *International Journal of Pharmaceutics* **600**: 120501.

[46] Xu, P., Li, J., Meda, A. et al. (2020). Development of a quantitative method to evaluate the printability of filaments for fused deposition modeling 3D printing. *International Journal of Pharmaceutics* **588**: 119760.

[47] Zhang, J., Feng, X., Patil, H. et al. (2017). Coupling 3D printing with hot-melt extrusion to produce controlled-release tablets. *International Journal of Pharmaceutics* **519** (1–2): 186–197.

[48] Gültekin, H.E., Tort, S., Tuğcu-Demiröz, F., and Acartürk, F. (2021). 3D printed extended release tablets for once daily use: an *in vitro* and *in vivo* evaluation study for a personalized solid dosage form. *International Journal of Pharmaceutics* **596**: 120222.

[49] Haryńska, A., Carayon, I., Kosmela, P. et al. (2020). A comprehensive evaluation of flexible FDM/FFF 3D printing filament as a potential material in medical application. *European Polymer Journal* **138**: 109958.

[50] Yu, X., Li, Y., and Li, L. (2015). Fracture mechanism of AZ31 magnesium alloy processed by equal channel angular pressing comparing three point bending test and tensile test. *Engineering Failure Analysis* **58**: 322–335.

[51] Alharbi, M., Kong, I., and Patel, V.I. (2020). Simulation of uniaxial stress–strain response of 3D-printed polylactic acid by non-linear finite element analysis. *Applied Adhesion Science* **8** (1): 5.

[52] Samaro, A., Janssens, P., Vanhoorne, V. et al. (2020). Screening of pharmaceutical polymers for extrusion-based additive manufacturing of patient-tailored tablets. *International Journal of Pharmaceutics* **586**: 119591.
[53] Tang, Z., Fujimoto, K., and Okazaki, S. (2020). All-atom molecular dynamics study of impact fracture of glassy polymers. II: Microscopic origins of stresses in elasticity, yielding, and strain hardening. *Polymer* **207**: 122908.
[54] Yang, Y., Wang, H., Xu, X., and Yang, G. (2021). Strategies and mechanisms to improve the printability of pharmaceutical polymers Eudragit® EPO and Soluplus®. *International Journal of Pharmaceutics* **599**: 120410.
[55] Cao, S., Yilmaz, E., and Song, W. (2019). Fiber type effect on strength, toughness and microstructure of early age cemented tailings backfill. *Construction and Building Materials* **223**: 44–54.
[56] Ramachandramoorthy, R., Schwiedrzik, J., Petho, L. et al. (2019). Dynamic plasticity and failure of microscale glass: rate-dependent ductile-brittle-ductile transition. *Nano Letters* **19** (4): 2350–2359.
[57] Peng, Z., Zhao, H., and Li, X. (2021). New ductile fracture model for fracture prediction ranging from negative to high stress triaxiality. *International Journal of Plasticity* **145**: 103057.
[58] Aho, J., Boetker, J.P., Baldursdottir, S., and Rantanen, J. (2015). Rheology as a tool for evaluation of melt processability of innovative dosage forms. *International Journal of Pharmaceutics* **494**: 623–642.
[59] Bhagia, S., Bornani, K., Agarwal, R. et al. (2021). Critical review of FDM 3D printing of PLA biocomposites filled with biomass resources, characterization, biodegradability, upcycling and opportunities for biorefineries. *Applied Materials Today* **24**: 101078.
[60] Weng, Z., Zhou, Y., Lin, W. et al. (2016). Structure-property relationship of nano enhanced stereolithography resin for desktop SLA 3D printer. Part A: Composites. *Applied Science and Manufacturing* **88**: 234–242.
[61] Liravi, F., Salarian, M., Dal Castel, C. et al. (2019). High-speed material jetting additive manufacturing of silicone structures: mechanical characterization. *Progress in Additive Manufacturing* **4** (4): 479–495.
[62] Liu, J., Zhao, Z., Wang, W. et al. (2019). Brittle-ductile transition in uniaxial compression of polymer glasses. *Journal of Polymer Science, Part B: Polymer Physics* **57** (12): 758–770.
[63] Bonilla, J.C., Erturk, M.Y., and Kokini, J.L. (2020). Understanding the role of gluten subunits (LMW, HMW glutenins and gliadin) in the networking behavior of a weak soft wheat dough and a strong semolina wheat flour dough and the relationship with linear and non-linear rheology. *Food Hydrocolloids* **108**: 106002.
[64] Song, H.Y., Faust, L., Son, J. et al. (2020). Small and medium amplitude oscillatory shear rheology of model branched polystyrene (PS) melts. *Polymers* **12** (2): 365.
[65] Ghorbani, M., Ai, J., Nourani, M.R. et al. (2017). Injectable natural polymer compound for tissue engineering of intervertebral disc: *in vitro* study. *Materials Science and Engineering C* **80**: 502–508.
[66] Elbadawi, M., Gustaffson, T., Gaisford, S., and Basit, A.W. (2020). 3D printing tablets: predicting printability and drug dissolution from rheological data. *International Journal of Pharmaceutics* **590**: 119868.
[67] Reddy, S.M.M., Dorishetty, P., Augustine, G. et al. (2017). A low-molecular-weight gelator composed of pyrene and fluorene moieties for effective charge transfer in supramolecular ambidextrous gel. *Langmuir* **33** (47): 13504–13514.
[68] Townsend, A.K. and Wilson, H.J. (2018). Small- and large-amplitude oscillatory rheometry with bead–spring dumbbells in Stokesian dynamics to mimic viscoelasticity. *Journal of Non-Newtonian Fluid Mechanics* **261**: 136–152.
[69] Boetker, J., Water, J.J., Aho, J. et al. (2016). Modifying release characteristics from 3D printed drug-eluting products. *European Journal of Pharmaceutical Sciences* **90**: 47–52.
[70] Ilyés, K., Kovács, N.K., Balogh, A. et al. (2019). The applicability of pharmaceutical polymeric blends for the fused deposition modelling (FDM) 3D technique: material considerations–printability–process modulation, with consecutive effects on in vitro release, stability and degradation. *European Journal of Pharmaceutical Sciences* **129**: 110–123.

[71] Isreb, A., Baj, K., Wojsz, M. et al. (2019). 3D printed oral theophylline doses with innovative 'radiator-like' design: impact of polyethylene oxide (PEO) molecular weight. *International Journal of Pharmaceutics* **564**: 98–105.

[72] Solanki, N.G., Gumaste, S.G., and Serajuddin, A.T.M. (2019). Effects of surfactants on itraconazole-hydroxypropyl methylcellulose acetate succinate solid dispersion prepared by hot melt extrusion. II: Rheological analysis and extrudability testing. *Journal of Pharmaceutical Sciences* **108** (9): 3063–3073.

[73] Marreto, R.N., Cardoso, G., Dos Santos Souza, B. et al. (2020). Hot melt-extrusion improves the properties of cyclodextrin-based poly(pseudo)rotaxanes for transdermal formulation. *International Journal of Pharmaceutics* **586**: 119510.

[74] Aho, J., Edinger, M., Botker, J. et al. (2016). Oscillatory shear rheology in examining the drug-polymer interactions relevant in hot melt extrusion. *Journal of Pharmaceutical Sciences* **105** (1): 160–167.

[75] Airey, G.D., Grenfell, J.R.A., Apeagyei, A. et al. (2016). Time dependent viscoelastic rheological response of pure, modified and synthetic bituminous binders. *Mechanics of Time-Dependent Materials* **20** (3): 455–480.

[76] Bertolino, M., Battegazzore, D., Arrigo, R., and Frache, A. (2021). Designing 3D printable polypropylene: material and process optimisation through rheology. *Additive Manufacturing* **40**: 101944.

[77] Monschke, M., Kayser, K., and Wagner, K.G. (2020). Processing of polyvinyl acetate phthalate in hot-melt extrusion-preparation of amorphous solid dispersions. *Pharmaceutics* **12** (4): 337.

[78] Solanki, N., Gupta, S.S., and Serajuddin, A.T.M. (2018). Rheological analysis of itraconazole-polymer mixtures to determine optimal melt extrusion temperature for development of amorphous solid dispersion. *European Journal of Pharmaceutical Sciences* **111**: 482–491.

[79] Bochmann, E.S., Üstüner, E.E., Gryczke, A., and Wagner, K.G. (2017). Predicting melt rheology for hot-melt extrusion by means of a simple T_g-measurement. *European Journal of Pharmaceutics and Biopharmaceutics* **119**: 47–55.

[80] Chen, Q., Han, L., Ren, J. et al. (2020). 4D printing via an unconventional fused deposition modeling route to high-performance thermosets. *ACS Applied Materials and Interfaces* **12** (44): 50052–50060.

[81] Lai, C.Q., Markandan, K., Luo, B. et al. (2021). Viscoelastic and high strain rate response of anisotropic graphene-polymer nanocomposites fabricated with stereolithographic 3D printing. *Additive Manufacturing* **37**: 101721.

[82] Jiang, F., Wörz, A., Romeis, M., and Drummer, D. (2022). Analysis of UV-assisted direct ink writing rheological properties and curing degree. *Polymer Testing* **105**: 107428.

[83] Zawaski, C.E., Wilts, E.M., Chatham, C.A. et al. (2019). Tuning the material properties of a water-soluble ionic polymer using different counterions for material extrusion additive manufacturing. *Polymer* **176**: 283–292.

[84] Azad, M.A., Olawuni, D., Kimbell, G. et al. (2020). Polymers for extrusion-based 3D printing of pharmaceuticals: a holistic materials–process perspective. *Pharmaceutics* **12**: 124.

[85] Garcia Cucalon, L., Kaseer, F., Arámbula-Mercado, E. et al. (2019). The crossover temperature: significance and application towards engineering balanced recycled binder blends. *Road Materials and Pavement Design* **20** (6): 1391–1412.

[86] Sadeghi-Mehr, A., Raudsepp, P., Brüggemann, D.A. et al. (2018). Dynamic rheology, microstructure and texture properties of model porcine meat batter as affected by different cold-set binding systems. *Food Hydrocolloids* **77**: 937–944.

[87] Barba, D., Arias, A., and Garcia-Gonzalez, D. (2020). Temperature and strain rate dependences on hardening and softening behaviours in semi-crystalline polymers: application to PEEK. *International Journal of Solids and Structures* **182–183**: 205–217.

[88] Yong, A.X.H., Sims, G.D., Gnaniah, S.J.P. et al. (2017). Heating rate effects on thermal analysis measurement of T_g in composite materials. *Advanced Manufacturing: Polymer and Composites Science* **3** (2): 43–51.

[89] Xie, R., Weisen, A.R., Lee, Y. et al. (2020). Glass transition temperature from the chemical structure of conjugated polymers. *Nature Communications* **11** (1): 893.

[90] Lipińska, M. and Imiela, M. (2019). Morphology, rheology and curing of (ethylene-propylene elastomer/hydrogenate acrylonitrile-butadiene rubber) blends reinforced by POSS and organoclay. *Polymer Testing* **75**: 26–37.

[91] Okwuosa, T.C., Stefaniak, D., Arafat, B. et al. (2016). A lower temperature FDM 3D printing for the manufacture of patient-specific immediate release tablets. *Pharmaceutical Research* **33** (11): 2704–2712.
[92] Gavan, A., Sylvester, B., Porfire, A. et al. (2022). NIR spectroscopy for monitoring of the critical manufacturing steps and quality attributes of paliperidone prolonged release tablets. *Journal of Molecular Structure* **1247**: 120501.
[93] Farquharson, A., Gladding, Z., Ritchie, G. et al. (2021). Drug content uniformity: quantifying loratadine in tablets using a created Raman excipient spectrum. *Pharmaceutics* **13** (3): 1–10.
[94] Riolo, D., Piazza, A., Cottini, C. et al. (2018). Raman spectroscopy as a PAT for pharmaceutical blending: advantages and disadvantages. *Journal of Pharmaceutical and Biomedical Analysis* **149**: 329–334.
[95] Peeters, E., Tavares da Silva, A.F., Toiviainen, M. et al. (2016). Assessment and prediction of tablet properties using transmission and backscattering Raman spectroscopy and transmission NIR spectroscopy. *Asian Journal of Pharmaceutical Sciences* **11** (4): 547–558.
[96] Trenfield, S.J., Awad, A., Goyanes, A. et al. (2018). 3D printing pharmaceuticals: drug development to frontline care. *Trends in Pharmacological Sciences* **39**: 440–451. Elsevier Ltd.
[97] Edinger, M., Iftimi, L.D., Markl, D. et al. (2019). Quantification of inkjet-printed pharmaceuticals on porous substrates using Raman spectroscopy and near-infrared spectroscopy. *AAPS PharmSciTech* **20** (5): 207.
[98] Bernardi, C., Nogueira, L., Orange, F. et al. (2021). The stereomicroscope and scanning electron microscope comparison of false starts bones produced by an oscillating autopsy saw. *Forensic Science International* **324**: 110816.
[99] Guo, Y., Zhang, J., Hu, W. et al. (2021). Shape-programmable liquid crystal elastomer structures with arbitrary three-dimensional director fields and geometries. *Nature Communications* **12** (1): 5936.
[100] Hausmann, M.K., Rühs, P.A., Siqueira, G. et al. (2018). Dynamics of cellulose nanocrystal alignment during 3D printing. *ACS Nano* **12** (7): 6926–6937.
[101] Anusuyadevi, P.R., Shanker, R., Cui, Y. et al. (2021). Photoresponsive and polarization-sensitive structural colors from cellulose/liquid crystal nanophotonic structures. *Advanced Materials* **33** (36): 2101519.
[102] Frelinger, S.N., Ledvina, M.D., Kyle, J.R., and Zhao, D. (2015). Scanning electron microscopy cathodoluminescence of quartz: principles, techniques and applications in ore geology. *Ore Geology Reviews* **65** (P4): 840–852.
[103] Cecchini, P., D'Aloisio, R., de Giacinto, C. et al. (2020). Scanning electron microscopy study of different one-piece foldable acrylic intraocular lenses after injection through microincisional cataract surgery cartridges. *International Ophthalmology* **40** (2): 369–376.
[104] Ferreira-Nunes, R., Cunha-Filho, M., Gratieri, T., and Gelfuso, G.M. (2021). Follicular-targeted delivery of spironolactone provided by polymeric nanoparticles. *Colloids and Surfaces B: Biointerfaces* **208**: 112101.
[105] Pang, S., Zhang, X., Zhang, X., and Lu, Y. (2019). A magnification-continuous calibration method for SEM-based nanorobotic manipulation systems. *Review of Scientific Instruments* **90** (5): 053706.
[106] Townsend, P., Larsson, E., Karlson, T. et al. (2021). Stochastic modelling of 3D fiber structures imaged with X-ray microtomography. *Computational Materials Science* **194**: 110433.
[107] Landis, E.N. and Keane, D.T. (2010). X-ray microtomography. *Materials Characterization* **61**: 1305–1316.
[108] da Silva, Í.B. (2018). X-ray computed microtomography technique applied for cementitious materials: a review. *Micron* **107**: 1–8.
[109] Markl, D., Zeitler, J.A., Rasch, C. et al. (2017). Analysis of 3D prints by X-ray computed microtomography and terahertz pulsed imaging. *Pharmaceutical Research* **34** (5): 1037–1052.
[110] Gioumouxouzis, C.I., Katsamenis, O.L., Bouropoulos, N., and Fatouros, D.G. (2017). 3D printed oral solid dosage forms containing hydrochlorothiazide for controlled drug delivery. *Journal of Drug Delivery Science and Technology* **40**: 164–171.
[111] Kissi, E.O., Nilsson, R., Nogueira, L.P. et al. (2021). Influence of drug load on the printability and solid-state properties of 3D-printed naproxen-based amorphous solid dispersion. *Molecules* **26** (15): 4492.

[112] Novikova, A., Markl, D., Zeitler, J.A. et al. (2018). A non-destructive method for quality control of the pellet distribution within a MUPS tablet by terahertz pulsed imaging. *European Journal of Pharmaceutical Sciences* **111**: 549–555.

[113] Pei, C., Lin, H., Markl, D. et al. (2018). A quantitative comparison of in-line coating thickness distributions obtained from a pharmaceutical tablet mixing process using discrete element method and terahertz pulsed imaging. *Chemical Engineering Science* **192**: 34–45.

[114] Markl, D., Zeitler, A., Rades, T. et al. (2018). Toward quality assessment of 3D printed oral dosage forms. *Journal of 3D Printing in Medicine* **2** (1): 27–33.

[115] Al-Sharabi, M., Markl, D., Vivacqua, V. et al. (2021). Terahertz pulsed imaging as a new method for investigating the liquid transport kinetics of α-alumina powder compacts. *Chemical Engineering Research and Design* **165**: 386–397.

[116] Patil, M.R., Ganorkar, S.B., Patil, A.S., and Shirkhedkar, A.A. (2020). Terahertz spectroscopy: encoding the discovery, instrumentation, and applications toward pharmaceutical prospectives. *Critical Reviews in Analytical Chemistry* **52** (2): 343–355.

[117] Maurer, L. and Leuenberger, H. (2009). Terahertz pulsed imaging and near infrared imaging to monitor the coating process of pharmaceutical tablets. *International Journal of Pharmaceutics* **370** (1–2): 8–16.

[118] Alves-Lima, D., Song, J., Li, X. et al. (2020). Review of terahertz pulsed imaging for pharmaceutical film coating analysis. *Sensors (Switzerland)* **20**: 1441.

[119] Russe, I.S., Brock, D., Knop, K. et al. (2012). Validation of terahertz coating thickness measurements using X-ray microtomography. *Molecular Pharmaceutics* **9** (12): 3551–3559.

[120] Gueche, Y.A., Sanchez-Ballester, N.M., Cailleaux, S. et al. (2021). Selective laser sintering (Sls), a new chapter in the production of solid oral forms (SOFs) by 3D printing. *Pharmaceutics* **13**: 1212.

[121] Fanous, M., Bitar, M., Gold, S. et al. (2021). Development of immediate release 3D-printed dosage forms for a poorly water-soluble drug by fused deposition modeling: study of morphology, solid state and dissolution. *International Journal of Pharmaceutics* **599**: 120417.

[122] Markl, D., Strobel, A., Schlossnikl, R. et al. (2018). Characterisation of pore structures of pharmaceutical tablets: a review. *International Journal of Pharmaceutics* **538**: 188–214.

[123] Goyanes, A., Kobayashi, M., Martínez-Pacheco, R. et al. (2016). Fused-filament 3D printing of drug products: microstructure analysis and drug release characteristics of PVA-based caplets. *International Journal of Pharmaceutics* **514** (1): 290–295.

[124] Schlumberger, C. and Thommes, M. (2021). Characterization of hierarchically ordered porous materials by physisorption and mercury porosimetry: a tutorial review. *Advanced Materials Interfaces* **8**: 200218.

[125] Vargas-Osorio, Z., Luzardo-Álvarez, A., Piñeiro, Y. et al. (2019). Three-dimensional hybrid mesoporous scaffolds for simvastatin sustained delivery with *in vitro* cell compatibility. *ACS Omega* **4** (3): 5496–5508.

[126] Nguyen HGT, Horn JC, Bleakney M, Siderius DW, Espinal L. (2019). Understanding material characteristics through signature traits from helium pycnometry. *Langmuir* **35** (6): 2115–2122.

9

Adoption of 3D Printing in Pharmaceutical Industry

Thomas Kipping

Life Science, Process Solutions, Upstream & Process Materials R&D, Merck Life Science KGaA, Darmstadt, Germany

The pharmaceutical industry is challenged by varying local manufacturing capacities, supply chain resilience, and faster development timelines, creating a high demand for novel manufacturing concepts. Continuous manufacturing, as well as integrated smart manufacturing approaches, are on the rise resulting in the creation of smart factories driving the industry 4.0 approaches. All new manufacturing approaches are highly data driven and integrate relevant control elements within their production lines [1].

3D printing provides an unprecedented manufacturing flexibility paired with new opportunities for resource saving production. As a highly digitalised technology, it generates a massive amount of data that can be integrated to self-regulating feedback loops [2]. But how can the pharmaceutical industry leverage the full potential of this rapidly growing technology?

In contrast to classical tablet manufacturing, less process steps are required to define the final dosage form. Also, the amount of drug substance for the early clinical phase can be drastically reduced, as the required formulation development part can be optimised. Internal data shows that up to 50% of active pharmaceutical ingredients (API) required during phase I–III trials can be reduced by applying dedicated 3D-printing technologies. Another aspect is the reduction of the duration required for formulation development and clinical trial supply throughout phases I–III. Considering an average time of 2–3 years, this time can be drastically reduced, allowing streamlined development and faster market entry.

3D Printing of Pharmaceutical and Drug Delivery Devices: Progress from Bench to Bedside, First Edition.
Edited by Dimitrios A. Lamprou, Dennis Douroumis and Sheng Qi.

Looking at other industrial sectors can give the first hint of how powerful additive manufacturing can impact industrial environments; especially the configuration of supply chains is already massively affected. For the classical additive manufacturing industry, the overexpressed scenario is the economy of one, which was presented by Petrick and Simpson in 2013 [3]. According to this model, not only the design of a product will be massively affected but also the way it will be manufactured. The system will move away from the traditional production approach involving an exchange of companies at certain development stages to a more holistic approach. To successfully manage this transition, the concept 'requires collaborative innovation between materials suppliers, product designers, and product producers at a level never before seen.'

Identifying additional value in 3D-printed products is critical for driving the evolution of this technology to compensate for critical unit manufacturing costs. In the plastics industry, 3D printing is often in direct competition with injection moulding technology, where high tooling costs and maintenance for expensive machines are critical cost drivers. Especially small batch production can be more feasible with 3D-printing approaches to avoid high upfront investments [4]. Looking deeper into the competitive range of 3D printing, especially the rather complex design parts at lower production rates, 3D printing can be competitive with injection moulding in production ranges between 50 and 5,000 units.

Similar approaches need to be considered for the pharmaceutical industry when estimating early prototyping or supply of clinical trial material. As many technologies for 3D-printed medications are still in early development stages, the full market potential is hard to predict.

There are mainly two companies on the market that are currently able to implement additive manufacturing in a large-scale Good Manufacturing Practice (GMP) compliant environment. Understanding their strategies is important to see the future potential of additive manufacturing.

Aprecia is one of the first innovators when it comes to implementing 3D printing on an industrial scale. They took a pharmaceutical approach to scale that had been widely considered as a rapid prototyping technology and focused their initial development on orodispersible dosage forms. An orodispersible form of levetiracetam was considered beyond the formulation capabilities of both soft compression and freeze-drying formulation technologies. This finally led to the approval of Spritam®. Understanding the way to get there is an important aspect in understanding the potential created. Required key actions are summarised below. It becomes obvious that a close contact with regulatory bodies and administrations is important to allow a successful evaluation of the technology.

Regulatory Strategy (Spritam®):

January 2008 – FDA Lunch & Learn Introducing 3DP for Pharmaceuticals

- PowerPoint Presentation
- Video (3DP Simulation)

March 2013 – Pre-NDA Type B Meeting with FDA for Levetiracetam

- Neurology Products Division
- Director: Russell (Rusty) Katz
- Video (December 2012)

May 2014 – Pre-NDA Type B Meeting with FDA for Levetiracetam

- Neurology Products Division
- Director: William (Billy) Dunn (Acting)

July 2014 – FDA Teleconference to Discuss Comments on Proposed Labelling
October 2014 – NDA filed for Spritam®
June 2015 – Participated in USP Nomenclature & Labelling Expert Committee Meeting
July 2015 – NDA for Spritam® Approved
September 2015 – FDA Meeting on Dosage Form Designation for Spritam®
February 2016 – Final Labelling approved for Spritam®
2015 Q4 to 2016 Q1 – Post-Approval Training for FDA Reviewers (2 separate sessions)
March 2016 – Introduction of Aprecia, 3DP, & Spritam® to Local FDA Field Office at BA Facility
June 2017 – PAS filed for BA site transfer from EW
October 2017 – PAS approved for BA site transfer from EW

The Technology Behind:

Elevating binder jetting 3DP technology to pharmaceutical mass production requires scaling and validating of the platform for commercial production of safe and reliable medication formulated in a novel alternative dosage form. Technological differentiation was achieved by unlocking new capabilities in formulation development, previously limited by available conventional technologies such as soft compression and freeze drying (high drug load, etc.).

In this early stage, it was important to rely on existing conventional GRAS excipients that are common to tableting, which was also a good differentiator to other 3D-printing technologies at the time. The next step was to advance technology capabilities and provide a versatile and flexible platform including the control of liquid and heat exposure to APIs, the ability to incorporate upstream functional particle technologies, and to enhance process efficacy by increasing volumes and waste reduction. Not having a commercial-scale production technology was a major pain point for the pharmaceutical industry when it came to alternative 3DP technologies. Alternative 3DP technologies were and are primarily focused on distributed point-of-care or low-volume custom manufacturing.

9.1 Partnering and Growing

Growing the platform is essential for long-term success. Collaborations with partners from the pharmaceutical industry as well as individual technology providers leads to the development of novel dosage forms with extended functionalities, whilst permanently expanding formulation capabilities and manufacturing efficiency of the technology. Latest developments include in-cavity printing, which provides new capabilities for NCEs and allows a faster adoption by industry for use in Research and Development (R&D). Next-generation technologies improve the material sparing capabilities. The major area of focus for pharma partners lies in proving the capability in pre-clinical/phase 1 formulation work. The top-level business model and strategy is based on being a specialty CDMO and technology innovator that can provide accelerated drug development timelines by rapid prototyping and development, reduced API quantities, and accelerated regulatory pathways, such as for orphan and rare diseases where no platform switch is ever required.

Melt-based Technology Concepts:

Triastek was founded in 2015 as the first 3D-printing pharmaceutical company in China. The company bases their offering on building a novel 3D-printing pharmaceutical technology platform with its comprehensive proprietary technologies encompassing dosage form design, digital pharmaceutical product development, and intelligent manufacturing.

The company developed a unique 3D-printing technology, melt extrusion deposition (MED), that is different from the existing powder binding or fused deposition modelling (FDM). MED can create complex geometric structures of multiple materials and does not need additional post-processing steps. MED also circumvents filament preparation and high process temperature, as in FDM. A key differentiation is a multiple nozzle-array design to produce the products at commercial-scale while maintaining high precision.

Process Analytical Technology (PAT):

In-line monitoring of multiple unit operations ensures the product quality in real time. Triastek is currently evaluating the integration of PAT into the 3D-printing system, which aims to conduct various inspection and monitoring during the process. Powered with various embedded sensors and inspection tools, the system is expected to inspect every tablet for quality control and real-time corrections with feedback loop. Each tablet is given a unique quick response (QR) code, which is traceable locally and remotely.

9.2 Regulatory Strategy

With its proprietary Melt Extrusion Deposition (MED®) 3D-printing technology, the company is the first Chinese pharmaceutical company to be accepted into the US FDA's Emerging Technology Program (ETP). This programme was initiated by the Center for Drug Evaluation and Research (CDER) as a response to the rising technical and regulatory challenges during implementation of innovative manufacturing approaches. As part of the FDA, the CDER's Office of Pharmaceutical Quality ensures the safety and effectiveness of drugs. As part of the ETP programme, industry representatives can meet with Emerging Technology Team (ETT) members to discuss, identify, and resolve potential technical and regulatory issues regarding the development and implementation of novel technologies prior to a regulatory submission [5].

Triastek is relying on a holistic approach when it comes to dedicated customer projects:

Drug delivery: Sophisticated structures with various internal geometric shapes and compartments to control release profiles.

Identified key aspects for target performance are mainly linked on release modification:

- control release rate
- control release onset time
- gastric retention
- control release site
- API/kinetics combinations
- low soluble drug substances.

9.2.1 Product Development

Triastek developed an individual formulation development approach matching their technology platform. This concept, named 3D Printing Formulation by Design (3DFbD®) (Figure 9.1), is intended to simplify formulation development by enhancing its development speed and predictability. In contrast to traditional formulation development drug release profiles, it can be predetermined by the parameters of tablet 3D structure and the excipients processing parameters.

9.2.2 Manufacturing

Pharmaceutical printer and printing systems are applied to achieve digital intelligent manufacturing. The fully automation enables a continuous manufacturing concept. Process analytical technologies (PAT) tools and relevant feedback controls are important features that are complemented by manufacturing robots.

Control, Fault Detection and Intervention:

Like many continuous manufacturing (CM) systems, minor malfunctions may trigger the entire system to stop. Therefore, the implementation of design concepts into the 3D-printing system can minimise the downtime and recovery time by introducing modular concepts for hardware design, and various sensors for system condition monitoring, as well as controls for passing and stopping an individual malfunctioning printing nozzle when needed.

Combining Additive – and Continuous Manufacturing Approaches:

Currently, FDA-approved CM manufacturing lines are mainly focusing on immediate release drug products. Adding a technology that can provide multiple release kinetics can be of high interest to further advance this field. Various release characteristics can be modified including release rate, mode, duration, and onset time.

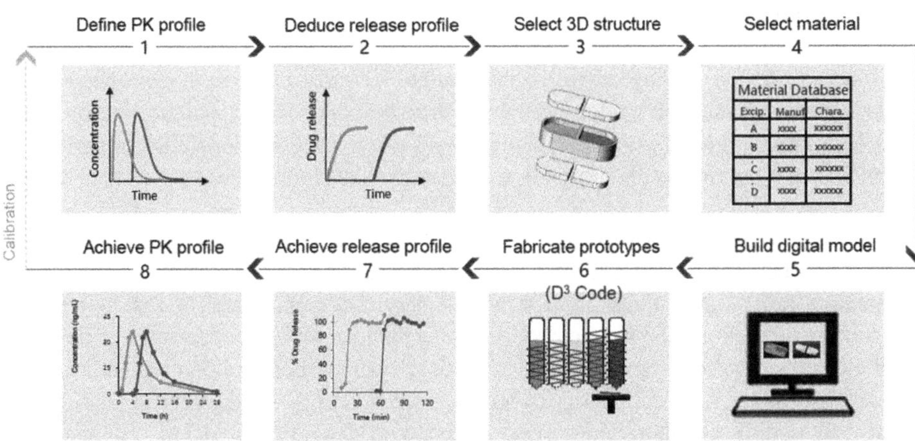

Figure 9.1 *Illustration of the 3D-printing formulation by design approach. (Source: 3DFbD® developed by Triastek).*

9.3 Business Model

Triastek is positioned as a 3D-printing pharmaceutical technology platform company, and its business model is divided into two directions for parallel development.

9.3.1 In-House Pipeline Products

Using MED® 3D-printing technology, Triastek has built a product pipeline including new formulations and new combination product candidates intended to meet specific clinical needs and thereby improving drug therapy outcomes. Triastek has submitted three preIND requests to the US FDA. Amongst these three preIND submissions, one has completed IND application and received IND clearance, and the other two have received the responses with agreement to the proposal (IND application is in preparation). All submissions are related to the 505b2 pathway.

At the stage of product clinical development, the strategy is to license out the market right to the originator or other multinational pharmaceutical companies (MNCs) and the licensee will commercialise the products, while Triastek is responsible for product development and manufacturing. Currently, first-in-house products include five products, and the sales of corresponding original products are close to or more than 1 billion US dollars respectively, which are the blockbuster drugs of multinational pharmaceutical companies. The API patents of these products will expire between 2025 and 2028, which creates the time window of new generation product development. With unique PK, clinical efficacy, and patents coverage, 3D-printed drugs can greatly extend the lifecycle of compounds and still secure the exclusive market right after the expiration of the original compound patent.

9.3.2 Co-Development

In addition to the development of in-house products, Triastek is a Partnership Development and Manufacturing Organization (PDMO), collaborating with leading multinational and Chinese pharmaceutical companies to explore a variety of application scenarios for the MED® 3D-printing technology, such as developing formulations to provide solution for the poor water soluble new chemical entities as well as extending the lifecycle of approved/developed products. Through these collaborations, Triastek provides a significant value proposition for partner companies through the development of novel, unique, and competitive pharmaceutical products in combating the challenges of developability of new compound and fine-tuning the PK profile in early product development to shorten the time for human study.

This kind of collaboration model is divided into research projects (such as feasibility studies) and product co-development. In the co-development model, Triastek is responsible for formulation R&D and manufacturing, while the partner is responsible for clinical research and commercialisation. The payment model relies on down/upfront payments and development milestones at different stages of product development. Furthermore, production management fees and royalties of net sales after the product goes on the market can be included.

The latest public collaboration programme is performed with Sperogenix Therapeutics, a China specialty company focusing on the development and commercialisation of rare diseases, seeking to advance 3D-printed orphan drugs. The collaboration will jointly explore new formulation and therapy options of 505b2 product T22, one of Triastek's pipeline products, for patients with pulmonary hypertension (PAH). According to the co-development agreement, Triastek is responsible for the R&D and production of T22, while Sperogenix is responsible for clinical trials, registration, and commercialisation in the collaboration area.

Looking at the personalised medication sector, FabRX is one of the most active companies in the area. FabRx is the first pharmaceutical company to focus on the personalised medicine market using 3D printing. 3D printing is the most suitable technology to produce personalisation of dose medicines for patients or to produce small batches for early clinical trial supply. The company has developed the first desktop pharmaceutical 3D printer M3DIMAKER™ and corresponding software to install this technology in hospitals and pharmacies. First clinical studies with the printing system were published in 2019 where a 3D printer was used to prepare personalised medicines in a hospital [6]. The goal is to utilise the 3D printer and software to prepare medicines using ink from different pharmaceutical companies close to the patient at the dispensing point.

9.4 Regulatory Strategy

As FabRx is a UK-based company, the technology can currently be applied following the local compounding or pharmaceutical special regulation. However, the regulation is changing and will be also adapted to 3D-printing technologies. In the UK, there will be a new regulatory framework for products supplied at the point-of-care. Driven by the Medicines and Healthcare Products Regulatory Agency (MHRA), the Point-of-Care (POC) framework is intended to complement existing manufacturing and supply arrangements, which will bring new therapies to patients [7]. FabRx is actively engaged and in contact with regulatory bodies in the EU and US to reach similar regulations in other countries.

9.5 Partnering and Growing

FabRx has a holistic approach when it comes to technology development. All proprietary knowledge is created in-house, ranging from hardware including 3D printers and quality control methods to software and application of AI technologies for printing optimisation and formulation development. Collaborations with hospitals are key to demonstrate the success of the technology concept, currently providing more than eight ongoing clinical studies. Partnerships with large pharma companies as well as startups are in place. Manufacturing partners are established, and the company already has initiated the expansion from the UK to the EU, US, and Japan.

9.6 Business Model and Strategy

FabRx is providing the technology to make personalised medicine a reality. The company is following a holistic approach based on the three cornerstones of 3D printing, hardware, software, and formulation. Partnerships with big pharma companies are part of the key strategy to assure a successful implementation.

9.6.1 Closing Remarks

The industrial implementation of pharmaceutical 3D printing is rapidly evolving. The key concepts are now differentiating and extending their application space. Besides the discussed rather matured technology concepts, selective laser sintering approaches as well as various melt-based approaches are on the rise. Advanced manufacturing concepts also provide new opportunities for the entire pharmaceutical supply chain. Process parameters of evolving technologies need to be fully understood and, in many cases, also place specific demands for drug manufacturers as well as for excipient providers. The holistic understanding of polymer requirements and respective particle design is key to enable successful development and assure broad industrial implementation. Close interdisciplinary collaborations between regulatory bodies, engineers, formulators, and excipient providers are required to lay the technological foundation of new 3D-printing manufacturing concepts. Defining differential focus areas at an early stage of development is key to assure a successful long-term implementation and pave the way for extended industrial applications at later stages. 3D printing will reshape our view on manufacturing concepts and supply strategies. The journey has just begun.

References

[1] Arden, N.S., Fisher, A.C., Tyner, K. et al. (2021). Industry 4.0 for pharmaceutical manufacturing: preparing for the smart factories of the future. *International Journal of Pharmaceutics* **602**: 120554.

[2] Elbadawi, M., McCoubrey, L.E., Gavins, F.K.H. et al. (2021). Disrupting 3D printing of medicines with machine learning. *Trends in Pharmacological Sciences* **42** (9): 745–757.

[3] Petrick, I.J. and Simpson, T.W. (2013). 3D printing disrupts manufacturing: how economies of one create new rules of competition. *Research-Technology Management* **56** (6): 12–16.

[4] Chan, H.K., Griffin, J., Lim, J.J. et al. (2018). The impact of 3D printing technology on the supply chain: manufacturing and legal perspectives. *International Journal of Production Economics* **205**: 156–162.

[5] CDER Emerging Technology Program Available from: https://www.fda.gov/about-fda/center-drug-evaluation-and-research-cder/emerging-technology-program (accessed 22 May 2023).

[6] Goyanes, A., Madla, C.M., Umerji, A. et al. (2019). Automated therapy preparation of isoleucine formulations using 3D printing for the treatment of MSUD: first single-centre, prospective, crossover study in patients. *International Journal of Pharmaceutics* **567**: 118497.

[7] Consultation on Point of Care manufacturing Available from: https://www.gov.uk/government/consultations/point-of-care-consultation/consultation-on-point-of-care-manufacturing (accessed 22 May 2023).

10

Clinical Benefits of 3D Printing in Healthcare

Atheer Awad[1,2], *Iria Seoane-Viaño*[2,3], *Abdul W. Basit*[2,4,5] *and Alvaro Goyanes*[2,4,5,6]

[1] *Department of Clinical, Pharmaceutical and Biological Sciences, University of Hertfordshire, Hatfield, UK*
[2] *UCL School of Pharmacy, University College London, London, UK*
[3] *Department of Pharmacology, Pharmacy and Pharmaceutical Technology, Paraquasil Group (GI-2109), Faculty of Pharmacy, iMATUS and Health Research Institute of Santiago de Compostela, Universidade de Santiago de Compostela, Santiago de Compostela, Spain*
[4] *FABRX Ltd., Henwood House, Henwood Ashford, Kent, UK*
[5] *FABRX Artificial Intelligence, Carretera de Escairón, Currelos (O Saviñao), Spain*
[6] *Departamento de Farmacología, Farmacia y Tecnología Farmacéutica, I+D Farma Group (GI-1645) Facultad de Farmacia, iMATUS and Health Research Institute of Santiago de Compostela, Universidade de Santiago de Compostela, Santiago de Compostela, Spain*

10.1 Introduction

The convergence of cutting-edge technologies with pharmaceutical manufacturing has the potential to usher in a new era of digital healthcare. Advanced non-invasive diagnostic tools, such as medical imaging, combined with innovative drug monitoring strategies (e.g., artificial intelligence tools or biosensors), provide valuable digital insights into the needs of individual patients. This wealth of information enables the production of medications customised to meet those unique requirements [1, 2]. However, traditional approaches centred around mass manufacturing of standardised dosage forms (e.g., capsule filling and

3D Printing of Pharmaceutical and Drug Delivery Devices: Progress from Bench to Bedside, First Edition.
Edited by Dimitrios A. Lamprou, Dennis Douroumis and Sheng Qi.

direct compaction [3]) are ill-suited for the era of personalised medicine. The adaptability of these methods for dose personalisation is costly and resource-intensive.

Three-dimensional (3D) printing, an additive manufacturing technology that constructs structures layer-by-layer, has emerged as the preferred platform of choice for manufacturing tailored medications, often referred to as 'Printlets™' [4]. To date, 3D printing has found extensive application in preclinical and clinical research, enabling fabrication of dosage forms with diverse sizes and doses [5, 6], varied geometries [7, 8], multiple routes of administration [9, 10], and even for different animal species [11]. The advantages offered by 3D printing in healthcare extend beyond these capabilities (Figure 10.1).

The ability to create customised formulations also holds great potential for the early stages of drug development, including preclinical studies, first-in-human (FIH) trials, and Phase I–II clinical trials. In these initial phases, new drug candidates are evaluated in animal models (pre-clinical) to assess efficacy, safety toxicology, and pharmacokinetic behaviour among other factors, before advancing to early human studies and subsequent clinical trials [12–16]. Integrating 3D printers into clinical settings would enable the automated production of small batches of pharmaceutical forms [17–20]. Moreover, 3D printing allows for the development of patient-friendly formulations, such as chewable or flavoured dosage forms, which improve medication compliance, particularly in geriatric and paediatric populations [5, 21, 22]. Customised drug release profiles, ranging from immediate to controlled release, can also be achieved, along with targeted delivery to specific regions of the gastrointestinal (GI) tract [23–25].

The portability of 3D printing makes it well-suited for manufacturing medicines in remote and hard-to-reach areas, including war and disaster zones, low-income countries, and potentially even future space missions. This technology enhances accessibility to vital medications in regions with limited resources [26]. In the front-line healthcare settings, such as community pharmacies and hospitals, 3D printing has the potential to reduce waiting times for compounded medications [10], as well as facilitate the production of personalised therapies, featuring tailored shapes, dosages, and sizes. Furthermore, being a digital technology, 3D printing seamlessly integrates other advanced technologies, such as artificial intelligence, medical imaging, and diagnostic tools [1, 27]. For instance, the results of point-of-care tests could be transmitted to a clinician for review and generation of a personalised prescription.

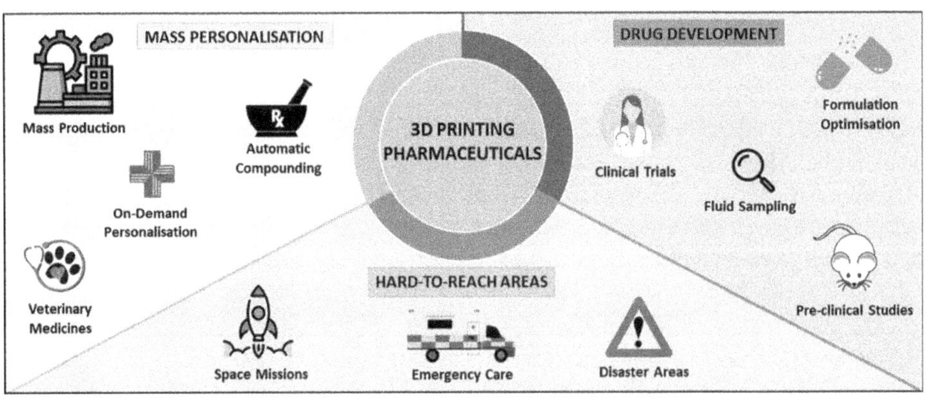

Figure 10.1 Pharmaceutical opportunities of 3D printing. (Source: Reproduced with permission from [4].)

The prescription is then sent to the pharmacy, where medications or medical devices with precise drug dosages are produced using 3D printing [28–30].

This chapter provides a concise overview of the five primary 3D printing technologies employed for pharmaceutical applications: binder jetting, vat photopolymerization, powder bed fusion, material jetting, and material extrusion [31–33]. Additionally, a comprehensive exploration of the applications and advancements of these 3D printing techniques in the preclinical and clinical settings, along with their advantages, limitations, and the current regulatory landscape, will also be presented.

10.2 3D Printing Technologies

According to the classification by the American Society of Testing Materials (ASTM), there are seven main categories of 3D printing [34]: binder jetting, vat photopolymerization, powder bed fusion, material jetting, material extrusion, directed energy deposition, and sheet lamination. However, within the realm of pharmaceutical development, the exploration primarily focuses on five key technologies.

10.2.1 Binder Jetting

Binder jet printing involves depositing a layer of powder onto a building platform (powder bed) and spraying a liquid binding solution over the powder via a nozzle. The wetted powder particles adhere together, solidifying the layers [35]. A notable application of this technology is ZipDose, developed by Aprecia Pharmaceuticals, which was utilised to manufacture the first FDA-approved 3D-printed tablet (Spritam®) [36]. Spritam® is a fast-dissolving tablet containing up to 1,000 mg of levetiracetam, designed to dissolve in the mouth within 11 seconds. This technology enables the production of highly porous, high drug load tablets [37–39].

10.2.2 Vat Photopolymerization

Vat photopolymerization encompasses various technologies, including digital light processing (DLP), two-photon polymerization (2PP), continuous liquid interface production (CLIP), volumetric and stereolithography (SLA). SLA is the most commonly used technique for pharmaceutical manufacturing. In vat photopolymerization, a vat of liquid resin is selectively solidified using a laser or other light source [40]. This technology provides the necessary high resolution for manufacturing drug delivery devices with intricate structures [8, 9, 26, 41]. However, challenges exist regarding the toxicity of photosensitive polymeric materials [42] and unwanted chemical reactions between drugs and resins can also occur [43].

10.2.3 Powder Bed Fusion

Powder bed fusion utilises a laser to selectively draw a specific pattern on a powder bed, bonding the powder particles together. This category includes selective laser sintering (SLS), multijet fusion (MJF), direct metal laser sintering/selective laser melting (DMLS/SLM), and electron beam melting (EBM) technologies [44]. The SLS technique is the most widely employed for pharmaceutical purposes [45], and has been used to fabricate Printlets with complex structures [46, 47] and special patterns [48].

10.2.4 Material Jetting

Material jetting, or in particular inkjet printing, involves the deposition of liquid droplets of materials on a surface, which could be made of, for example, edible paper [49], or simply be the build plate itself [50]. The main inkjet dispensing systems are piezoelectric and thermal inkjet printing [51]. Prominent material jetting techniques include drop-on-demand (DoD), where drops spontaneously solidify, and nanoparticle jetting (NPJ), where the drops are cured or fused using UV light or a heat source [52–54].

10.2.5 Material Extrusion

10.2.5.1 Fused Deposition Modelling

Fused deposition modelling (FDM) entails extruding a melted polymer filament through a heated nozzle and depositing it layer-by-layer onto a build plate [55]. Filaments are commonly produced using hot-melt extrusion (HME) [56, 57]. This technology enables the creation of hollow structures and objects with various geometrical shapes, such as cubes, spheres, and pyramids [58]. Drug release patterns from the 3D-printed structures can be modified by adjusting the infill percentage (i.e., the amount of materials incorporated within the structure, and ranges between 0%-completely hollow and 100%-completely filled) [59], to achieve immediate [60, 61] or controlled drug release [24, 62]. However, high temperatures required for melting and extrusion can pose challenges for thermolabile or heat-sensitive drugs [63, 64].

10.2.5.2 Semi-Solid Extrusion

In semi-solid extrusion (SSE), the printing involves extruding a paste, wax or gel from a syringe-like system to create a 3D structure. The material solidifies through cooling, solvent evaporation, or UV light curing [65]. SSE printing does not require high temperatures, making it suitable for printing thermolabile drugs [66]. It is highly recommended for the production of chewable and patient-friendly dosage forms [5, 67] and lipidic formulations intended for drug release upon contact with body temperature [10].

10.2.5.3 Direct Powder Extrusion

Direct powder extrusion (DPE) eliminates the step of creating a filament using HME. Instead, powdered material can be directly printed using a small hot-melt extruder within a printhead nozzle [68]. This approach reduces the amount of drug and excipients required and eliminates the need for HME, resulting in benefits such as reduced drug development and optimisation timelines [69, 70].

These technologies offer diverse capabilities and advantages for pharmaceutical manufacturing, paving the way for personalised medicine and the production of complex dosage forms tailored to specific patient needs.

10.3 Preclinical Applications of 3D Printing

During early-phase drug development, preclinical studies in animals are crucial to assess the efficacy, safety, toxicity, and pharmacokinetic behaviour of new drug candidates. These studies involve administering new molecule across a wide dose range [71, 72]. Rodents, being easy to handle, cost-effective, and physiologically similar to humans, are commonly used as animal models in preclinical studies [73]. Typically, the new drug candidate is administered in these animal models using liquid formulations, such as suspensions, or administered within small capsules (e.g., size 9 capsules) [13]. However, these methods are labour-intensive, may present solubility challenges, and do not facilitate the preparation of formulations with a wide dosage range. Given the high failure rate in the drug development process, it is crucial to have a technology that enables the rapid production of dosage forms to identify suitable dose ranges and molecules for further advancement into FIH clinical trials [74].

3D printing provides a flexible manufacturing platform for producing Printlets in a short timeframe, making it ideal for rapid preclinical progression at low cost. The technology allows the production of small batches of prints with a wide variety of doses on demand. In preclinical settings, 3D printing enables the preparation of dosage forms with different excipients or combinations thereof, facilitating the testing of factors such as compatibility and interactions with the drug. During the late stages of FIH trials, the ability to produce Printlets immediately prior to administration can reduce the length of storage duration, mitigating some stability issues and minimising the need for long-term stability studies [66].

The high degree of design freedom offered by 3D printing enables the creation of dosage forms with geometries and sizes tailored to the selected preclinical animal model. For example, small capsule-shaped devices and small pellets have already been 3D printed, showing potential application for the oral drug administration to rodents [17, 75]. Additionally, small 3D-printed suppositories have already been used in a preclinical study to test a formulation intended for human use [18]. The decreasing cost and size of 3D printers, along with the creation of more user-friendly interfaces, make 3D printing an affordable technology that can be easily integrated into preclinical or laboratory environments. Researchers and lab staff can manufacture small batches of Printlets on demand, ready for use in preclinical studies. To date, various 3D-printed formulations have been evaluated in preclinical animal models, including oral drug products for immediate or modified drug release, rectal dosage forms, and 3D-printed devices targeting specific regions of the GI tract.

10.3.1 Immediate and Modified Release Oral Printlets

The pharmacokinetic behaviour of some 3D-printed oral drug products has been investigated in animal studies, demonstrating their potential in achieving immediate and modified release profiles. For instance, 3D-printed tablets of warfarin, a drug with a narrow therapeutic index often requiring splitting of commercial tablets for correct dosage, were prepared using FDM in a size and dose suitable for oral administration to rats [20]. A comparison of the pharmacokinetic behaviour between the 3D tablets and the drug solution revealed a superior profile for the Printlets due to sustained drug release. FDM was also utilised to fabricate a two-compartment dosage unit loaded with rifampicin and isoniazid, two anti-tuberculosis drugs, enabling the physical isolation and controlled release of each drug (Figure 10.2) [76]. The separation of the drugs prevented their simultaneous release in the stomach, avoiding potential interactions. *In vivo* release profiles confirmed the sequential release of rifampicin following by a slower and more delayed release of isoniazid from the 3D-printed dosage form.

The potential applications of 3D printing also extend to the oral or localised delivery of peptides and other biological drugs. In one study, 3D-printed capsules sensitive to pressure were fabricated using FDM with filaments composed of extruded Eudragit RS powder, a polymer commonly used in enteric coatings [77]. These pressure-sensitive capsules, loaded with the octapeptide octreotide, were designed to break at a specific area of the GI tract. Comparisons of their performance with traditional enteric coated tablets and gelatin capsules were conducted in Beagle dogs, revealing similar octreotide bioavailability for the

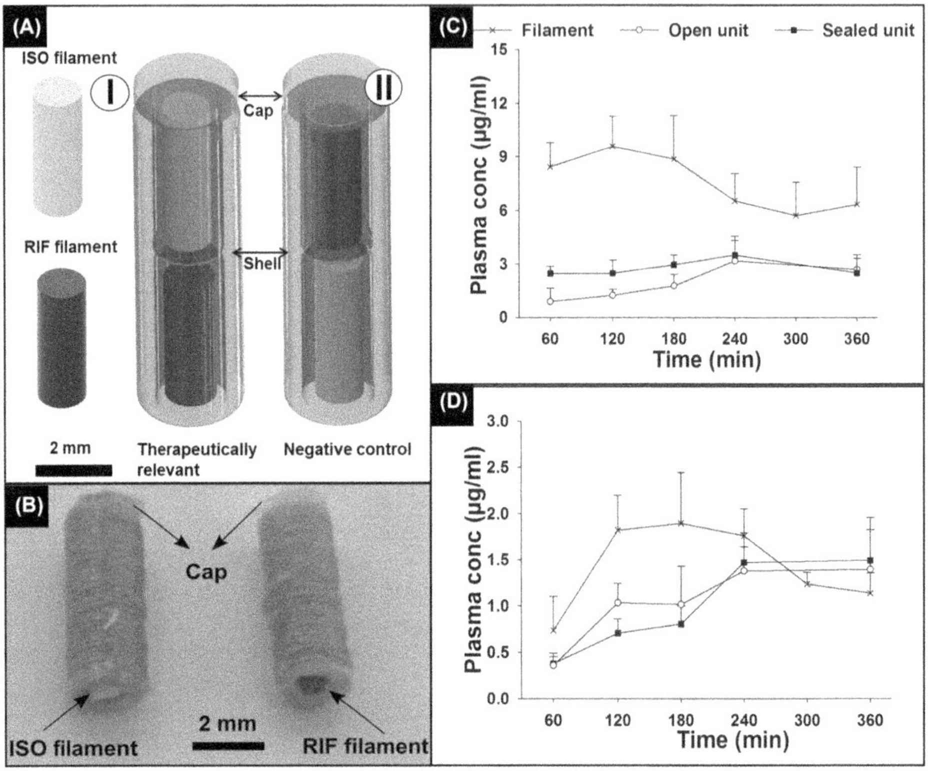

Figure 10.2 (A) Schematic representation of (I) isoniazid and rifampicin filaments prepared by HME and (II) the final dual-compartmental dosage units. (B) Images of the 3D-printed dual-compartmental dosage units. Plasma concentration-time profiles of (D) isoniazid and (E) rifampicin in fasted rats after oral administration of filaments and dual compartmental dosage units with and without sealing. (Source: Reproduced from [76] with permission from Elsevier.)

pressure-sensitive capsules. Another study in Beagle dogs focused on modifying the regional absorption of the drug lamivudine by varying the wall thickness of 3D-printed capsules [19]. The approach resulted in different release profiles, allowing evaluation of regional drug absorption in the animal model. In a study utilising PET/CT medical imaging, the transit of 3D-printed capsules through the GI tract was visualised. Four polymer-based capsules, printed using FDM and filled with the radiotracer [^{18}F]FDG, were tracked to evaluate their intestinal behaviour after administration to rats (Figure 10.3) [75]. Interestingly, the capsules did not empty from the stomach of the rats, despite being manufactured in sizes recommended for rodent administration (e.g., 9 and 9h sizes). Subsequent studies clarified that the size of the capsule and especially the use of anaesthetic agents strongly influence the gastric emptying of the devices [13].

By manipulating certain parameters such as the number of shells and the infill percentage, immediate and sustained release Printlets can be obtained [78]. Drug filaments loaded with diltiazem and drug-loaded and drug-free filaments, for FDM were prepared using HME. By alternating the use of drug-free and drug-loaded filaments, both pulsatile and chronologically controlled release tablets were created, yielding *in vivo* release profiles in rats similar to *in vitro* results. The ability to combine more than two drugs in a single dosage form is another

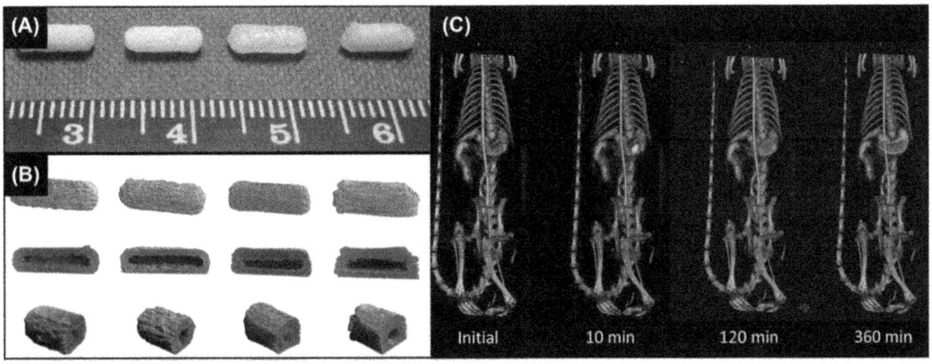

Figure 10.3 (A) Images and (B) micro-CT scans of FDM 3D-printed size 9 capsules composed of (from left to right) Kollicoat IR, Klucel EF, Aqualon N7, and Aquasolve-LG. (C) Gastrointestinal tracking of the Klucel EF device using PET/CT imaging at different time points. (Source: Reproduced from [75] with permission from Elsevier.)

advantage of 3D printing. For example, sustained release tablets containing three drugs (efavirenz, tenofovir disoproxil fumarate, and emtricitabine) used in human immunodeficiency virus (HIV) treatment were successfully produced using SSE 3D printing [79]. A preclinical *in vivo* study conducted in pigs showed enhanced bioavailability of the drugs compared to the marketed formulation, attributed to the use of humic acid-polyquaternium 10 (HA-PQ10), which improved drug solubility through hydrogen bonding.

Gastroretentive systems that prolong formulation gastric residence time have been developed using 3D printing. Commercial tablets or capsules can be included within the gastroretentive devices, as demonstrated in studies utilising FDM [80]. In one study, a sustained-release tablet of acyclovir was incorporated into a gastro-floating shell, which was 3D printed and administered to Beagle dogs (Figure 10.4) [25]. X-ray imaging revealed that

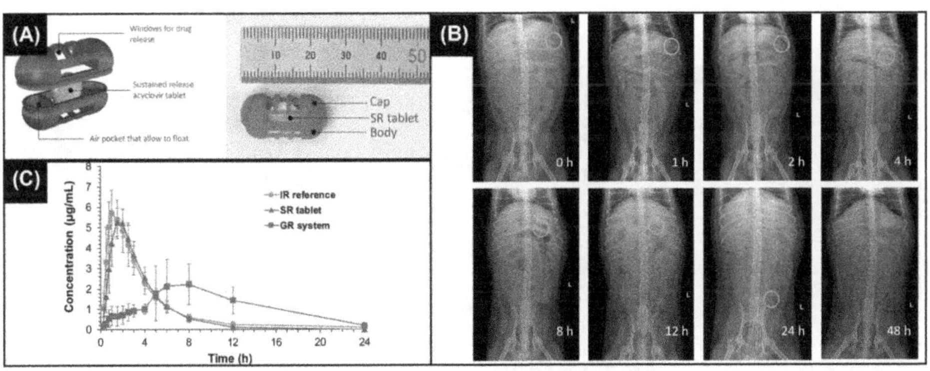

Figure 10.4 (A) (Left) 3D design and (right) image of a gastroretentive device composed of an FDM 3D-printed gastro-floating shell with an acyclovir tablet inside it. (B) Abdominal X-ray images tracking the position of the device following oral administration. (C) In vivo pharmacokinetics of the drug after oral administration of the gastroretentive system (GR) or immediate-release (IR) and sustained-release (SR) alone. (Source: (A), (B) and (C) Reproduced with permission from [25]/PLOS/CC by 4.0.)

the formulation remained in the stomach for at least 10 to 12 hours. Another study employed FDM to create gastroretentive devices using different infill percentages and shell numbers, loading the drug domperidone onto the filaments [81]. The devices exhibited a prolonged release of the drug and remained in the stomach of rabbits for at least 10 hours, as visualised through X-ray imaging. Gastroretentive devices hold promise in the treatment requiring sustained blood levels of the drug over an extended period of time. For instance, an ultra-long-acting drug delivery capable of sustaining release of the antimalarial drug ivermectin for at least one week was fabricated using 3D printing and tested in a pig animal model [82].

Moreover, 3D printing has been applied to the development of personalised biomedical electronics, enabling new therapeutic functionalities and advanced personalised diagnostics [2]. Instead of performing invasive procedures to implant long-term resident electronic devices, these devices can be ingested and retained in the GI tract for several days or even weeks. A 3D-printed wireless gastroretentive electronic device tested in pigs demonstrated controlled drug release and maintained *in vivo* wireless communication for a significant period, with a gastric residence period of 36 day [83].

10.3.2 3D-Printed Drug Delivery Devices for Other Routes of Administration

In addition to the oral route, 3D printing has found applications in preclinical studies involving rectal dosage forms. Early efforts in this area focused on creating moulds for rectal or vaginal suppositories rather than directly printing them [84]. However, more recent advancements have enabled the direct printing of lipid-based suppositories without the need for moulds, achieved through 3D printing [10, 85]. Various drugs have been tested using this approach, including tacrolimus [85], tofacitinib citrate, and budesonide, with investigations into the possibility of combining multiple drugs within a single suppository [86].

In a preclinical study, previously developed 3D-printed lipid-based tacrolimus suppositories [10] were adapted in size and dose for their administration to rats with induced experimental colitis (Figure 10.5) [18]. Tacrolimus is an immunosuppressant drug used in the treatment of ulcerative colitis. The objective of the study was to evaluate the remission of the disease in the experimental animal model following the administration of the 3D-printed rectal dosage forms. PET/CT medical imaging [87] was employed to assess

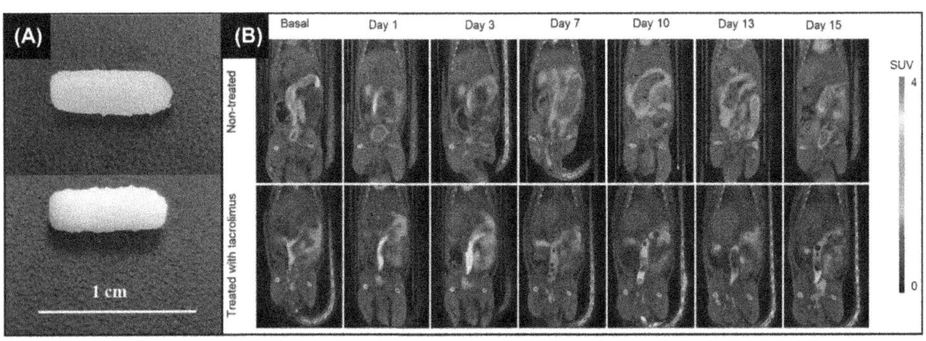

Figure 10.5 (A) SSE 3D-printed lipid-based tacrolimus suppositories adapted in size and dose for administration to rats. (B) Fused PET/CT images over time of rats receiving the tacrolimus suppositories, and non-treated animals. Experimental colitis was induced on day 3. The metabolic activity is coded on a colour scale ranging from blue (low [^{18}F]FDG uptake) to red (high [^{18}F]FDG uptake). (Source: [18]/MDPI/CC BY 4.0.)

disease progression before and after suppository administration, confirming disease remission from day 7 compared to the non-treated group.

More recently, this approach was applied to rectal biological delivery, specifically formulating infliximab into a suppository using SSE [88]. The promising results from this study demonstrate the potential of 3D printing as a novel platform for fabricating dosage forms containing biopharmaceuticals. This approach addresses patient compliance issues often associated with injectables and aims to meet the unmet needs of patients suffering from acute severe ulcerative colitis.

10.4 Clinical Applications of 3D Printing

Thus far, various types of 3D-printed formulations have been produced, including orally disintegrating tablets [89, 90] and films [91–94], chewable tablets [5, 67, 95], capsule shells [96, 97], suppositories [10, 18, 85–88], as well as intrauterine [98–102] and intravesical devices [9, 103], hearing aids [104], and punctal plugs [8].

10.4.1 Personalised Medications

In recent years, there has been a growing interest in exploring 3D printing as a novel method for the production of personalised medications tailored to the individual needs of patients. 3D printing is well suited for creating and dispensing small or custom batches of individualised Printlets, with specific doses, dosage forms, drug(s) content, and even aesthetics, right at the point-of-care, such as in a clinic or pharmacy.

The use of 3D printing offers particular advantages in formulating drugs with narrow therapeutics indices (i.e., drugs with small differences between their therapeutic and toxic doses), ensuring their safe use [26, 105]. It also addresses the specific dosing and formulation requirements of geriatric and paediatric patients, whose needs may differ significantly from those of standard adult patients. For example, ageing can lead to visual or swallowing difficulties, requiring specific considerations in medication formulation packaging [106]. Similarly, medications for young children need to be tailored to their physical characteristics and pharmacokinetics [106, 107]. Currently, these specialised doses and formulations are not always readily available, and the practice of splitting or crushing tablets to achieve the desired doses is associated with risks of errors and inaccuracies, or even dose dumping [108–111]. Tailored drug products can also be prepared using extemporaneous compounding; however, since this is performed manually, it is laborious and time-consuming.

The digital and automatic nature of 3D printing makes it a suitable alternative enabling the rapid and flexible preparation of patient-specific drug products. For instance, SSE 3D printing can be used to produce subdivided Printlets of spironolactone and hydrochlorothiazide, offering an alternative to split tablets (Figure 10.6) [112]. Studies have shown that these Printlets meet the requirements for mass variation, drug content, and content uniformity specified in the European Pharmacopeia. Furthermore, the Printlets exhibit superior aesthetics compared to split tablets, making them more appealing to young patients.

In 2019, the first clinical study on 3D printing of pharmaceuticals was conducted, focusing on the personalised fabrication of chewable isoleucine Printlets for children with maple syrup urine disease (MSUD), a rare and severe metabolic disorder [5]. The Printlets, created in different sizes, flavours, and colours (Figure 10.7A), were compared with conventional capsules prepared through extemporaneous compounding. The study monitored isoleucine blood

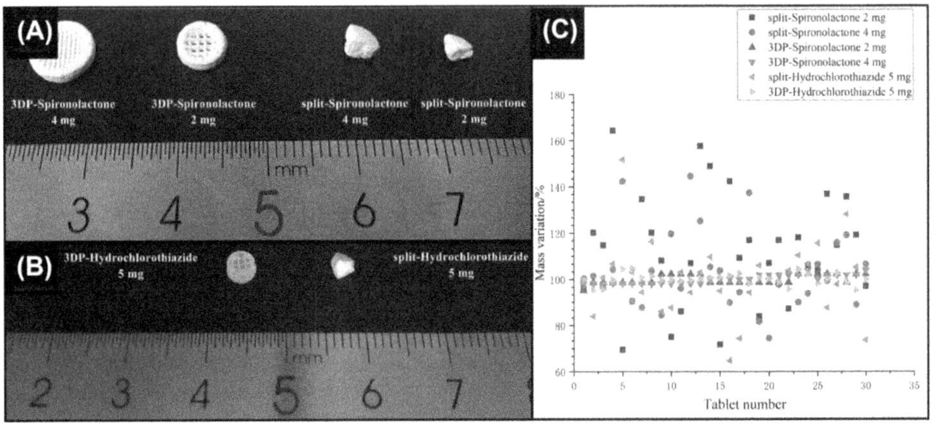

Figure 10.6 Images of (A) spironolactone (2 and 4 mg) and (B) hydrochlorothiazide (5 mg) Printlets, compared with split tablets containing the same doses. (C) Mass variation for 3D-printed subdivided tablets and split commercial tablets. (Source: Reproduced with permission from [112], Elsevier.)

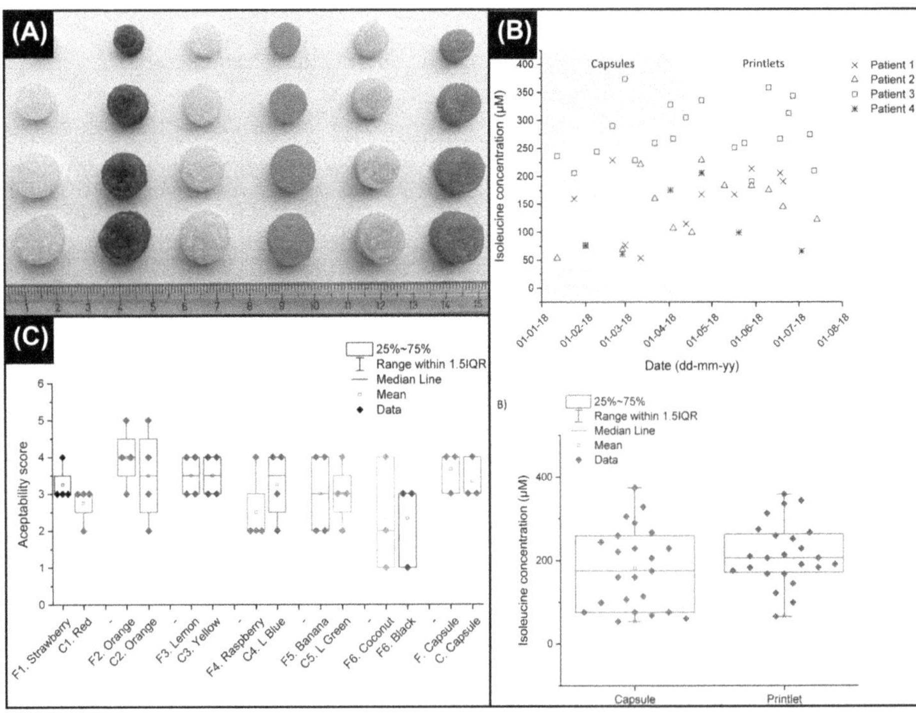

Figure 10.7 (A) Image of chewable Printlets in different flavours, colours, and with different doses of isoleucine. (B) (top) Isoleucine blood levels of the patients during the study and (bottom), isoleucine blood levels and mean values for Printlets and capsules during the study. (C) Patient reported outcomes scores for the flavours (F.) and colours (C.) of the chewable Printlets and the capsule. (Source: Reproduced with permission from [5], Elsevier.)

levels and patient acceptability to both treatments were monitored over a period of six months. The 3D-printed Printlets demonstrated higher accuracy in achieving the targeted isoleucine blood concentrations compared to the capsules (Figures 10.7B and 10.7C), which showed variability. Additionally, the Printlets were well accepted by the children, with certain flavours and colours being preferred (e.g., orange-orange and lemon-yellow).

Apart from oral dosage forms, 3D printing has also been employed to fabricate customised drug delivery devices with various designs and release properties. Examples include personalised mouthguards containing clobetasol propionate as a model drug [113] and bespoke orthodontic retainers incorporating chonidine hydrochloride (Figure 10.8) [114]. These devices exhibited specific drug release profiles, meeting the individualised treatment needs of patients.

10.4.2 Improved Acceptability and Medication Compliance

3D printing enables the engineering of drug products with specific shapes, sizes, or formulation characteristics, thereby enhancing patient acceptability and medication compliance. Various patient-friendly formulations, such as orodispersible tablets [89, 90], films [91–94], and chewable jelly-like [5, 95] or chocolate-based [67] Printlets, have been created using 3D printing techniques. By offering visually appealing and pleasant-tasting medications, these formulations ensure that patients receive therapies consistently, reducing side effects and minimising the need for hospitalisations.

10.4.2.1 Paediatric Patients

In the case of paediatric patients, 3D printing offers significant advantages by allowing the fabrication of dosage forms with tailored flavours and colours, and appealing shapes. This customisation addresses the issue of unpleasant taste, which is a common barrier to medication compliance among young patients [115]. For example, using the FDM 3D printing technology, chewable Printlets in different shapes such as hearts, bottles, bears, rings, and

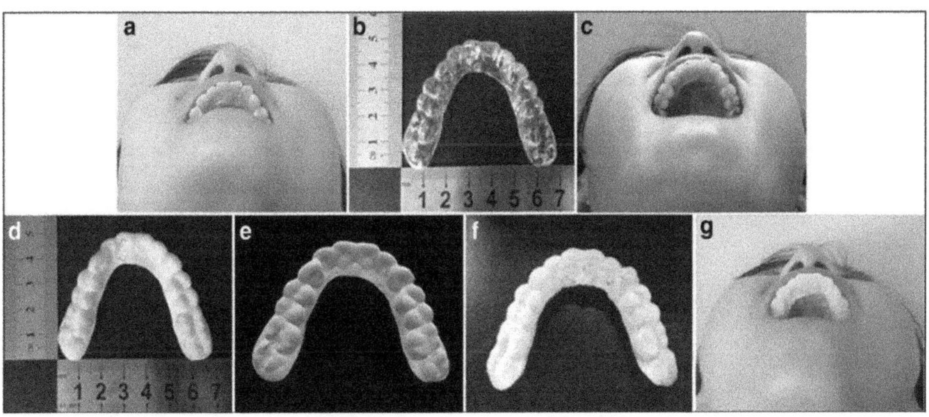

Figure 10.8 *Design and preparation process of 3D-printed orthodontic retainers containing clonidine hydrochloride (CH): (a) appearance of the upper teeth of the volunteer; (b) moulded orthodontic retainer; (c) moulded orthodontic retainer worn by the volunteer; (d) film-coated moulded orthodontic retainer; (e) computer image of the orthodontic retainer; (f) CH–loaded 3D-printed retainer; and (g) 3D-printed orthodontic retainer worn by the volunteer [114]. (Source: Reproduced with permission from Springer Nature.)*

lions have been created with taste-masking properties and improved palatability [116]. Additionally, gelatin-based Lego™ brick-shaped formulations [117] and chocolate-based Printlets [67] have been successfully fabricated, further demonstrating the versatility of 3D printing in creating patient-friendly formulations for paediatric use.

Acceptability studies have been conducted to assess the preference of children (aged 4–11 years) for Printlets produced using different 3D printing technologies [118]. The visual assessment involved comparing placebo Printlets made using FDM, SSE, DLP, or SLS 3D printing (Figure 10.9A). Initially, DLP Printlets were found to be the most visually preferred (preferred by 61.7% of the participants), followed by SLS Printlets, while the FDM and SSE Printlets scored lower (Figure 10.9B). However, when children were

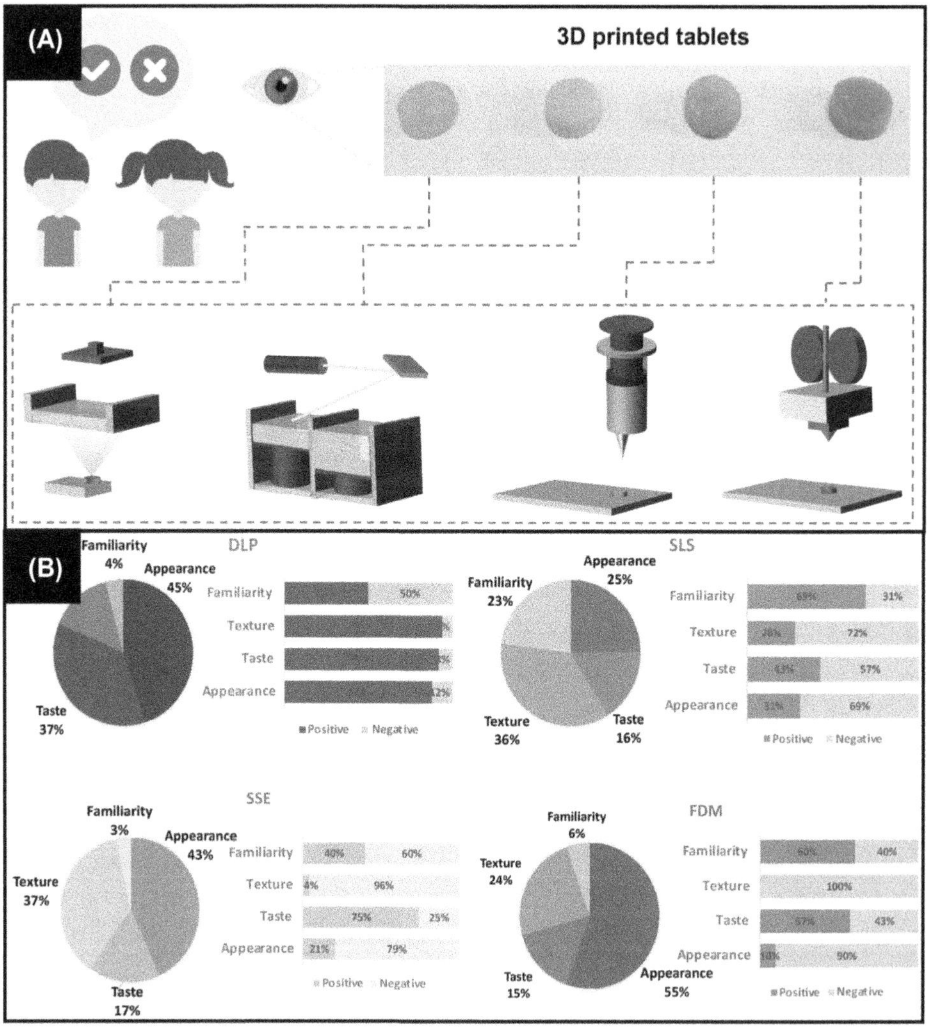

Figure 10.9 *(A) Outline of the patient acceptability study using placebo Printlets fabricated using (from left to right) DLP, SLS, SSE, and FDM 3D printing. (B) Summary of visual description data for the Printlets based on familiarity, appearance, perceived taste, and texture (DLP, n = 244; SLS, n = 170; SSE, n = 125; FDM, n = 92). (Source: [118]/MDPI/CC BY 4.0.)*

informed that SSE Printlets were chewable, their preferences shifted in favour of the SSE Printlets, highlighting the importance of formulation type in the acceptability of a treatment.

10.4.2.2 Adult and Geriatric Patients

In the case of adult and geriatric patients, 3D printing offers the potential to generate formulations with unique properties that cannot be achieved through traditional mass manufacturing processes. Studies have shown that the acceptability of 3D-printed medications is influenced by factors such as shape, size, and colour (Figure 10.10) [7, 119]. Patients tend to prefer dosage forms that resemble conventional formulations, such as capsules and discs, due to familiarity. However, novel shapes like torus (i.e., donut-like shape) Printlets have also been found to be well accepted, indicating the potential for exploring unconventional shapes in practice.

For visually impaired patients, 3D printing has been utilised to create orally disintegrating Printlets [48] and intraoral films with Braille or Moon patterns on their surface [120] (Figure 10.11). These tactile patterns enable patients to identify medications, especially when they are taken out of their original packaging. The patterns on the films and Printlets were successfully read by visually impaired volunteers, demonstrating the feasibility and safety of this approach.

Although still in its early stages and yet to undergo human trials, the utilisation of SSE warfarin orodispersible films presents a promising alternative to extemporaneously compounded oral warfarin powder sachets [92]. These films offer several advantages,

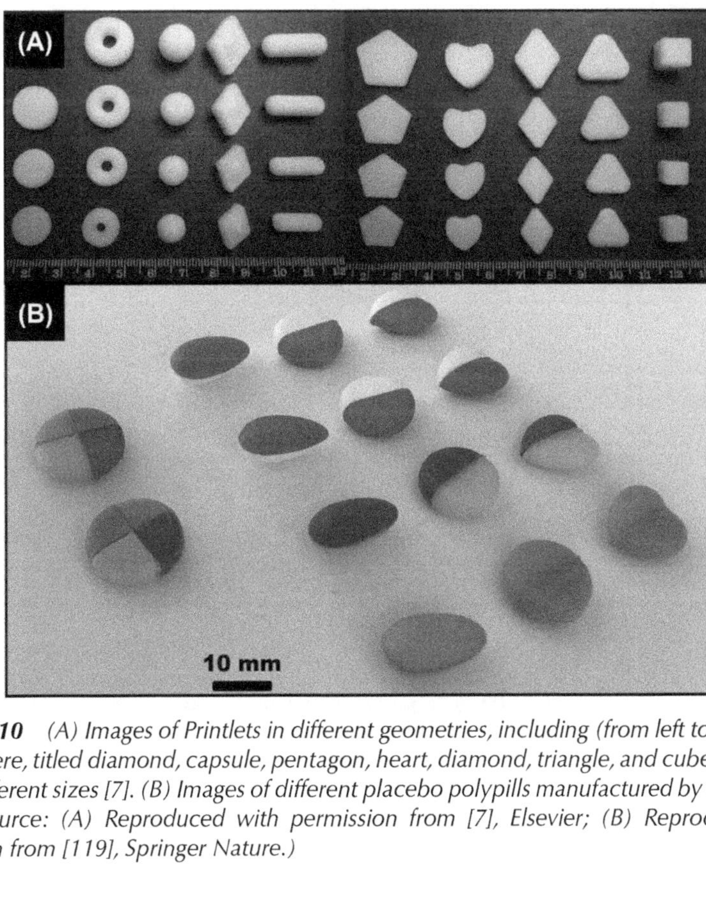

Figure 10.10 (A) Images of Printlets in different geometries, including (from left to right) disc, torus, sphere, titled diamond, capsule, pentagon, heart, diamond, triangle, and cube, presented in four different sizes [7]. (B) Images of different placebo polypills manufactured by 3D printing [119]. (Source: (A) Reproduced with permission from [7], Elsevier; (B) Reproduced with permission from [119], Springer Nature.)

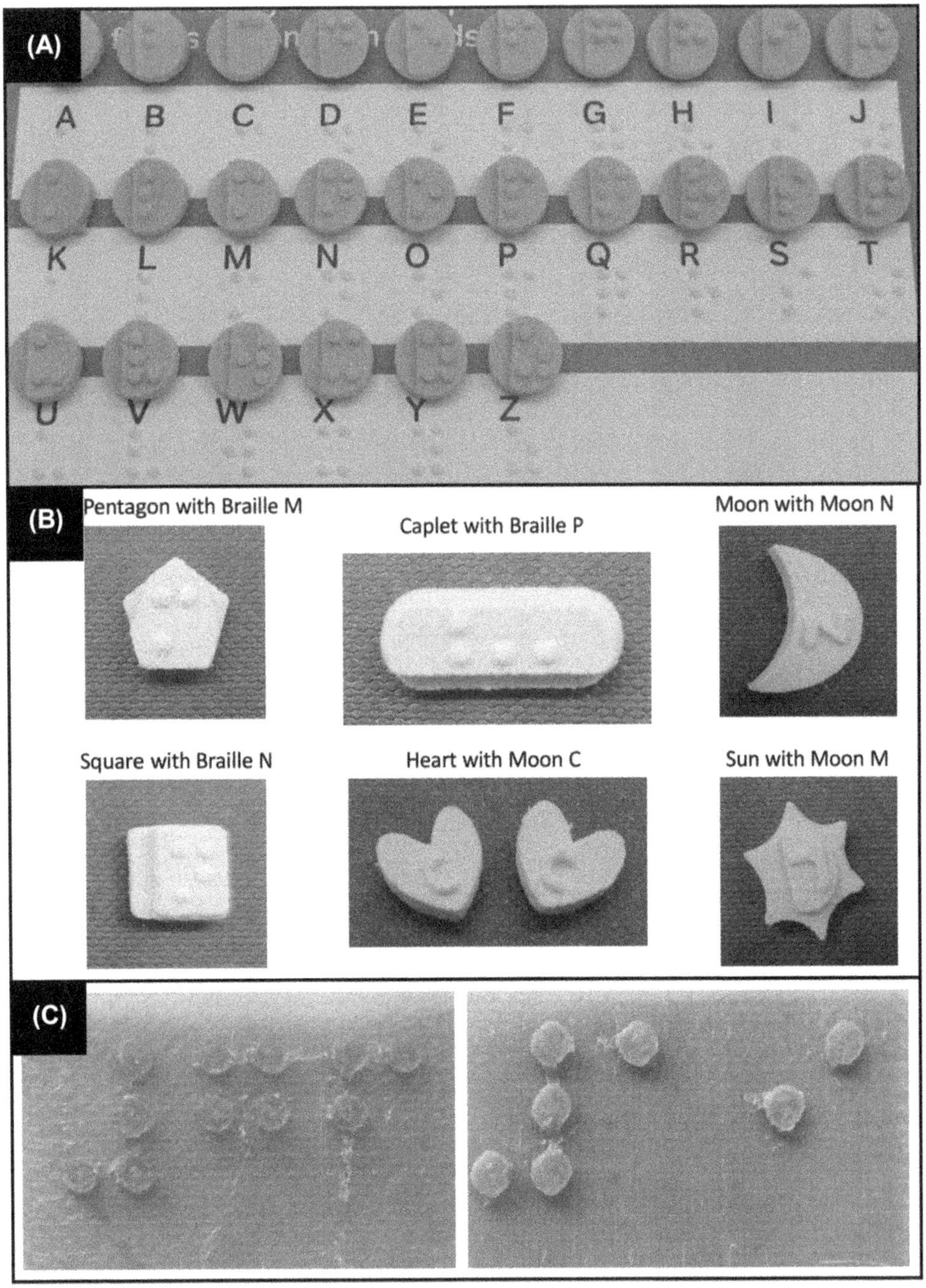

Figure 10.11 *Images of (A) SLS Printlets containing all the Braille alphabet on their surface; (B) SLS Printlets with different shapes and containing Braille or Moon patterns [48]; and (C) FDM films with Braille patterns, containing different dosage strengths [120]. (Source: (A),(B) [48] /MDPI/CC BY 4.0; (C) [120], Reproduced with permission from Elsevier.)*

including enhanced safety and identification features. By incorporating quick response (QR) codes within the films, dosage form information can be easily accessed, minimising the risk of administration errors. Notably, the 3D-printed films have demonstrated superior performance in delivering accurate drug doses compared to traditional powder sachets. Furthermore, their ability to be consumed without the need for water provides is an additional convenience and ease of use for patients.

Polypharmacy (i.e., the concurrent use of five or more medications), a common practice among elderly patients, poses challenges to medication compliance due to high pill burden and complex treatment regimens [121–123]. 3D printing offers a solution by allowing the design of polypills, referred to as polyprintlets, that combine multiple drugs or drug doses in a single dosage form based on the individual patient's treatment regimen. Studies have demonstrated that various iterations can be produced [17, 124–126]. As an example, highly modular polypill capsules with tailored release profiles were fabricated using FDM 3D printing combined with hot-filling [28]. The polypills can be designed with different internal arrangements to achieve immediate or delayed drug release profiles, offering more control over the drug release characteristics.

10.4.3 Mass Manufacturing

Beyond small-scale manufacturing, 3D printing has the potential to revolutionise mass manufacturing of drug products. An example of successful mass manufacturing using the 3D printing technology is Spritam® (Aprecia Pharmaceuticals), the world's only FDA-approved 3D-printed drug product [127]. Spritam® utilises the Zipdose® technology, which involves scaled-up binder jetting 3D printing. While personalisation is not the primary goal of this technology, Spritam® is produced with multiple fixed doses of the anti-epileptic drug, levetiracetam. The advantage lies in the generation of highly porous tablets that disintegrate in the mouth, allowing high drug loading of up to 1,000 mg. Comparative studies have shown comparable drug plasma concentrations between Spritam® and immediate-release reference formulations, with improved acceptability and swallowability observed for Spritam® [128] .

10.4.4 Decentralised On-Demand Fabrication

Traditionally, pharmaceutical production has been centralised due to the bulkiness of manufacturing equipment. However, the portability of most desktop 3D printing platforms now enables decentralised fabrication of dosage forms in various settings, including clinics, pharmacies, remote areas, and even at a patient's home [26, 129]. This decentralisation opens up opportunities for on-demand fabrication, particularly in low- and middle-income countries, disaster zones, and space. As an example, NASA is considering the use of 3D-printed prosthetics for treating space injuries [130] or for creating medical supplies in orbit [131]. More recently, a 3D printer that functions using a smartphone's screen has been suggested as an enabler for medicine production at the point-of-care [26].

10.4.5 Veterinary Applications

The benefits of 3D printing extend beyond human medicine and can be applied to veterinary medicine as well. In the case of household pets (e.g., dogs, cats, rabbits, or guinea pigs) medication dosing is typically determined based on age or body weight, which varies depending on the drug and treatment indication. The required doses are generally obtained by splitting commercial formulations and mixing them with food or directly administering into the animal's mouth with a syringe. However, similar to humans, animals may refuse to swallow unpalatable formulations, not to mention the high risk of errors associated with this approach [132, 133]. 3D printing offers a safer and more reproducible way to fabricating veterinary formulations, which can be done in veterinary clinics or even at home [11]. This approach reduces the risk of errors and improves medication administration for animals.

10.5 Challenges, Regulatory View and Future Applications

The combination of 3D printing with other digital health technologies, such as biosensors, robots, artificial intelligence, digital prescriptions, and virtual and augmented reality, is transitioning the pharmaceutical sector into a new digital era (Figure 10.12) [1, 27]. This innovative healthcare model envisions a decentralised platform where personalised medicines are automatically produced based on remote monitoring data. However, several regulatory, quality, and technical challenges need to be addressed for the widespread adoption and acceptance of this model in the pharmaceutical sector.

One of the pivotal concerns currently is the lack of clear and comprehensive regulatory and legal guidance. Establishing a safe environment for the use of technologies like 3D printing while patient data and identities is of paramount importance. Regulatory agencies have been urged to develop a *Meaningful Regulation*, that covers both hardware and software aspects and oversees the use of digital technologies in healthcare. For on-demand 3D printing of dosage forms and medical devices, traditional regulations may not be directly

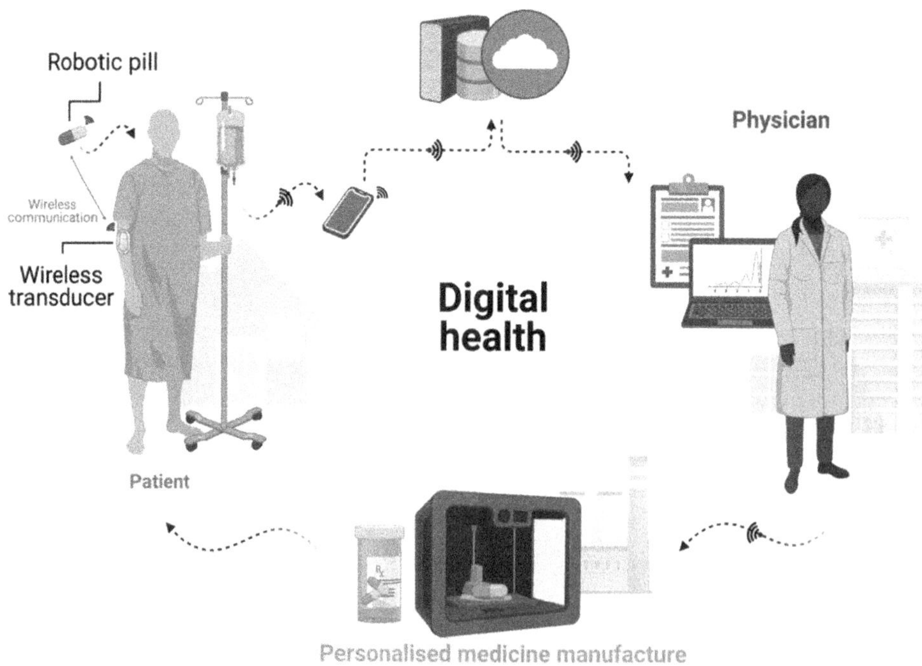

Figure 10.12 The virtual cycle of digital health. (Source: Reproduced with permission from [1].)

applicable, necessitating adaptable and versatile guidelines. Additionally, the final drug products must comply with regulatory standards for dose uniformity and end-product consistency [3]. Ensuring that the 3D printers used adhere to current Good Manufacturing Practice (GMP) requirements is also essential [134, 135].

The classification of 3D printing as a pharmaceutical production process or a form of extemporaneous compounding is a topic of ongoing discussion. Current data tends to support the notion that it should be regarded as a form of compounding. To ensure patient safety and product reproducibility, specific considerations must be taken into account when implementing 3D printing in clinics. Traditional quality control (QC) measures are unsuitable due to their destructive nature, time-consuming processes, and labour-intensive nature. Non-destructive analytical techniques such as near infrared (NIR) and Raman spectroscopy have been suggested as viable alternatives [136, 137]. To enhance traceability and ensure drug quality across the supply chain, the use QR codes and data matrices has been proposed as an additional track-and-trace measure [138, 139]. Implementation of this concept could provide patients with tailored information regarding medication dosing.

The rapid evolution of digital technologies poses a challenge for regulatory agencies to keep up with the constant changes in the landscape. Regular revisions and modifications to legislations are necessary to ensure they remain up to date with the latest technological advancements. Furthermore, while the pharmaceutical industry is known for its reluctance to change, regulatory bodies must strike a balance between guidelines that protect patients and healthcare practitioners and allowing the adoption of advanced technologies like 3D printing. This requires a multidisciplinary discussion involving various pharmaceutical stakeholders, including clinicians, patients, researchers, and pharmaceutical companies. Once the full potential of 3D printing is realised, it is envisioned that it will play a pivotal role in a closed-loop system of continuous monitoring, diagnosis, and on-demand dispensing based on digital outputs, thereby making healthcare more efficient and accessible worldwide.

10.6 Conclusion

The pharmaceutical industry is undergoing a significant transformation in the wake of the latest advancements in digital technologies. The integration of novel cutting-edge tools such as 3D printing is poised to revolutionise the way medications are designed and manufactured, ushering in a new era of personalised on-demand treatments. This groundbreaking approach holds tremendous potential for expediting the drug development process while simultaneously maintaining therapeutic efficacy and reducing adverse effects and hospitalisations.

One of the key advantages of 3D printing is its affordability and portability, which make it highly suitable for use in clinics, pharmacies, and even patient homes. This accessibility empowers healthcare providers to produce customised medications at the point-of-care, tailoring treatments to individual patients' needs. Furthermore, the ability to rapidly iterate and refine drug formulations using the 3D printing technology opens up new avenues for innovation and personalised medicine.

However, to fully realise the potential of 3D printing in the pharmaceutical sector, it is imperative to establish robust regulations and stringent quality control measures. Clear guidelines need to be developed to ensure the safety, efficacy, and consistency of 3D-printed drugs. Regulatory agencies must keep pace with the rapid advancements in digital technologies, regularly revising and updating legislation to adapt to the evolving landscape.

Collaboration among various stakeholders, including regulatory bodies, pharmaceutical companies, clinicians, researchers, and patients, is crucial in shaping this new pharmaceutical model. By fostering multidisciplinary discussions and forging partnerships, the industry can strike a balance between patient safety and the utilisation of advanced technologies, enabling the seamless integration of 3D printing into healthcare systems.

Looking ahead, the future of pharmaceuticals holds great promise. The convergence of digital technologies and 3D printing is poised to create a paradigm shift in drug development and delivery. A new pharmaceutical model, characterised by personalised and on-demand treatments, is within reach. With the right regulations, quality control measures, and collaborative efforts, we can unlock the full potential of 3D printing, making healthcare more efficient, accessible, and patient-centric.

References

[1] Awad, A., Trenfield, S.J., Pollard, T.D. et al. (2021). Connected healthcare: improving patient care using digital health technologies. *Advanced Drug Delivery Reviews* **178**: 113958.
[2] Ghosh, U., Ning, S., Wang, Y., and Kong, Y.L. (2018). Addressing unmet clinical needs with 3D printing technologies. *Advanced Healthcare Materials* **7** (17): 1800417.
[3] Awad, A., Trenfield, S.J., and Basit, A.W. (2021). Chapter 19 – Solid oral dosage forms. In: *Remington*, 23e (ed. A. Adejare), 333–358. Academic Press.
[4] Seoane-Viaño, I., Trenfield, S.J., Basit, A.W., and Goyanes, A. (2021). Translating 3D printed pharmaceuticals: from hype to real-world clinical applications. *Advanced Drug Delivery Reviews* **174**: 553–575.
[5] Goyanes, A., Madla, C.M., Umerji, A. et al. (2019). Automated therapy preparation of isoleucine formulations using 3D printing for the treatment of MSUD: first single-centre, prospective, crossover study in patients. *International Journal of Pharmaceutics* **567**: 118497.
[6] Boniatti, J., Januskaite, P., Fonseca, L.B.D. et al. (2021). Direct powder extrusion 3D printing of praziquantel to overcome neglected disease formulation challenges in paediatric populations. *Pharmaceutics* **13** (8): 1114.
[7] Goyanes, A., Scarpa, M., Kamlow, M. et al. (2017). Patient acceptability of 3D printed medicines. *International Journal of Pharmaceutics* **530** (1–2): 71–78.
[8] Xu, X., Awwad, S., Diaz-Gomez, L. et al. (2021). 3D printed punctal plugs for controlled ocular drug delivery. *Pharmaceutics* **13** (9): 1421.
[9] Xu, X., Goyanes, A., Trenfield, S.J. et al. (2021). Stereolithography (SLA) 3D printing of a bladder device for intravesical drug delivery. *Materials Science and Engineering C: Materials for Biological Applications* **120**: 111773.
[10] Seoane-Viaño, I., Ong, J.J., Luzardo-Álvarez, A. et al. (2021). 3D printed tacrolimus suppositories for the treatment of ulcerative colitis. *Asian Journal of Pharmaceutical Sciences* **16** (1): 110–119.
[11] Sjöholm, E., Mathiyalagan, R., Rajan Prakash, D. et al. (2020). 3D-printed veterinary dosage forms: a comparative study of three semi-solid extrusion 3D printers. *Pharmaceutics* **12** (12): 1239.
[12] Madla, C.M., Gavins, F.K.H., Merchant, H.A. et al. (2021). Let's talk about sex: differences in drug therapy in males and females. *Advanced Drug Delivery Reviews* **175**: 113804.
[13] Gómez-Lado, N., Seoane-Viaño, I., Matiz, S. et al. (2020). Gastrointestinal tracking and gastric emptying of coated capsules in rats with or without sedation using CT imaging. *Pharmaceutics* **12** (1): 81.
[14] Ku, M.S. and Dulin, W. (2012). A biopharmaceutical classification-based Right-First-Time formulation approach to reduce human pharmacokinetic variability and project cycle time from first-in-human to clinical proof-of-concept. *Pharmaceutical Development and Technology* **17** (3): 285–302.
[15] Seoane-Viaño, I., Gómez-Lado, N., Lázare-Iglesias, H. et al. (2019). Evaluation of the therapeutic activity of melatonin and resveratrol in inflammatory bowel disease: a longitudinal PET/CT study in an animal model. *International Journal of Pharmaceutics* **572**: 118713.

[16] Bendicho-Lavilla, C., Seoane-Viaño, I., Otero-Espinar, F.J., and Luzardo-Álvarez, A. (2021). Fighting type 2 diabetes: formulation strategies for peptide-based therapeutics. *Acta Pharmaceutica Sinica B* **12** (2): 621–636.
[17] Awad, A., Fina, F., Trenfield, S.J. et al. (2019). 3D printed pellets (miniprintlets): a novel, multi-drug, controlled release platform technology. *Pharmaceutics* **11** (4): 148.
[18] Seoane-Viaño, I., Gómez-Lado, N., Lázare-Iglesias, H. et al. (2020). 3D printed tacrolimus rectal formulations ameliorate colitis in an experimental animal model of inflammatory bowel disease. *Biomedicines* **8** (12): 563.
[19] Smith, D., Kapoor, Y., Hermans, A. et al. (2018). 3D printed capsules for quantitative regional absorption studies in the GI tract. *International Journal of Pharmaceutics* **550** (1): 418–428.
[20] Arafat, B., Qinna, N., Cieszynska, M. et al. (2018). Tailored on demand anti-coagulant dosing: an *in vitro* and *in vivo* evaluation of 3D printed purpose-designed oral dosage forms. *European Journal of Pharmaceutics and Biopharmaceutics* **128**: 282–289.
[21] Al-Metwali, B. and Mulla, H. (2017). Personalised dosing of medicines for children. *Journal of Pharmacy and Pharmacology* **69** (5): 514–524.
[22] Rycerz, K., Stepien, K.A., Czapiewska, M. et al. (2019). Embedded 3D printing of novel bespoke soft dosage form concept for pediatrics. *Pharmaceutics* **11** (12): 630.
[23] Linares, V., Casas, M., and Caraballo, I. (2019). Printfills: 3D printed systems combining fused deposition modeling and injection volume filling. Application to colon-specific drug delivery. *European Journal of Pharmaceutics and Biopharmaceutics* **134**: 138–143.
[24] Ghanizadeh Tabriz, A., Nandi, U., Hurt, A.P. et al. (2021). 3D printed bilayer tablet with dual controlled drug release for tuberculosis treatment. *International Journal of Pharmaceutics* **593**: 120147.
[25] Shin, S., Kim, T.H., Jeong, S.W. et al. (2019). Development of a gastroretentive delivery system for acyclovir by 3D printing technology and its *in vivo* pharmacokinetic evaluation in Beagle dogs. *PloS One* **14** (5): e0216875–e.
[26] Xu, X., Seijo-Rabina, A., Awad, A. et al. (2021). Smartphone-enabled 3D printing of medicines. *International Journal of Pharmaceutics* **609**: 121199.
[27] Trenfield, S.J., Awad, A., McCoubrey, L.E. et al. (2022). Advancing pharmacy and healthcare with virtual digital technologies. *Advanced Drug Delivery Reviews* **182**: 114098.
[28] Pereira, B.C., Isreb, A., Isreb, M. et al. (2020). Additive manufacturing of a point-of-care 'polypill:' fabrication of concept capsules of complex geometry with bespoke release against cardiovascular disease. *Advanced Healthcare Materials* **9** (13): 2000236.
[29] Teo, A.Q.A., Dqk, N., Lee, P., and O'Neill, G.K. (2021). Point-of-care 3D printing: a feasibility study of using 3D printing for orthopaedic trauma. *Injury* **52** (11): 3286–3292.
[30] Papadakis, G., Pantazis, A.K., Ntogka, M. et al. (2019). 3D-printed point-of-care platform for genetic testing of infectious diseases directly in human samples using acoustic sensors and a smartphone. *ACS Sensors* **4** (5): 1329–1336.
[31] Basit, A.W. and Gaisford, S. (2018). *3D Printing of Pharmaceuticals*, 1e. Springer International Publishing. 10.1007/978-3-319-90755-0.
[32] Seoane-Viaño, I., Otero-Espinar, F.J., and Goyanes, Á. (2021). *Chapter 18 – 3D Printing of Pharmaceutical Products.* In: *Additive Manufacturing* (eds. J. Pou, A. Riveiro, and J.P. Davim), 569–597. Elsevier.
[33] Capel, A.J., Rimington, R.P., Lewis, M.P., and Christie, S.D.R. (2018). 3D printing for chemical, pharmaceutical and biological applications. *Nature Reviews Chemistry* **2** (12): 422–436.
[34] ASTM (2012). Standard terminology for additive manufacturing technologies. F2792 – 12a.
[35] Sen, K., Mehta, T., Sansare, S. et al. (2021). Pharmaceutical applications of powder-based binder jet 3D printing process: a review. *Advanced Drug Delivery Reviews* **177**: 113943.
[36] Aprecia Pharmaceuticals (2018). Technology: the future of pharmaceutical manufacturing. Available from: https://www.aprecia.com/technology (accessed 23 January 2020).
[37] Kozakiewicz-Latała, M., Nartowski, K.P., Dominik, A. et al. (2022). Binder jetting 3D printing of challenging medicines: from low dose tablets to hydrophobic molecules. *European Journal of Pharmaceutics and Biopharmaceutics* **170**: 144–159.
[38] Chang, S.Y., Li, S.W., Kowsari, K. et al. (2020). Binder-jet 3D printing of indomethacin-laden pharmaceutical dosage forms. *Journal of Pharmaceutical Sciences* **109** (10): 3054–3063.

[39] Infanger, S., Haemmerli, A., Iliev, S. et al. (2019). Powder bed 3D-printing of highly loaded drug delivery devices with hydroxypropyl cellulose as solid binder. *International Journal of Pharmaceutics* **555**: 198–206.
[40] Xu, X., Awad, A., Robles-Martinez, P. et al. (2021). Vat photopolymerization 3D printing for advanced drug delivery and medical device applications. *Journal of Controlled Release* **329**: 743–757.
[41] Economidou, S.N., Lamprou, D.A., and Douroumis, D. (2018). 3D printing applications for transdermal drug delivery. *International Journal of Pharmaceutics* **544** (2): 415–424.
[42] Carve, M. and Wlodkowic, D. (2018). 3D-printed chips: compatibility of additive manufacturing photopolymeric substrata with biological applications. *Micromachines* **9** (2): 91.
[43] Xu, X., Robles-Martinez, P., Madla, C.M. et al. (2020). Stereolithography (SLA) 3D printing of an antihypertensive polyprintlet: case study of an unexpected photopolymer-drug reaction. *Additive Manufacturing* **33**: 101071.
[44] Barakh Ali, S.F., Mohamed, E.M., Ozkan, T. et al. (2019). Understanding the effects of formulation and process variables on the printlets quality manufactured by selective laser sintering 3D printing. *International Journal of Pharmaceutics* **570**: 118651.
[45] Awad, A., Fina, F., Goyanes, A. et al. (2020). 3D printing: principles and pharmaceutical applications of selective laser sintering. *International Journal of Pharmaceutics* **586**: 119594.
[46] Fina, F., Goyanes, A., Madla, C.M. et al. (2018). 3D printing of drug-loaded gyroid lattices using selective laser sintering. *International Journal of Pharmaceutics* **547** (1–2): 44–52.
[47] Yang, Y., Xu, Y., Wei, S., and Shan, W. (2021). Oral preparations with tunable dissolution behavior based on selective laser sintering technique. *International Journal of Pharmaceutics* **593**: 120127.
[48] Awad, A., Yao, A., Trenfield, J.S. et al. (2020). 3D printed tablets (printlets) with braille and moon patterns for visually impaired patients. *Pharmaceutics* **12** (2): 172.
[49] Buanz, A.B., Saunders, M.H., Basit, A.W., and Gaisford, S. (2011). Preparation of personalized-dose salbutamol sulphate oral films with thermal ink-jet printing. *Pharmaceutical Research* **28** (10): 2386–2392.
[50] Cader, H.K., Rance, G.A., Alexander, M.R. et al. (2019). Water-based 3D inkjet printing of an oral pharmaceutical dosage form. *International Journal of Pharmaceutics* **564**: 359–368.
[51] Scoutaris, N., Alexander, M.R., Gellert, P.R., and Roberts, C.J. (2011). Inkjet printing as a novel medicine formulation technique. *Journal of Controlled Release* **156** (2): 179–185.
[52] Içten, E., Purohit, H.S., Wallace, C. et al. (2017). Dropwise additive manufacturing of pharmaceutical products for amorphous and self-emulsifying drug delivery systems. *International Journal of Pharmaceutics* **524** (1): 424–432.
[53] Evans, S.E., Harrington, T., Rodriguez Rivero, M.C. et al. (2021). 2D and 3D inkjet printing of biopharmaceuticals: a review of trends and future perspectives in research and manufacturing. *International Journal of Pharmaceutics* **599**: 120443.
[54] He, Y., Foralosso, R., Trindade, G.F. et al. (2020). A reactive prodrug ink formulation strategy for inkjet 3D printing of controlled release dosage forms and implants. *Advanced Therapeutics* **3** (6): 1900187.
[55] Awad, A., Gaisford, S., and Basit, A.W. (2018). Fused deposition modelling: advances in engineering and medicine. In: *3D Printing of Pharmaceuticals* (ed. A.W. Basit and S. Gaisford), 107–132. Cham: Springer International Publishing.
[56] Goyanes, A., Buanz, A.B., Basit, A.W., and Gaisford, S. (2014). Fused-filament 3D printing (3DP) for fabrication of tablets. *International Journal of Pharmaceutics* **476** (1–2): 88–92.
[57] Tan, D.K., Maniruzzaman, M., and Nokhodchi, A. (2018). Advanced pharmaceutical applications of hot-melt extrusion coupled with fused deposition modelling (FDM) 3D printing for personalised drug delivery. *Pharmaceutics* **10** (4).
[58] Goyanes, A., Martinez, P.R., Buanz, A. et al. (2015). Effect of geometry on drug release from 3D printed tablets. *International Journal of Pharmaceutics* **494** (2): 657–663.
[59] Dos Santos, J., Deon, M., da Silva, G.S., and Beck, R.C.R. (2021). Multiple variable effects in the customisation of fused deposition modelling 3D-printed medicines: a design of experiments (DoE) approach. *International Journal of Pharmaceutics* **597**: 120331.
[60] Eleftheriadis, G.K., Ritzoulis, C., Bouropoulos, N. et al. (2019). Unidirectional drug release from 3D printed mucoadhesive buccal films using FDM technology: *in vitro* and *ex vivo* evaluation. *European Journal of Pharmaceutics and Biopharmaceutics* **144**: 180–192.

[61] Fanous, M., Bitar, M., Gold, S. et al. (2021). Development of immediate release 3D-printed dosage forms for a poorly water-soluble drug by fused deposition modeling: study of morphology, solid state and dissolution. *International Journal of Pharmaceutics* **599**: 120417.
[62] Đuranović, M., Obeid, S., Madžarević, M. et al. (2020). Paracetamol extended release FDM 3D printlets: evaluation of formulation variables on printability and drug release. *International Journal of Pharmaceutics* **592**: 120053.
[63] Goyanes, A., Buanz, A.B., Hatton, G.B. et al. (2015). 3D printing of modified-release aminosalicylate (4-ASA and 5-ASA) tablets. *European Journal of Pharmaceutics and Biopharmaceutics* **89**: 157–162.
[64] Kollamaram, G., Croker, D.M., Walker, G.M. et al. (2018). Low temperature fused deposition modeling (FDM) 3D printing of thermolabile drugs. *International Journal of Pharmaceutics* **545** (1): 144–152.
[65] Seoane-Viaño, I., Januskaite, P., Alvarez-Lorenzo, C. et al. (2021). Semi-solid extrusion 3D printing in drug delivery and biomedicine: personalised solutions for healthcare challenges. *Journal of Controlled Release* **332**: 367–389.
[66] Aita, I.E., Breitkreutz, J., and Quodbach, J. (2020). Investigation of semi-solid formulations for 3D printing of drugs after prolonged storage to mimic real-life applications. *European Journal of Pharmaceutical Sciences* **146**: 105266.
[67] Karavasili, C., Gkaragkounis, A., Moschakis, T. et al. (2020). Pediatric-friendly chocolate-based dosage forms for the oral administration of both hydrophilic and lipophilic drugs fabricated with extrusion-based 3D printing. *European Journal of Pharmaceutical Sciences* **147**: 105291.
[68] Goyanes, A., Allahham, N., Trenfield, S.J. et al. (2019). Direct powder extrusion 3D printing: fabrication of drug products using a novel single-step process. *International Journal of Pharmaceutics* **567**: 118471.
[69] Ong, J.J., Awad, A., Martorana, A. et al. (2020). 3D printed opioid medicines with alcohol-resistant and abuse-deterrent properties. *International Journal of Pharmaceutics* **579**: 119169.
[70] Fanous, M., Gold, S., Muller, S. et al. (2020). Simplification of fused deposition modeling 3D-printing paradigm: feasibility of 1-step direct powder printing for immediate release dosage form production. *International Journal of Pharmaceutics* **578**: 119124.
[71] Agoram, B.M. (2009). Use of pharmacokinetic/ pharmacodynamic modelling for starting dose selection in first-in-human trials of high-risk biologics. *British Journal of Clinical Pharmacology* **67** (2): 153–160.
[72] Nair, A., Morsy, M.A., and Jacob, S. (2018). Dose translation between laboratory animals and human in preclinical and clinical phases of drug development. *Drug Development Research* **79** (8): 373–382.
[73] Hatton, G.B., Yadav, V., Basit, A.W., and Merchant, H.A. (2015). Animal farm: considerations in animal gastrointestinal physiology and relevance to drug delivery in humans. *Journal of Pharmaceutical Sciences* **104** (9): 2747–2776.
[74] Awad, A., Goyanes, A., Basit, A.W. et al. (2023). A review of state-of-the-art on enabling additive manufacturing processes for precision medicine. *Journal of Manufacturing Science and Engineering* **145** (1): 010802.
[75] Goyanes, A., Fernández-Ferreiro, A., Majeed, A. et al. (2018). PET/CT imaging of 3D printed devices in the gastrointestinal tract of rodents. *International Journal of Pharmaceutics* **536** (1): 158–164.
[76] Genina, N., Boetker, J.P., Colombo, S. et al. (2017). Anti-tuberculosis drug combination for controlled oral delivery using 3D printed compartmental dosage forms: from drug product design to *in vivo* testing. *Journal of Controlled Release* **268**: 40–48.
[77] Berg, S., Krause, J., Björkbom, A. et al. (2020). *In vitro* and *in vivo* evaluation of 3D printed capsules with pressure triggered release mechanism for oral peptide delivery. *Journal of Pharmaceutical Sciences* **110** (10): 228–238.
[78] Kadry, H., Al-Hilal, T.A., Keshavarz, A. et al. (2018). Multi-purposable filaments of HPMC for 3D printing of medications with tailored drug release and timed-absorption. *International Journal of Pharmaceutics* **544** (1): 285–296.
[79] Siyawamwaya, M., du Toit, L.C., Kumar, P. et al. (2019). 3D printed, controlled release, tritherapeutic tablet matrix for advanced anti-HIV-1 drug delivery. *European Journal of Pharmaceutics and Biopharmaceutics* **138**: 99–110.

[80] Charoenying, T., Patrojanasophon, P., Ngawhirunpat, T. et al. (2020). Fabrication of floating capsule-in- 3D-printed devices as gastro-retentive delivery systems of amoxicillin. *Journal of Drug Delivery Science and Technology* **55**: 101393.
[81] Chai, X., Chai, H., Wang, X. et al. (2017). Fused deposition modeling (FDM) 3D printed tablets for intragastric floating delivery of domperidone. *Scientific Reports* **7** (1): 2829.
[82] Bellinger, A.M., Jafari, M., Grant, T.M. et al. (2016). Oral, ultra-long-lasting drug delivery: application toward malaria elimination goals. *Science Translational Medicine* **8** (365): 365ra157.
[83] Kong, Y.L., Zou, X., McCandler, C.A. et al. (2019). 3D-printed gastric resident electronics. *Advanced Materials Technologies* **4** (3): 1800490.
[84] Tagami, T., Ito, E., Hayashi, N. et al. (2020). Application of 3D printing technology for generating hollow-type suppository shells. *International Journal of Pharmaceutics* **589**: 119825.
[85] Chatzitaki, A.-T., Tsongas, K., Tzimtzimis, E.K. et al. (2021). 3D printing of patient-tailored SNEDDS-based suppositories of lidocaine. *Journal of Drug Delivery Science and Technology* **61**: 102292.
[86] Awad, A., Hollis, E., Goyanes, A. et al. (2023). 3D printed multi-drug-loaded suppositories for acute severe ulcerative colitis. *International Journal of Pharmaceutics: X* **5**: 100165.
[87] Seoane-Viaño, I., Gómez-Lado, N., Lázare-Iglesias, H. et al. (2018). Longitudinal PET/CT evaluation of TNBS-induced inflammatory bowel disease rat model. *International Journal of Pharmaceutics* **549** (1–2): 335–342.
[88] Awad, A., Goyanes, A., Orlu, M. et al. (2023). 3D printed infliximab suppositories for rectal biologic delivery. *International Journal of Pharmaceutics: X* **5**: 100176.
[89] Allahham, N., Fina, F., Marcuta, C. et al. (2020). Selective laser sintering 3D printing of orally disintegrating printlets containing ondansetron. *Pharmaceutics* **12** (2): 110.
[90] Fina, F., Madla, C.M., Goyanes, A. et al. (2018). Fabricating 3D printed orally disintegrating printlets using selective laser sintering. *International Journal of Pharmaceutics* **541** (1): 101–107.
[91] Musazzi, U.M., Selmin, F., Ortenzi, M.A. et al. (2018). Personalized orodispersible films by hot melt ram extrusion 3D printing. *International Journal of Pharmaceutics* **551** (1): 52–59.
[92] Öblom, H., Sjöholm, E., Rautamo, M., and Sandler, N. (2019). Towards printed pediatric medicines in hospital pharmacies: comparison of 2D and 3D-printed orodispersible warfarin films with conventional oral powders in unit dose sachets. *Pharmaceutics* **11** (7): 334.
[93] Sjöholm, E. and Sandler, N. (2019). Additive manufacturing of personalized orodispersible warfarin films. *International Journal of Pharmaceutics* **564**: 117–123.
[94] Jamróz, W., Kurek, M., Łyszczarz, E. et al. (2017). 3D printed orodispersible films with Aripiprazole. *International Journal of Pharmaceutics* **533** (2): 413–420.
[95] Tabriz, A.G., Fullbrook, D.H., Vilain, L. et al. (2021). Personalised tasted masked chewable 3D printed fruit-chews for paediatric patients. *Pharmaceutics* **13** (8): 1301.
[96] Nober, C., Manini, G., Carlier, E. et al. (2019). Feasibility study into the potential use of fused-deposition modeling to manufacture 3D-printed enteric capsules in compounding pharmacies. *International Journal of Pharmaceutics* **569**: 118581.
[97] Melocchi, A., Parietti, F., Maccagnan, S. et al. (2018). Industrial development of a 3D-printed nutraceutical delivery platform in the form of a multicompartment HPC capsule. *AAPS PharmSciTech* **19**: 3343–3354.
[98] Tiboni, M., Campana, R., Frangipani, E., and Casettari, L. (2021). 3D printed clotrimazole intravaginal ring for the treatment of recurrent vaginal candidiasis. *International Journal of Pharmaceutics* **596**: 120290.
[99] Farmer, Z.-L., Utomo, E., Domínguez-Robles, J. et al. (2021). 3D printed estradiol-eluting urogynecological mesh implants: influence of material and mesh geometry on their mechanical properties. *International Journal of Pharmaceutics* **593**: 120145.
[100] Arany, P., Papp, I., Zichar, M. et al. (2021). Manufacturing and examination of vaginal drug delivery system by FDM 3D printing. *Pharmaceutics* **13** (10).
[101] Fu, J., Yu, X., and Jin, Y. (2018). 3D printing of vaginal rings with personalized shapes for controlled release of progesterone. *International Journal of Pharmaceutics* **539** (1): 75–82.
[102] Januszewicz, R., Mecham, S.J., Olson, K.R., and Benhabbour, S.R. (2020). Design and characterization of a novel series of geometrically complex intravaginal rings with digital light synthesis. *Advanced Materials Technologies* **5** (8): 2000261.

[103] Melocchi, A., Inverardi, N., Uboldi, M. et al. (2019). Retentive device for intravesical drug delivery based on water-induced shape memory response of poly(vinyl alcohol): design concept and 4D printing feasibility. *International Journal of Pharmaceutics* **559**: 299–311.

[104] Vivero-Lopez, M., Xu, X., Muras, A. et al. (2021). Anti-biofilm multi drug-loaded 3D printed hearing aids. *Materials Science and Engineering C* **119**: 111606.

[105] Alomari, M., Mohamed, F.H., Basit, A.W., and Gaisford, S. (2015). Personalised dosing: printing a dose of one's own medicine. *International Journal of Pharmaceutics* **494** (2): 568–577.

[106] Breitkreutz, J. and Boos, J. (2007). Paediatric and geriatric drug delivery. *Expert Opinion on Drug Delivery* **4** (1): 37–45.

[107] Kearns, G.L., Abdel-Rahman, S.M., Alander, S.W. et al. (2003). Developmental pharmacology: drug disposition, action, and therapy in infants and children. *New England Journal of Medicine* **349** (12): 1157–1167.

[108] Habib, W.A., Alanizi, A.S., Abdelhamid, M.M., and Alanizi, F.K. (2014). Accuracy of tablet splitting: comparison study between hand splitting and tablet cutter. *Saudi Pharmaceutical Journal* **22** (5): 454–459.

[109] Peek, B.T., Al-Achi, A., and Coombs, S.J. (2002). Accuracy of tablet splitting by elderly patients. *Journal of the American Medical Association* **288** (4): 451–452.

[110] McDevitt, J.T., Gurst, A.H., and Chen, Y. (1998). Accuracy of tablet splitting. *Pharmacotherapy: The Journal of Human Pharmacology and Drug Therapy* **18** (1): 193–197.

[111] Hill, S., Varker, A.S., Karlage, K., and Myrdal, P.B. (2009). Analysis of drug content and weight uniformity for half-tablets of 6 commonly split medications. *Journal of Managed Care Pharmacy* **15** (3): 253–261.

[112] Zheng, Z., Lv, J., Yang, W. et al. (2020). Preparation and application of subdivided tablets using 3D printing for precise hospital dispensing. *European Journal of Pharmaceutical Sciences* **149**: 105293.

[113] Liang, K., Carmone, S., Brambilla, D., and Leroux, J.C. (2018). 3D printing of a wearable personalized oral delivery device: a first-in-human study. *Science Advances* **4** (5): eaat2544.

[114] Jiang, H., Fu, J., Li, M. et al. (2019). 3D-printed wearable personalized orthodontic retainers for sustained release of clonidine hydrochloride. *AAPS PharmSciTech* **20** (7): 260.

[115] Mennella, J.A., Roberts, K.M., Mathew, P.S., and Reed, D.R. (2015). Children's perceptions about medicines: individual differences and taste. *BMC Pediatrics* **15** (1): 1–6.

[116] Scoutaris, N., Ross, S.A., and Douroumis, D. (2018). 3D printed 'starmix' drug loaded dosage forms for paediatric applications. *Pharmaceutical Research* **35** (2): 34.

[117] Mathias, D., Snider, C., Hicks, B., and Ranscombe, C. (2019). Accelerating product prototyping through hybrid methods: coupling 3D printing and LEGO. *Design Studies* **62**: 68–99.

[118] Januskaite, P., Xu, X., Ranmal, S.R. et al. (2020). I spy with my little eye: a paediatric visual preferences survey of 3D printed tablets. *Pharmaceutics* **12** (11): 1100.

[119] Fastø, M.M., Genina, N., Kaae, S., and Kälvemark Sporrong, S. (2019). Perceptions, preferences and acceptability of patient designed 3D printed medicine by polypharmacy patients: a pilot study. *International Journal of Clinical Pharmacy* **41** (5): 1290–1298.

[120] Eleftheriadis, G.K. and Fatouros, D.G. (2021). Haptic evaluation of 3D-printed Braille-encoded intraoral films. *European Journal of Pharmaceutical Sciences* **157**: 105605.

[121] Merchant, H.A., Liu, F., Orlu Gul, M., and Basit, A.W. (2016). Age-mediated changes in the gastrointestinal tract. *International Journal of Pharmaceutics* **512** (2): 382–395.

[122] Hajjar, E.R., Cafiero, A.C., and Hanlon, J.T. (2007). Polypharmacy in elderly patients. *American Journal of Geriatric Pharmacotherapy* **5** (4): 345–351.

[123] Murray, M.D. and Kroenke, K. (2001). Polypharmacy and medication adherence: small steps on a long road. *Journal of General Internal Medicine* **16** (2): 137.

[124] Khaled, S.A., Burley, J.C., Alexander, M.R. et al. (2015). 3D printing of tablets containing multiple drugs with defined release profiles. *International Journal of Pharmaceutics* **494** (2): 643–650.

[125] Goyanes, A., Wang, J., Buanz, A. et al. (2015). 3D printing of medicines: engineering novel oral devices with unique design and drug release characteristics. *Molecular Pharmaceutics* **12** (11): 4077–4084.

[126] Rowe, C., Katstra, W., Palazzolo, R. et al. (2000). Multimechanism oral dosage forms fabricated by three dimensional printing™. *Journal of Controlled Release* **66** (1): 11–17.

[127] Aprecia Pharmaceuticals (2016). Z-FREE Zipdose technology: the world's first 3DP dosage form. Available from: https://www.aprecia.com/technology/zipdose (accessed 23 January 2020).

[128] Boudriau, S., Hanzel, C., Massicotte, J. et al. (2016). Randomized comparative bioavailability of a novel three-dimensional printed fast-melt formulation of levetiracetam following the administration of a single 1000-mg dose to healthy human volunteers under fasting and fed conditions. *Drugs in R&D* **16** (2): 229–238.

[129] Sagandira, C.R., Siyawamwaya, M., and Watts, P. (2020). 3D printing and continuous flow chemistry technology to advance pharmaceutical manufacturing in developing countries. *Arabian Journal of Chemistry* **13** (11): 7886.

[130] Rahman, A. (2019). NASA fund backs 3D printing in space for treating astronaut injuries like mallet finger. NS Medical Devices. Available from: https://www.nsmedicaldevices.com/news/nasa-3d-printing-space-medical-devices (accessed 23 January 2020).

[131] Saunders, S. (2017). Astronauts 3D print the first medical supplies in space, which can also teach us more about healthcare on earth. 3DPrint.com. Available from: https://3dprint.com/162241/3d-print-medical-supplies-in-space (accessed 23 January 2020).

[132] Davidson, G. (2017). Veterinary compounding: regulation, challenges, and resources. *Pharmaceutics* **9** (1): 5.

[133] Thamm, D.H. and Gustafson, D.L. (2020). Drug dose and drug choice: optimizing medical therapy for veterinary cancer. *Veterinary and Comparative Oncology* **18** (2): 143–151.

[134] Melocchi, A., Briatico-Vangosa, F., Uboldi, M. et al. (2021). Quality considerations on the pharmaceutical applications of fused deposition modeling 3D printing. *International Journal of Pharmaceutics* **592**: 119901.

[135] Awad, A., Fina, F., Goyanes, A. et al. (2021). Advances in powder bed fusion 3D printing in drug delivery and healthcare. *Advanced Drug Delivery Reviews* **174**: 406–424.

[136] Trenfield, S.J., Xian Tan, H., Goyanes, A. et al. (2020). Non-destructive dose verification of two drugs within 3D printed polyprintlets. *International Journal of Pharmaceutics* **577**: 119066.

[137] Trenfield, S.J., Goyanes, A., Telford, R. et al. (2018). 3D printed drug products: non-destructive dose verification using a rapid point-and-shoot approach. *International Journal of Pharmaceutics* **549** (1): 283–292.

[138] Trenfield, S.J., Xian Tan, H., Awad, A. et al. (2019). Track-and-trace: novel anti-counterfeit measures for 3D printed personalized drug products using smart material inks. *International Journal of Pharmaceutics* **567**: 118443.

[139] Edinger, M., Bar-Shalom, D., Sandler, N. et al. (2018). QR encoded smart oral dosage forms by inkjet printing. *International Journal of Pharmaceutics* **536** (1): 138–145.

11

Regulatory Aspects of 3D-Printed Medicinal Products

Maria Malamatari, Ka-Wai Wan and Fotios Baxevanis

Medicines & Healthcare Products Regulatory Agency, London, UK

11.1 Introduction

Three-dimensional printing (3D printing or sometimes referred to as additive manufacturing), in the context of pharmaceutical application, involves the deposition of the active drug substance and excipients in a layer-by-layer manner for the construction of a pharmaceutical dosage form or drug delivery system with the use of a 3D printer equipped with an appropriate computer-aided design (CAD) software.

This chapter aims to highlight the relevant regulatory consideration associated with the manufacture of pharmaceutical products using 3D-printing (3DP) technologies. It is not intended to provide a comprehensive list of regulatory requirements for the satisfactory approval of a 3DP device or a 3DP-printed pharmaceutical product. Consideration relating to the 3DP device, the use of computer simulations, or computer-aided design (CAD) of the product are not extensively discussed in this chapter. The focus of this chapter is mainly on the manufacturing process and controls of the 3DP pharmaceutical product intended for oral delivery, although it is acknowledged that different regulatory aspects for medical implants, stents, vaginal, or topical delivery systems, such as microneedles and pessaries, as well as 3D bioprinting of tissues and organs warrant further consideration, as well as the use of 3DP technology for medical device purposes.

3D Printing of Pharmaceutical and Drug Delivery Devices: Progress from Bench to Bedside, First Edition.
Edited by Dimitrios A. Lamprou, Dennis Douroumis and Sheng Qi.

11.2 Current Regulatory Framework

Even though additive manufacturing (AM) is a rapidly developing technology, there are no separate regulatory pathways for the regulation of products manufactured using 3DP. Despite their unique nature, the chemistry, manufacture, and control (CMC) aspects are regulated in line with the regulatory and procedural guidance which governs conventional medicines and devices in the APAC/EMEA/LAD/NA areas.

In the US, the FDA regulates 3DP medicinal products and devices via the Centre for Biologics Evaluation and Research (CBER), Centre for Devices and Radiological Health (CDRH), and the Centre for Drug Evaluation and Research (CDER) [1], depending on the product type and intended use [2]. In December 2017, the FDA published a guidance for industry and for the FDA staff on the technical considerations for additive-manufactured medical devices [3]. The scope of the document was to provide guidance on the design, development, characterisation, and testing of medical devices, at least one element of which is manufactured using an AM process. The guidance does not intend to address all regulatory requirements for such devices but rather to provide an insight into the FDA's rationale in the evaluation of such products and complement the existing device-specific regulatory guidance. It is noted that the use of biological components incorporated in 3DP devices is not addressed by the guideline.

Although several AM devices have been granted 510(k) FDA clearance [4], only one 3DP product, the antiepileptic Spritam® (levetiracetam), has been so far approved by the FDA [5]. In the US, marketing authorisation applications for medicinal products are submitted via the 505(b)(1), 505(b)(2), and 505(j) regulatory pathways, depending on whether: a) they concern new drug substances supported by full studies on quality, safety, and efficacy; b) a combination of full investigations and studies not conducted by or for the applicant and the applicant has not obtained the right to refer to; or c) generic drugs [6].

Spritam® (levetiracetam) is an orodispersible tablet for the treatment of epilepsy licensed to Aprecia Pharmaceuticals. As stated above, it was the first medicinal product manufactured by 3DP technology to receive a marketing authorisation by the US FDA. Spirtam® tablets are manufactured using the ZipDose technology, a platform developed by Aprecia which uses binder jetting 3DP, to create a porous, rapidly disintegrating formulation, without a compression step, formed in a one-piece layer [7, 8]. Spritam tablets are appropriate for people with swallowing difficulties as the tablets disintegrate in the mouth [9], which makes formulations using this technology ideal for the paediatric population.

In 2018, Health Canada published its first draft on a Guidance document on Supporting Evidence for Implantable Medical Devices Manufactured by 3DP, the final version of which was adopted in 2019. The guidance provides an overview of the 3DP process and information on the extent of data required as evidence on aspects like device design parameters, safety and effectiveness standards, labelling, and process validation to support the licensing of Class III and Class IV applications for devices manufactured by AM processes [10].

In June 2020, the UK regulator, Medicines and Healthcare Products Regulatory Agency (MHRA), published a Guidance on 3DP of medical device or component parts during the coronavirus (COVID-19) pandemic. Depending on the 3DP product being characterised as a device or a personal protective equipment (PPE), the requirements of the Medical Devices Directive/Regulations or PPE Regulation 2016/425 respectively must be met [11]. To support increased manufacture of products at the point where a patient receives it, the MHRA is developing a new point-of-care (POC) regulatory framework which 'seeks to balance regulatory requirements for the control of these products to ensure levels of safety equivalent to current products while avoiding unnecessary regulatory barriers by adapting regulatory requirement' [12]. 3DP products are included within the product range covered by the POC regulatory framework. Between August and September 2021, the MHRA carried out a consultation inviting businesses, individuals, or representative bodies to respond to specific questions raised. The consultation survey closed on 23 September 2021 and its conclusions will be used to identify regulatory changes that are required to support POC manufacturing.

In the EU, no unique legal framework for the regulation of 3DP medicinal products and medical devices is in place. 3DP medical devices are regulated by the EU Regulation on Medical Devices 2017/745′ (MDR) or the EU Regulation on *In Vitro* Diagnostic Medical Devices 2017/746′ (IVDR), which came into place in 2022 and replaced the existing Medical Devices Directive (93/42/EEC) (MDD) and the Active Implantable Medical Devices Directive (90/385/EEC) (AIMDD). In the EU, like all medicinal products, medicines manufactured using AM are licensed based on the EC pharmaceutical Regulations (e.g., Regulation (EC) No. 726/2004), the provisions of the EC pharmaceutical Directives (e.g., 2001/83/EC and its amending Directives), the Notice to Applicants, European Medicines Agency (EMA) guidelines, and also according to each Member State's national guidelines, policy, and best practice. The quality, safety, and efficacy of all medicinal products, including those using 3DP technology for such products to be licensed in the EU, should be demonstrated. The EMA's Committee for Medicinal Products for human use (CHMP), in collaboration with the national competent authorities of each Member State, have developed a series of scientific guidelines to facilitate the interpretation of the quality, safety, and efficacy principles as described in the Community Directives [13]. The Guidelines cover quality, clinical efficacy and safety, and non-clinical aspects with the Common Technical Document (eCTD) structure followed where possible. The eCTD is the standard format for submitting regulatory information (e.g., applications, amendments, supplements, and reports) to the concerned Heath Authorities.

The compilation also includes other related documents, such as reflection papers, public statements, and questions and answers. EMA guidelines are harmonised through the International Council on Harmonisation (ICH) of Technical Requirements for Registration of Pharmaceuticals for Human Use, the scope of which since its formation in 1990 has been to bring together the regulatory authorities and pharmaceutical industry to discuss scientific and technical aspects of pharmaceuticals [14]. In terms of their quality, 3DP medicinal products are expected to follow all stability requirements for the drug substance and product as stated in the ICH Q1A-1F, with analytical methods appropriately described and validated (ICH Q2), impurities identified and qualified where appropriate (ICH Q3), and specifications appropriate to the drug substance and type of formulation (ICH Q6A, B). ICH Q8-10 have set a new quality paradigm where a risk-based enhanced approach based on the knowledge of product development and quality risk management throughout the product's lifecycle is followed.

11.3 Quality Aspects of 3D-Printed Medicinal Products

The required standards for a safe, effective, and medicinal product of appropriate quality for the treatment and prevention of diseases remain the same, regardless of the technology used to produce the medicinal product. Therefore, the intended use of the pharmaceutical product and targeted condition/treatment regimen must be considered for any 3DP medicinal products. It is recognised that a risk-based regulatory approach is applied for any new technology when limited experience is available for the innovation. Nevertheless, the perceived challenges in regulatory uncertainty should not be a deterrent for advancing pharmaceutical technologies to help enhance patient care.

The principle of Quality by Design (QbD) and knowledge of the design space are still applicable for 3DP pharmaceuticals as in the case of conventional oral dosage forms. Process understanding and monitoring are considered critical for the success of 3DP in pharmaceuticals, particularly when the quality of the product is built into the system during manufacturing. The ability to control the required critical process parameters (CPPs) at any time during the production process, in order to meet the necessary specification or critical quality attributes (CQAs) that define the Quality Target Product Profiles (QTTPs), is especially critical for point-of-care and personalised medicine manufacture, where the sample size of the 'production batch' may be significantly smaller than the commercial batch size (e.g., 100,000 tablets) of a conventional immediate-release dosage form often tested in process validation studies. Regardless of whether the proposed 3DP product manufacturing is made as a batch process or continuous manufacturing, the application of QbD in 3DP product is particularly important.

The 3DP technologies that have been mainly used to produce drug delivery systems are fused deposition modelling (FDM), stereolithography (SLA, aka vat photopolymerisation), selective laser sintering (SLS), and binder jetting [15]. The majority of research articles on 3DP drug products focus on extrusion-based techniques such as the FDM, while binder jetting, SLA, and SLS are used by approximately 10%, 2.5%, and 3% of developments as per the research articles, respectively [16, 17]. The main reasons for the popularity of FDM on 3DP of pharmaceuticals are its low cost, ability to fabricate hollow objects, and its ease of use.

As fundamentally different processes are used for the 3DP of pharmaceuticals, the material attributes and process parameters that may affect the drug product critical quality attributes (CQAs) will differ significantly depending upon the 3DP process used. As per the ICH Q8(R2) guideline, fishbone (Ishikawa) diagrams are useful risk assessment tools, as they can aid to identify potential variables which may have an impact on CQAs. A fishbone diagram for 3DP of pharmaceuticals by FDM is given in Figure 11.1.

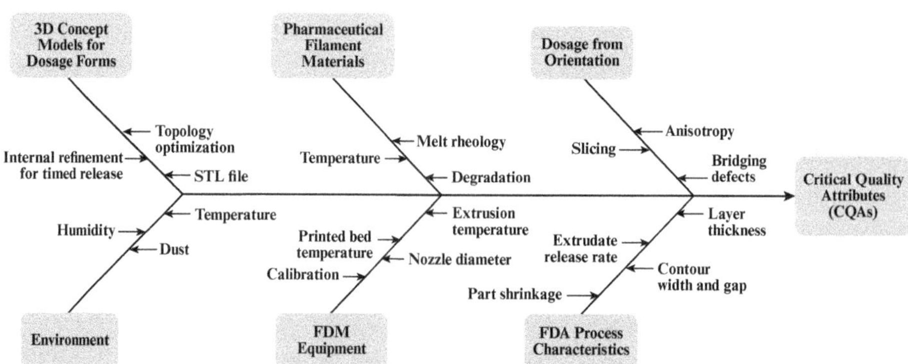

Figure 11.1 *Fishbone diagram for 3DP of pharmaceuticals by FDM. (Source: Adapted from [18].)*

It is essential to understand the CPPs for the production of 3DP dosage forms and have appropriate controls of these parameters. For instance, in conventional pharmaceutical manufacturing of tablets, milling, granulation (wet or dry), and compression are carefully validated within the necessary processing and control range. Similarly for FDM, the diameter and temperature of the extruder, thickness of the tablet to be printed per layer/geometry, printing and extrusion speed, ratio of each additive employed, and the required density of the produced tablet are all considered critical. The proposed prototype of the 3DP manufacturing process is generally based on the robust pharmaceutical developmental data gathered. Reference to ICH guideline Q8 (R2) on Pharmaceutical Development is recommended.

Quality should be built-in or by design. In particular, the quality by testing approach is not possible when the batch size of a 3DP product is relatively small, or where a bespoke or patient-specific formulation is adopted or frequently adjusted. The potential of real-time release of the product and continuous real-time quality assurance depend principally on the ability of the system to monitor and adjust the operating conditions within the optimal range in order to maintain the required CQAs and consistency in the production process by minimising variability and thus mitigate the risks of manufacturing a defective product. By adopting a robust pharmaceutical quality management system, the risks of any activities or materials associated with the production of the product using the innovative manufacturing process should be considered, as recommended in the ICH guideline Q10 on pharmaceutical quality systems.

To achieve this, a robust assessment of all variables should be undertaken and the variability of the process should be managed and controlled within the established design space. An appropriate process analytical technology (PAT), perhaps integrated real-time PAT into the DP technology, following QbD principle, is therefore useful to support AM. The capability of having at-line, on-line, or in-line analysis during the manufacturing process would enable development of a continuous and real-time product release that meets the required finished product specifications.

To align with the concept of QbD and PAT, regulatory considerations of the Good Manufacturing Practice (GMP) aspects were made a decade ago. The guideline on parametric release/real-time release testing (RTRT), Annex 15 and 17 of the EU GMP guidelines as well as validation guidelines specifically relating to continuous validation, ICH Q8, Q9, and Q10, were revised accordingly [19], which are also relevant for consideration for AM.

How could the finished product specifications be met simultaneously to facilitate real-time release? Table 11.1 illustrates some of the commonly exploited techniques for analysis of the required specification parameters for an oral dosage form manufactured using a conventional process. Many of these analytical methods are considered as destructive techniques employed following sampling of a specific batch of product.

However, non-destructive techniques that allow simultaneous testing during production are desirable for 3DP pharmaceutical manufacture. In view of this, spectroscopic analytical methods have been exploited as PAT tools for the characterisation of 3DP materials. This includes infrared (IR), near-infrared (NIR), and Raman spectroscopy. IR spectroscopy is an absorption technique based on the measurement of bond vibrations, while Raman is based on scattering. There are different types of IR spectroscopy, such as Fourier-Transform Infrared Spectroscopy (FTIR) and Attenuated Total Reflectance (ATR), each with its own mechanism of measurement and limitations. In general, the IR spectrum can confirm the identity and concentration of the substance of interest in the sample. Therefore, it could

Table 11.1 *Commonly employed analytical techniques for testing of the finished drug product.*

Test parameters	Typical test methods	Additional note/Reasons for testing
Physical attributes of the product (size, shape, colour)	Calibrated measurer	Information would also need to appear in the SmPC
Identification	Fourier-Transform Infrared Spectroscopy (FTIR), HPLC	Comparison with validated standard
Assay	HPLC, ultraviolet (UV) spectrometer	Usually 95–105% specification limit
Impurities	HPLC	Drug substance monographs should be considered and justification for specified and unspecified impurities to be provided
Polymorphism	X-ray powder diffractometry (XRPD)	Provide definitive θ values for a particular polymorph. To investigate any changes in the solid state of the drugs/excipients prior to, during, and after printing, as well as upon storage
	Differential scanning calorimetry (DSC) and thermogravimetric analysis (TGA)	To study the plasticising effect of the drugs on the filaments. To study crystallinity of the drug and miscibility with the extrudates. To study thermal stability of the drugs and excipients (polymers) during the manufacturing processing (e.g., via hot melt extrusion)
Microbiological burden	As per the pharmacopoeial methods (e.g., Ph. Eur. chapters 2.6.12 and 2.6.13)	Validation for the chosen methods is required to confirm suitability of the method for the concerned product
Disintegration	Disintegration tester (pharmacopoeial)	As per pharmacopoeial requirement
Dissolution / rate of drug release	Dissolution tester (pharmacopoeial). e.g., USP Type II apparatus	As per pharmacopoeial requirement
Uniformity of dosage units (content uniformity or mass variation)	Weighing balance (mass)/HPLC (content)	As per pharmacopoeial requirement (2.9.40)
Tablet hardness or tensile strength (depending on the nature of the dosage form)	Tablet hardness tester/mechanical strength tester	To determine the tablet crushing strength or tensile strength
Tablet friability	Tablet friability tester	Friability – dependent on particle size of the binder used (if using binder-jet technology)
Porosity/density	Scanning electron microscopy (SEM)	Particularly if this attribute is considered critical for the intended release profile of the 3D-printed product

be used for the qualitative and quantitative analysis of a 3DP pharmaceutical product during manufacture (e.g., for assay and identification purposes). Nevertheless, it is sensitive to water, which limits the use of the technique for compounds or products with high water content, or where water-based inks are employed in the manufacturing process [20]. ATR-FTIR can also be used to study the interactions between the drugs and the selected excipients, justifying the proposed formulation of the 3DP product during development [21]. Similar to IR spectroscopy, NIR and Raman spectroscopy are useful quantification tools for both the active substance and excipients, although NIR spectroscopy is also sensitive to water. By coupling with chemical imaging (NIR-CI and Raman-CI) and microscopy (e.g., confocal Raman microscopy), the distribution and concentration of the active substance can be visualised [22, 23], providing invaluable information as a control testing method to help investigate variation of the drug substance distribution in the 3DP product. Indeed, the extensive applications of these spectroscopic techniques for the characterisation of pharmaceutical products have been well established [24].

Thermal analysis, along with X-ray powder diffractions (XRPD) analysis, is useful to reveal information about the solid state (i.e., crystallinity/amorphous nature) of the drugs and excipients prior to, during, and after 3DP. It is anticipated that these analyses are performed during pharmaceutical development for the prototype design, but it may not be easily implemented online and simultaneously performed during production as a PAT tool.

Other analytical techniques that one may expect to apply during development may include rheological and morphological characterisation of the product using a rheometer and scanning electron microscopy (SEM), which would provide information about the surface properties and physical characteristics of the polymer filaments formed during manufacturing of the 3DP. More sophisticated methods such as X-ray micro-computed tomography (Micro-CT) have also been used in the past to study the density and porosity of the 3DP product, allowing a better understanding of the tablet internal structure [25]. To help understand the composition of the 3DP product, advanced NIR hyperspectral imaging, which is considered as a non-destructive method, has also been employed [26].

Decisions on whether these characterisation techniques are employed for quality control testing depend on the understanding of the required CQAs of the finished 3DP product. For instance, if a precise control of drug release is necessary for the intended dosage form, then specific control of the internal geometric structure is likely to be required as part of the quality control test strategy and should be included in the finished product specifications. 3DP tablets or other oral dosage forms containing various internal geometric structures may be seen as a platform for tailored drug delivery system for one drug to another. However, there is still a need to optimise the required pharmacokinetics for different drugs with different clinical needs. Therefore, relevant validation for the different active substances with any one particular 3DP dosage form to achieve the necessary drug release profile and clinical outcomes is still considered essential.

As discussed above, it is useful to implement tools that are capable of real-time analysis to help minimise analytical burden at the end of product characterisation. Spectroscopic techniques that enable online monitoring for the quantification, polymorphism, and distribution of the active substance/excipients, water content, and thickness of the 3DP product could replace end-product testing; for example, assay, uniformity of content/dosage unit, dimension measurement, and possibly drug disintegration and dissolution. Since the success of utilising these analytical tools, whether qualitatively or quantitatively, rely on correlation with calibrated standards, appropriate method validation in accordance with, for instance, ICH Q2(R2), is anticipated.

It should be borne in mind that it is not just the active substance that requires identification/quantification, but colourants, which are often employed in the 3DP products, would also need to be identified and controlled in the finished product specifications [27]. Excipients used in the manufacturing process must comply with the relevant pharmacopeial specifications, where applicable, which would provide quality reassurance of the product.

In general, relevant guidance on analytical method validation should be followed. Any changes to the physicochemical properties of the active(s) and raw materials during processing should be thoroughly considered and appropriately managed to meet the required control specifications and critical quality attributes.

11.4 3D-Printed Paediatric Medicinal Products

The 'Paediatric Regulation' entered into force in the EU in January 2007 [28]. The aim of this regulation is to encourage the development and accessibility of age-appropriate paediatric medicines [29]. As highlighted in a ten-year report published by the EC, the implementation of the 'Paediatric Regulation' has already started to bear fruit regarding the development of paediatric medicines, as an increase in paediatric trials, authorised medicines, and product information has been observed [30]. The increase in paediatric research along with a positive agenda of actions will ensure the reduction of the off-label use of adult medicines in the paediatric population over time.

Despite the progress made, administration of medicines to children remains a challenge for both the healthcare professionals, parents, and caregivers [31]. The main reason for this is the lack of age-appropriate formulations, especially for very young children (i.e., under 2 years old). This has led to either patient refusal of the medication or the need for manipulation of adult oral medication with a prevalence of more than 30% in both outpatient and inpatient settings [32]. Manipulation of adult oral medication for children (e.g., crushing tablets, opening capsules, taste masking, cutting patches, etc.) is time-consuming, increases the risk of errors (e.g., dose calculation errors), and its impact on the efficacy and safety of the drug is unknown [33].

3DP technologies have been identified as a promising approach for paediatric drug development [34]. The main reason for this is the dose individualisation that can be attained by printing technologies, a property particularly suited for paediatric drug development, given the heterogenicity of the patient population and the need for relatively small batch sizes. Apart from dose flexibility, the capabilities of 3DP can be linked and modulate several of the key attributes of a paediatric quality target product profile (pQTPP), as proposed by Walsh et al. [35]. Specifically, 3DP could modulate attributes such as the target release profile, dosage form, dosage strength, patient acceptability, stability and storage conditions, manufacturing, and patient access.

Several oral dosage forms for the paediatric population have been produced by 3DP, including tablets, orodispersible tablets, chewable tablets, and orodispersible films [36, 37]. Eduardo et al. [38] prepared paediatric orodispersible printlets of hydrochlorothiazide, using semi-solid micro-extrusion 3DP process. The printing surface material and the first printing layer were found to be critical process parameters influencing the quality attributes of the printlets, namely the printlet appearance, dimensions, weight, and disintegration time. Upon optimisation of the process parameters, printlets with reduced size and high drug load were printed, meeting the requirements of Ph. Eur. regarding mass, content uniformity, disintegration time (i.e., NMT 3 minutes), and dissolution. Printing techniques such as the semi-solid extrusion were also found to produce orodispersible warfarin films with acceptable characteristics and superior uniformity of dosage units compared to the conventional manufacturing method for compounding oral powders in unit-dose sachets [39].

FDM has also been employed to produce 3DP minitablets with diameters between 1.5 mm and 4.00 mm for paediatric use. Dose individualisation and different release profiles were achieved by varying the size of the minitablet or the polymer used [40]. However, it should be mentioned that the minitablets with the smallest diameters exhibited uneven appearance and higher coefficient of variation of mass variation, indicating that production of small tablets or low-dose formulations is challenging as variation of the percentage of the incorporated drug, due to printing defects, is more pronounced.

Patient acceptability is an important consideration in the development of medicines for children as it 'is likely to have a significant impact on patient adherence and, consequently, on the safety and efficacy of a medicinal product' [29]. Patient acceptability depends on several aspects, including palatability, swallowability, and appearance. 3DP technologies have the ability for taste masking, fabrication of solid oral dosage forms with different shapes, sizes, and colours; aspects that can be correlated with the patient's acceptability regarding picking and swallowing [41]. Based on visual preferences survey completed by children 4–11 years old, printlets produced by different 3DP technologies scored differently regarding the preference, indicating that appearance, perceived taste, texture, and familiarity are important parameters that need to be considered when building the pQTPP of 3DP drug products [42]. Currently, limited evidence (but promising) has been obtained on the acceptability of 3DP formulations by children, especially when chronic administration of the medicine is required [43]. Thus, the acceptability of a 3DP paediatric formulation is important to be assessed using either analytical methods, taste/smell sensors (e.g., electronic torque), or taste panels, with children being the most suitable panel for taste assessment of paediatric formulations [44].

It should be highlighted that as stated in the Guideline on pharmaceutical development of medicines for paediatric use, 'the attractiveness of a paediatric medicinal products should be carefully balanced between the risk of inadequate patient acceptability and accidental intake [...]' and 'paediatric preparations must not become too attractive to children (candy like) as this is known to increase the rate of accidental poisoning' [29]. This is important for 3DP paediatric dosage forms, as a formulation which does not differentiate the medicinal product from confectionary and toys may be associated with increased attractiveness of the product to children, potentially leading to safety issues.

The importance of the safety of excipients in pharmaceutical development cannot be overemphasised. The Elixir Sulphonamide disaster in 1937 was one of the most serious poisoning of the twentieth century, where more than 100 patients, including many children, died due to diethylene glycol which was used as a solvent and sweetening agent in the formulation [45]. This incident, which led to the passage of the 1938 Federal, Drug and Cosmetic Act, also highlighted the different exposures and thus safety that excipients may have in adults compared to children. Thus, the selection of suitable excipients in paediatric medicinal products is a critical part of the pharmaceutical development.

In the context of 3DP paediatric formulations, the safety of excipients should be considered, and it should be justified. Excipients such as propylene glycol, ethanol, sorbitol, etc. that are likely to cause damage and side effects in the paediatric population should be avoided or included in line with the recommended limits. For example, acceptable daily intake limits, depending on the age of the child, have been set for propylene glycol, an excipient that has been used as a solvent and viscosity modifier in inkjet printing [46]. In case the use of solvents such as propylene glycol and ethanol cannot be avoided, it may be required to be demonstrated that such solvents are consistently removed through the manufacturing process (e.g., drying step) or that their residual content in the finished drug product is well controlled. Apart from the guidelines published by regulatory authorities and the qualitative composition of excipients in medicinal products currently authorised, the Safety and Toxicity of Excipients for Paediatrics (STEP) database can be used as a tool to assess the safety profile of excipients for paediatric formulations [47].

11.5 3D-Printed Systems With Tailored Release Profiles

3DP has been used to produce dosage forms with different release profiles (e.g., immediate, prolonged, delayed release, etc.). Specifically, more than 70% of the 3DP oral dosage prepared by FDM that have been reported in the literature exhibit modified release [48]. Factors that have been found to impact on the release profile include geometry, infill percentage, infill pattern, wall thickness, and the composition of the excipients used for 3DP formulations [49, 50]. It is clear then that sufficient knowledge should be acquired during pharmaceutical development on how each of these factors influence drug release. Based on this knowledge, satisfactory in-process controls should be set to ensure that the manufacturing process operates under control.

Currently, several polymers have been used for 3DP dosage forms [17]. The most significant polymer property for melt-based printing are the thermal properties and rheology of the melted polymer. For solvent-based printing, rheology and surface tension of printing inks are the key polymer properties [51]. From a pharmaceutics point of view, selection of a polymer for 3DP has been described as a balancing act between performance and printability [51]. However, from a regulatory point of view, the emphasis should be placed on performance and safety (Figure 11.2).

In line with the guideline on quality of oral modified release products, the release

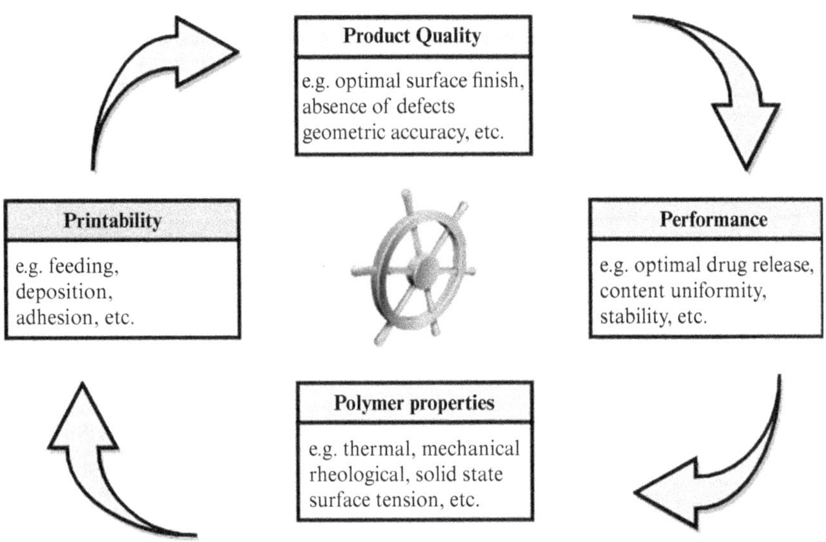

Figure 11.2 *Selecting a polymer for 3DP medicinal products is a balancing act between printability and performance. (Source: Adapted from [51].)*

mechanisms and kinetics of the modified-release dosage forms (e.g., erosion, diffusion, osmosis, etc.) should be well understood and reported. As part of the pharmaceutical development of these systems, it should be demonstrated that the drug release characteristics remain unchanged regardless of the between-subject and within-subject variability of physiological conditions [52]. Especially, focus should be given on food effect studies and concurrent alcohol intake, whenever relevant.

Concurrent ingestion of alcohol with modified-release formulations may lead to dose dumping (i.e., unintended, rapid release of the entire amount (or significant fraction) of drug contained in the dosage form). Dose dumping may lead to significant safety and efficacy concerns. Currently, only a few studies have investigated the potential occurrence of dose dumping in 3DP formulations. Skalická et al. used 3DP for coating tramadol formulations using polyvinyl alcohol (PVA) and hydroxypropyl methylcellulose (HPMC) as the coating materials [53]. From the polymers used, only HPMC was found effective to change the release rate of the drug substance (i.e., delayed release), but also could prevent the dose dumping incidence. Ong et al. produced alcohol-resistant and abused-deterrent tramadol tablets by using powder extrusion 3DP with HPMC polymers [54].

Most of the studies on 3DP of modified-release dosage forms have only been characterised by *in vitro* dissolution tests, while *in vivo* studies are sparse. Studies to investigate the efficacy, safety, biopharmaceutical, and pharmacokinetic properties of the modified-release products should be designed, conducted, and evaluated as part of the clinical development [55]. *In vitro–in vivo* correlations (IVIVCs) are highly recommended to be established. Validated IVIVCs would allow 'to predict, accurately and precisely, expected bioavailability characteristics from dissolution profile characteristics' [56]. Ultimately, depending on the level of correlation established, dissolution could act as a surrogate for bioequivalence studies facilitating changes that may take part in later stages of the clinical development or after the approval of the medicinal product.

11.6 Conclusions

Continual development on other unique features associated with 3D printing is likely to bring other opportunities and challenges to 3DP medicinal products/devices. It is recommended that the researchers or prospective marketing authorisation holders should discuss and engage with the relevant regulatory authorities to obtain specific feedback on the research development or marketing authorisation proposal. The Innovation Office at the Medicines and Healthcare Products Regulatory Agency (MHRA) is able to provide scientific and regulatory advice to support companies and research institutions [57]. Under certain conditions, the Innovative Licencing and Access Pathway (ILAP) may also be used as a route to innovative 3DP medicines approval and access [58].

Disclaimer

The views expressed in this article are the personal views of the authors and may not be understood or quoted as being made on behalf of or reflecting the position of the regulatory agencies or other organisations with which the authors are affiliated. **Conflict of Interest**: The authors are employees of the Medicines and Healthcare products Regulatory Agency, UK.

References

[1] Office of the Commissioner (2022). FDA overview organization chart. US Food and Drug Administration. Available from: https://www.fda.gov/about-fda/fda-organization-charts/fda-overview-organization-chart (accessed 3 May 2022).
[2] Di Prima, M., Coburn, J., Hwang, D. et al. (2015). Additively manufactured medical products: the FDA perspective. *3D Printing in Medicine* **2** (1): 1. Available from: http://dx.doi.org/10.1186/s41205-016-0005-9.
[3] Center for Devices and Radiological Health (2017). Technical considerations for additive manufactured medical devices. US Food and Drug Administration. Available from: https://www.fda.gov/media/97633/download (accessed 28 June 2023).
[4] 510(k) premarket notification print share E-mail. US Food and Drug Administration. Available from: https://www.accessdata.fda.gov/scripts/cdrh/cfdocs/cfPMN/pmn.cfm (acessefd 5 March 2022).
[5] Drugs@FDA: FDA-approved drugs. US Food and Drug Administration. Available from: https://www.accessdata.fda.gov/scripts/cder/daf/index.cfm?event=overview.process&ApplNo=207958 (accessed 5 March 2022).
[6] Centre for Drug Evaluation and Research (CDER) (1999). *Guidance for industry applications covered by section 505(b)(2)*. Available from: https://www.fda.gov/regulatory-information/search-fda-guidance-documents/applications-covered-section-505b2 (accessed 28 June 2023).
[7] Trofimiuk, M., Wasilewska, K., and Winnicka, K. (2019). How to modify drug release in paediatric dosage forms? Novel technologies and modern approaches with regard to children's population. *International Journal of Molecular Sciences* **20** (13): 3200. Available from: http://dx.doi.org/10.3390/ijms20133200 (accessed 28 June 2023).
[8] Li, X., Liang, E., Hong, X. et al. (2021). *In vitro* and *in vivo* bioequivalence study of 3D-printed instant-dissolving levetiracetam tablets and subsequent personalised dosing for Chinese children based on physiological pharmacokinetic modeling. *Pharmaceutics* **14** (1): 20. Available from: http://dx.doi.org/10.3390/pharmaceutics14010020.
[9] Home – SPRITAM® (levetiracetam) (2021). SPRITAM® (levetiracetam). Available from: https://www.spritam.com/#/patient/about-spritam/what-is-spritam (accessed 3 May 2022).
[10] Health Canada (2019). Guidance document – supporting evidence for implantable medical devices manufactured by 3D printing. Available from: https://www.canada.ca/en/health-canada/services/drugs-health-products/medical-devices/application-information/guidance-documents/3d-licensing-requirements/document.html (accessed 28 June 2023).
[11] Medicines and Healthcare products Regulatory Agency (2020). 3D printing (additive manufacturing) of medical devices or component parts during the coronavirus (COVID-19) pandemic. Available from: https://www.gov.uk/guidance/3d-printing-additive-manufacturing-of-medical-devices-or-component-parts-during-the-coronavirus-covid-19-pandemic (accessed 28 June 2023).
[12] Consultation on point of care manufacturing GOV.UK. Available from: https://www.gov.uk/government/consultations/point-of-care-consultation/consultation-on-point-of-care-manufacturing (accessed 30 May 2022).
[13] Scientific guidelines. European Medicine Agency. Available from: https://www.ema.europa.eu/en/human-regulatory/research-development/scientific-guidelines (accessed 3 May 2022).
[14] ICH Official web site: ICH. Ich.org. Available from: https://www.ich.org/page/quality-guidelines (accessed 4 May 2022).
[15] Rahman, Z., Barakh Ali, S.F., Ozkan, T. et al. (2018). Additive manufacturing with 3D printing: progress from bench to bedside. *The AAPS Journal* **20** (6): 101. Available from: http://dx.doi.org/10.1208/s12248-018-0225-6 (accessed 28 June 2023).
[16] Jamróz, W., Szafraniec, J., Kurek, M., and Jachowicz, R. (2018). 3D printing in pharmaceutical and medical applications: recent achievements and challenges. *Pharmaceutical Research* **35** (9): 176. Available from: http://dx.doi.org/10.1007/s11095-018-2454-x.
[17] Azad, M.A., Olawuni, D., Kimbell, G. et al. (2020). Polymers for extrusion-based 3D printing of pharmaceuticals: a holistic materials-process perspective. *Pharmaceutics* **12** (2): 124. Available from: http://dx.doi.org/10.3390/pharmaceutics12020124 (accessed 28 June 2023).
[18] Ozkan, T., Ienina, M., Barakh Ali, S.F. et al. (2018). Manufacturing engineering challenges of pharmaceutical 3D printing for on-demand drug delivery. *Engineering Technology Open Access Journal.* **1** (5): 555575. Available from: https://juniperpublishers.com/etoaj/pdf/ETOAJ.MS.ID.555575.pdf.

[19] EudraLex – Volume 4: Public Health. Available from: https://ec.europa.eu/health/medicinal-products/eudralex/eudralex-volume-4_el (accessed 4 May 2022).
[20] Wartewig, S. and Neubert, R.H.H. (2005). Pharmaceutical applications of Mid-IR and Raman spectroscopy. *Advanced Drug Delivery Reviews* **57** (8): 1144–1170. Available from: http://dx.doi.org/10.1016/j.addr.2005.01.022.
[21] Ehtezazi, T., Algellay, M., Islam, Y. et al. (2018). The application of 3D printing in the formulation of multilayered fast dissolving oral films. *Journal of Pharmaceutical Sciences* **107** (4): 1076–1085. Available from: http://dx.doi.org/10.1016/j.xphs.2017.11.019.
[22] Khorasani, M., Edinger, M., Raijada, D. et al. (2016). Near-infrared chemical imaging (NIR-CI) of 3D printed pharmaceuticals. *International Journal of Pharmaceutics* **515** (1–2): 324–330. Available from: http://dx.doi.org/10.1016/j.ijpharm.2016.09.075.
[23] Thabet, Y., Sibanc, R., and Breitkreutz, J. (2018). Printing pharmaceuticals by inkjet technology: proof of concept for stand-alone and continuous in-line printing on orodispersible films. *Journal of Manufacturing Processes* **35**: 205–215. Available from: http://dx.doi.org/10.1016/j.jmapro.2018.07.018.
[24] Ewing, A.V. and Kazarian, S.G. (2018). Recent advances in the applications of vibrational spectroscopic imaging and mapping to pharmaceutical formulations. *Spectrochim Acta Part A Molecular and Biomolecular Spectroscopy* **197**: 10–29. Available from: http://dx.doi.org/10.1016/j.saa.2017.12.055 (accessed 28 June 2023).
[25] Goyanes, A., Fina, F., Martorana, A. et al. (2017). Development of modified release 3D printed tablets (printlets) with pharmaceutical excipients using additive manufacturing. *International Journal of Pharmaceutics* **527** (1–2): 21–30. Available from: http://dx.doi.org/10.1016/j.ijpharm.2017.05.021.
[26] Vakili, H., Kolakovic, R., Genina, N. et al. (2015). Hyperspectral imaging in quality control of inkjet printed personalised dosage forms. *International Journal of Pharmaceutics* **483** (1–2): 244–249. Available from: http://dx.doi.org/10.1016/j.ijpharm.2014.12.034.
[27] Specifications and control tests on the finished product (1991). European Medicines Agency. Available from: https://www.ema.europa.eu/en/specifications-control-tests-finished-product#current-effective-version-section.
[28] Regulation (EC) No. 1901/2006 of the European Parliament and of the Council of 12 December 2006 on medicinal products for paediatric use and amending Regulation (EEC) No. 1768/92, Directive 2001/20/EC, Directive 2001/83/EC and Regulation (EC) No. 726/2004. **12**, 19. Available from: https://eur-lex.europa.eu/eli/reg/2006/1901/oj.
[29] Committee for Medicinal Products for Human Use (2013). Guideline on pharmaceutical development of medicines for paediatric use. Available from: https://www.ema.europa.eu/en/pharmaceutical-development-medicines-paediatric-use (accessed on 28th June 2023).
[30] European Commission (2017). State of paediatric medicines in the EU – 10 years of the EU paediatric regulation. Available from: https://ec.europa.eu/health/system/files/2017-11/2017_childrensmedicines_report_en_0.pdf.
[31] van der Vossen, A.C., Al-Hassany, L., Buljac, S. et al. (2019). Manipulation of oral medication for children by parents and nurses occurs frequently and is often not supported by instructions. *Acta Paediatrica* **108** (8): 1475–1481. Available from: http://dx.doi.org/10.1111/apa.14718.
[32] Venables, R., Batchelor, H., Hodson, J. et al. (2015). Determination of formulation factors that affect oral medicines acceptability in a domiciliary paediatric population. *International Journal of Pharmaceutics* **480** (1–2): 55–62. Available from: http://dx.doi.org/10.1016/j.ijpharm.2015.01.023.
[33] Richey, R.H., Shah, U.U., Peak, M. et al. (2013). Manipulation of drugs to achieve the required dose is intrinsic to paediatric practice but is not supported by guidelines or evidence. *BMC Pediatrics* **13** (1): 81. Available from: http://dx.doi.org/10.1186/1471-2431-13-81.
[34] World Health Organisation (2020). Innovative delivery systems for paediatric medicines technology landscape. Available from: https://www.who.int/publications/i/item/9789240008182.
[35] Walsh, J., Schaufelberger, D., Iurian, S. et al. (2021). Path towards efficient paediatric formulation development based on partnering with clinical pharmacologists and clinicians, a conect4children expert group white paper. British Journal of Clinical Pharmacology. Available from: http://dx.doi.org/10.1111/bcp.14989.
[36] Quodbach, J., Bogdahn, M., Breitkreutz, J. et al. (2021). Quality of FDM 3D printed medicines for pediatrics: considerations for formulation development, filament extrusion, printing process and printer design. *Therapeutics Innovative and Regulatory Science*. Available from: http://dx.doi.org/10.1007/s43441-021-00354-0.

[37] Lafeber, I., Ruijgrok, E.J., Guchelaar, H.-J., and Schimmel, K.J.M. (2022). 3D printing of pediatric medication: the end of bad tasting oral liquids? A scoping review. *Pharmaceutics* **14** (2): 416. Available from: https://www.mdpi.com/1999-4923/14/2/416 (accessed 4 May 2022).
[38] Diaz-Tores, E., Ana, S.-E., and Farina, J.B. (2021). A micro-extrusion 3D printing platform for fabrication of orodispersible printlets for pediatric use. *International Journal of Pharmaceutics* **605**: 120854. Available from: http://dx.doi.org/10.1016/j.ijpharm.2021.120854.
[39] Öblom, H., Sjöholm, E., Rautamo, M., and Sandler, N. (2019). Towards printed pediatric medicines in hospital pharmacies: comparison of 2D and 3D-printed orodispersible warfarin films with conventional oral powders in unit dose sachets. *Pharmaceutics* **11** (7): 334. Available from: http://dx.doi.org/10.3390/pharmaceutics11070334.
[40] Krause, J., Müller, L., Sarwinska, D. et al. (2021). 3D printing of mini tablets for pediatric use. *Pharmaceuticals (Basel)* **14** (2): 143. Available from: http://dx.doi.org/10.3390/ph14020143.
[41] Goyanes, A., Scarpa, M., Kamlow, M. et al. (2017). Patient acceptability of 3D printed medicines. *International Journal of Pharmaceutics* **530** (1–2): 71–78. Available from: http://dx.doi.org/10.1016/j.ijpharm.2017.07.064.
[42] Januskaite, P., Xu, X., Ranmal, S.R. et al. (2020). I spy with my little eye: a paediatric visual preferences survey of 3D printed tablets. *Pharmaceutics* **12** (11): 1100. Available from: http://dx.doi.org/10.3390/pharmaceutics12111100.
[43] Goyanes, A., Madla, C.M., Umerji, A. et al. (2019). Automated therapy preparation of isoleucine formulations using 3D printing for the treatment of MSUD: first single-centre, prospective, crossover study in patients. *International Journal of Pharmaceutics* **567**: 118497. Available from: http://dx.doi.org/10.1016/j.ijpharm.2019.118497.
[44] Committee for Medicinal Products for Human Use (2006). Reflection paper: formulations of choice for the paediatric population. Available from: https://www.ema.europa.eu/en/documents/scientific-guideline/reflection-paper-formulations-choice-paediatric-population_en.pdf (accessed on 28th June 2023).
[45] Woolf, A.D. (2022). Sulfanilamide (diethylene glycol) disaster: United States, 1937. In: *History of Modern Clinical Toxicology* (ed. A.D. Woof), 139–148. Elsevier.
[46] Genina, N., Janßen, E.M., Breitenbach, A. et al. (2013). Evaluation of different substrates for inkjet printing of rasagiline mesylate. *European Journal of Pharmaceutics and Biopharmaceutics* **85** (3 Pt B): 1075–1083. Available from: http://dx.doi.org/10.1016/j.ejpb.2013.03.017.
[47] Salunke, S., Brandys, B., Giacoia, G., and Tuleu, C. (2013). The STEP (safety and toxicity of excipients for paediatrics) database. Part 2: The pilot version. *International Journal of Pharmaceutics* **457** (1): 310–322. Available from: http://dx.doi.org/10.1016/j.ijpharm.2013.09.013.
[48] Brambilla, C.R.M., Okafor-Muo, O.L., Hassanin, H., and El Shaer, A. (2021). 3DP printing of oral solid formulations: a systematic review. *Pharmaceutics* **13** (3): 358. Available from: http://dx.doi.org/10.3390/pharmaceutics13030358.
[49] Goyanes, A., Wang, J., Buanz, A. et al. (2015). 3D printing of medicines: engineering novel oral devices with unique design and drug release characteristics. *Molecular Pharmaceuticals* **12** (11): 4077–4084. Available from: http://dx.doi.org/10.1021/acs.molpharmaceut.5b00510.
[50] Obeid, S., Madžarević, M., and Ibrić, S. (2021). Tailoring amlodipine release from 3D printed tablets: influence of infill patterns and wall thickness. *International Journal of Pharmaceutics* **610**: 121261. Available from: http://dx.doi.org/10.1016/j.ijpharm.2021.121261.
[51] Govender, R., Kissi, E.O., Larsson, A., and Tho, I. (2021). Polymers in pharmaceutical additive manufacturing: a balancing act between printability and product performance. *Advanced Drug Delivery Reviews* **177**: 113923. Available from: http://dx.doi.org/10.1016/j.addr.2021.113923.
[52] Committee for Medicinal Products for Human Use (2014). Quality of oral modified release products. Available from: https://www.ema.europa.eu/en/quality-oral-modified-release-products (accessed 28 June 2023).
[53] Skalická, B., Matzick, K., Komersová, A. et al. (2021). 3D-printed coating of extended-release matrix tablets: effective tool for prevention of alcohol-induced dose dumping effect. *Pharmaceutics* **13** (12): 2123. Available from: http://dx.doi.org/10.3390/pharmaceutics13122123.
[54] Ong, J.J., Awad, A., Martorana, A. et al. (2020). 3D printed opioid medicines with alcohol-resistant and abuse-deterrent properties. *International Journal of Pharmaceutics* **579**: 119169. Available from: http://dx.doi.org/10.1016/j.ijpharm.2020.119169.
[55] Committee for Medicinal Products for Human Use (2014). Guideline on the pharmacokinetic and clinical evaluation of modified release dosage forms. Available from: https://www.ema.

europa.eu/en/pharmacokinetic-clinical-evaluation-modified-release-dosage-forms (accessed 28 June 2023).
[56] US Food and Drug Administration (1997). Extended release oral dosage forms: development, evaluation, and application of *in vitro/in vi*vo correlations. Available from: https://www.fda.gov/media/70939/download (accessed 28 June 2023).
[57] MHRA Innovation Office (2014). GOV.UK. Available from: https://www.gov.uk/government/groups/mhra-innovation-office (accessed 4 May 2022).
[58] GOV.UK (2002). Innovative licensing and access pathway. Available from: https://www.gov.uk/guidance/innovative-licensing-and-access-pathway (accessed 30 May 2022).

Index

3D Printing of Pharmaceutical and Drug Delivery Devices: Progress from Bench to Bedside, First Edition.
Edited by Dimitrios A. Lamprou, Dennis Douroumis and Sheng Qi.